T0176897

MULTI-DIMENSIONAL IMAGING

MULTI-DIMENSIONAL IMAGING

Edited by

Bahram Javidi
University of Connecticut, USA

Enrique Tajahuerce
University Jaume I, Spain

Pedro Andrés
University of Valencia, Spain

IEEE PRESS

Library of Congress Cataloging-in-Publication Data applied for.

A catalogue record for this book is available from the British Library.

ISBN: 9781118449837

Typeset in 10/12pt TimesLTStd by Laserwords Private Limited, Chennai, India
Printed and bound in Singapore by Markono Print Media Pte Ltd

1   2014

For Bethany, Ariana, Darius, and Vida

In memory of our friend and colleague, Dr Fumio Okano

Contents

About the Editors

Bahram Javidi is the Board of Trustees Distinguished Professor at University of Connecticut.

He has been recognized by nine best paper awards, and major awards from professional societies, including fellowships of IEEE, OSA, EOS, and SPIE. In 2008, he received the Fellow Award from the John Simon Guggenheim Foundation. He has written over 870 publications, which have been cited 11 000 times according to the ISI Web of Knowledge (*h index* = 55). He has received the 2008 IEEE Donald G. Fink Prize Paper Award, the 2010 George Washington University's Distinguished Alumni Scholar Award, the 2008 SPIE Technology Achievement Award, and the 2005 SPIE Dennis Gabor Award in Diffractive Wave Technologies. In 2007, the Alexander von Humboldt Foundation awarded him the Humboldt Prize for Outstanding Scientists. He was the recipient of the (IEEE) Photonics Distinguished Lecturer Award in 2003–2005. He was awarded the Best Journal Paper Award from the *IEEE Transactions on Vehicular Technology* in 2002 and 2005. In 2003 he was selected, as one of the nation's top 160 engineers between the ages of 30–45 by the National Academy of Engineering (NAE), to be an invited speaker at The Frontiers of Engineering Conference. He is an alumnus of the Frontiers of Engineering of The NAE since 2003. He was a National Science Foundation Presidential Young Investigator and received The Engineering Foundation and the IEEE Faculty Initiation Awards. He is on the Editorial Board of the *Proceedings of the IEEE* journal (ranked number one in electrical engineering), and is on the Advisory Board of the *IEEE Photonics* journal. He was on the founding editorial board of the *IEEE Journal of Display*. In 2008, he was elected by the members to be on The Board of Directors of the SPIE. He received his BSc from George Washington University, and his PhD from the Pennsylvania State University.

Enrique Tajahuerce was born in Soria, Spain, in 1964. He received his PhD in Physics from
the University of Valencia (UV), Spain, in 1998. Dr Tajahuerce
was a researcher at the Technological Institute of Optics, Colour
and Imaging (AIDO) in Paterna, Spain, from 1989–1992. Since
1992 he has been member of the Physics Department in the Universitat Jaume I (UJI), in Castelló, Spain, where he is an Associate
Professor. He is currently Secretary of the Physics Department
and Deputy Director of the Institute of New Imaging Technologies
(INIT).

Dr Tajahuerce's research interests lie in the areas of diffractive
optics, digital holography, ultrafast optics, computational imaging,
and microscopy. He has co-authored more than 90 scientific publications, and over 140 communications in conference meetings
(35 of them by invitation). He is member of the SPIE, OSA, EOS,
and the Spanish Optical Society (SEDO). In 2008, Dr Tajahuerce
received the IEEE Donald G. Fink Prize Paper Award.

Pedro Andrés was born in Valencia, Spain, in 1954. He earned a PhD in physics/optics
from the University of Valencia (UV) in 1983. His thesis received
the 1984 Special Distinction awarded by the UV. Dr Andrés has
been full a Professor of Optics since 1994 at the UV. He acted as
the UV's Head of the Department of Optics from 1998–2006. He
was also the Director of both the PhD and the Masters Program in
the Faculty of Physics (UV) from 2008–2010.

His current research interests include static and dynamic diffractive optical elements, advanced imaging systems, microstructured
fibers, temporal imaging, and ultrafast optics. He has co-authored
more than 130 peer-reviewed papers. Two of these articles have
received more than 200 citations each. He also supervised 13 PhD
works (four of them received a Special Distinction awarded by the
University of Valencia).

Currently, Professor Andrés is an expert on the Board (Branch
Science) for the Evaluation of Faculty Members of Spanish
Universities, President of the Iberian-American Network for
Optics, Fellow of the OSA, elected member of the Board of
Directors of the European Optical Society (EOS), Past-President
of the Imaging Committee of the Spanish Optical Society
(SEDOPTICA), and Academic Mentor of the EOS Comunidad
Valenciana Student Club.

List of Contributors

Pedro Andrés, Department d'Òptica, Universitat de València, Spain

Yasuhiro Awatsuji, Division of Electronics, Kyoto Institute of Technology, Japan

Michal Baranek, Department of Optics, Palacky University Olomouc, Czech Republic

Vittorio Bianco, CNR, Istituto Nazionale di Ottica, Sezione di Napoli, Italy

Pere Clemente, GROC·UJI, Departament de Física, and Servei Central d'Instrumentació Científica, Universitat Jaume I, Spain

Vicent Climent, GROC·UJI, Departament de Física and Institut de Noves Tecnologies de la Imatge (INIT), Universitat Jaume I, Spain

Loïc Denis, Laboratoire Hubert Curien, Saint Etienne University, France

Christian Depeursinge, Institute of Microengineering, École Polytechnique Fédérale de Lausanne, Switzerland

Adrián Dorado, Department of Optics, University of Valencia, Spain

Frank Dubois, Microgravity Research Centre, Université Libre de Bruxelles, Belgium

Vicente Durán, GROC·UJI, Departament de Física and Institut de Noves Tecnologies de la Imatge (INIT), Universitat Jaume I, Spain

Michael T. Eismann, Air Force Research Laboratory, USA

Mercedes Fernández-Alonso, GROC·UJI, Departament de Física and Institut de Noves Tecnologies de la Imatge (INIT), Universitat Jaume I, Spain

Pietro Ferraro, CNR, Istituto Nazionale di Ottica, Sezione di Napoli, Italy

Andrea Finizio, CNR, Istituto Nazionale di Ottica, Sezione di Napoli, Italy

Thierry Fournel, Laboratoire Hubert Curien, Saint Etienne University, France

Corinne Fournier, Laboratoire Hubert Curien, Saint Etienne University, France

Javier Garcia, Departamento de Óptica, Universitat Valencia, Spain

Eran Gur, Department of Electrical Engineering and Electronics, Azrieli – College of Engineering, Israel

Tobias Haist, Institute für Technische Optik, University of Stuttgart, Germany

Malte Hasler, Institute für Technische Optik, University of Stuttgart, Germany

Yoshio Hayasaki, Center for Optical Research and Education (CORE), Utsunomiya University, Japan

Esther Irles, GROC·UJI, Departament de Física, Universitat Jaume I, Spain

Kazuyoshi Itoh,Graduate School of Engineering, Department of Material & Life Science, Osaka University, Japan and Science Technology Entrepreneurship Laboratory (e-square), Osaka University, Japan

Bahram Javidi, Department of Electrical and Computer Engineering, University of Connecticut, USA

Boaz Jessie Jackin, Center for Optical Research and Education, Utsunomiya University, Japan

Jesús Lancis, GROC·UJI, Departament de Física and Institut de Noves Tecnologias de la Imatge (INIT), Universitat Jaume I, Spain

Chun-Hea Lee, Industrial Design Department, Joongbu University, Korea

Daniel A. LeMaster, Air Force Research Laboratory, USA

Anabel LLavador, Department of Optics, University of Valencia, Spain

Massimiliano Locatelli, CNR, Istituto Nazionale di Ottica, Largo E. Fermi, Italy

Ahmed El Mallahi, Microgravity Research Centre, Université Libre de Bruxelles, Belgium

Pierre Marquet, Centre de Neurosciences Psychiatriques, Centre Hospitalier Universitaire Vaudois, Département de Psychiatrie, Switzerland and Brain Mind Institute, Institute of Microengineering, École Polytechnique Fédérale de Lausanne, Switzerland

Manuel Martínez-Corral, Department of Optics, University of Valencia, Spain

Lluís Martínez-León, GROC·UJI, Departament de Física and Institut de Noves Tecnologies de la Imatge (INIT), Universitat Jaume I, Spain

Amihai Meiri, Faculty of Engineering, Bar-Ilan University, Israel

Omel Mendoza-Yero, GROC·UJI, Departament de Física and Institut de Noves Tecnologies de la Imatge (INIT), Universitat Jaume I, Spain

Riccardo Meucci, CNR, Istituto Nazionale di Ottica, Largo E. Fermi, Italy

Lisa Miccio, CNR, Istituto Nazionale di Ottica, Sezione di Napoli, Italy

Christophe Minetti, Microgravity Research Centre, Université Libre de Bruxelles, Belgium

Gladys Mínguez-Vega, GROC·UJI, Departament de Física and Institut de Noves Tecnologies de la Imatge (INIT), Universitat Jaume I, Spain

Vicente Micó, Departamento de Óptica, University of Valencia, Spain

Wolfgang Osten, Institute für Technische Optik, University of Stuttgart, Germany

Yasuyuki Ozeki, Graduate School of Engineering, Department of Material & Life Science, Osaka University, Japan

Min-Chul Park, Sensor System Research Center, Korea Institute of Science and Technology, Korea

Melania Paturzo, CNR, Istituto Nazionale di Ottica, Sezione di Napoli, Italy

Anna Pelagotti, CNR, Istituto Nazionale di Ottica, Largo E. Fermi, Italy

Jorge Pérez-Vizcaíno, GROC·UJI, Departament de Física and Institut de Noves Tecnologies de la Imatge (INIT), Universitat Jaume I, Spain

Pasquale Poggi, CNR, Istituto Nazionale di Ottica, Largo E. Fermi, Italy

Eugenio Pugliese, CNR, Istituto Nazionale di Ottica, Largo E. Fermi, Italy

Yair Rivenson, Department of Electrical and Computer Engineering, Ben-Gurion University of the Negev, Israel

Joseph Rosen, Department of Electrical and Computer Engineering, Ben-Gurion University of the Negev, Israel

Genaro Saavedra, Department of Optics, University of Valencia, Spain

Yusuke Sando, Center for Optical Research and Education, Utsunomiya University, Japan

Mozhdeh Seifi, Laboratoire Hubert Curien, Saint Etienne University, France

Fernando Soldevila, GROC·UJI, Departament de Física, Universitat Jaume I, Spain

Jung-Young Son, Biomedical Medical Engineering Department, Konyang University, Korea

Wook-Ho Son, Content Platform Research Department, Electronics and Communication Technology Research Institute, Korea

Adrian Stern, Department of Electro-Optics Engineering, Ben-Gurion University of the Negev, Israel

Enrique Tajahuerce, GROC·UJI, Departament de Física and Institut de Noves Tecnologies de la Imatge (INIT), Universitat Jaume I, Spain

Koki Wakunami, Global Scientific Information and Computing Center, Tokyo Institute of Technology, Japan

Masahiro Yamaguchi, Global Scientific Information and Computing Center, Tokyo Institute of Technology, Japan

Toyohiko Yatagai, Center for Optical Research and Education, Utsunomiya University, Japan

Catherine Yourassowsky, Microgravity Research Centre, Université Libre de Bruxelles, Belgium

Zeev Zalevsky, Faculty of Engineering, Bar-Ilan University, Israel

Preface

Imaging sciences and engineering are rapidly evolving in many ways by encompassing more sensing modalities, display media, digital domains, and consumer products. This field of research and development is frenetically active in multiple scientific, innovative disciplines including those of materials, sensors, displays, algorithms, and applications. Today, the term "optical image" refers not only to the concept of image formation and its multiple analysis, reconstruction, and visualization techniques, but also to computer vision, terahertz frequencies and electromagnetic imaging, medical imaging, algorithms for processing of images, and three-dimensional image sensing, among many others.

In the last two decades, research into advanced imaging systems has made great progress. There are many new procedures in microscopy that overcome the classical resolution limit. The field has benefited from the astonishing results of computational imaging techniques. The advances in imaging through turbid and scattering media allow the achievement of images with good resolution, either from deep layers of tissue in living beings, or the cosmos through telescopes on Earth's surface. Optics in the life sciences incorporates new methods for non-invasive imaging of *in vivo* biological material and the tools to translate that knowledge and procedures for the study, diagnosis, and treatment of diseases. Sources of entangled photons in quantum imaging can provide high-quality images at a very low level of illumination. To all this, we must add many other rapidly evolving areas such as modern adaptive optics, imaging in nuclear medicine, optical tweezers that are opening new avenues for the study of single cells, the role of spatial light modulators in advanced imaging, and so on.

Recently, there have been rapid advances in imaging systems because of the introduction of various multi-dimensional imaging techniques, including digital holography, integral imaging, multiview, light field, multispectral imaging, polarimetric imaging, temporal multiplexing; development of new algorithms, such as those used for compressive sensing or computational imaging; and the application of new light sources, such as ultrashort lasers, laser diodes, super-continuum sources, and so on. In parallel to the development of new imaging techniques, there has been a great advance in image resolution by increasing the number of pixels of different detector arrays and reducing pixel size. It has been recognized that, in many situations, it is also very important to measure not only the spatial intensity distribution of the object, but also other useful dimensions of an image, such as spectral, polarization, optical phase, or three-dimensional structure, leading to the development of multi-dimensional imaging. As a result, there have been substantial multidisciplinary activities in the development of polarimetric cameras, multispectral sensors, holographic techniques, three-dimensional visualization devices, and so on, integrated with special pur- pose algorithms to produce multi-dimensional imaging systems for a variety of applications, including medical, defense and security, robotics, education, entertainment, environment, and manufacturing.

Given the great interest in multi-dimensional imaging research, development, and educa- tion, this book, entitled *Multi-dimensional Imaging* aims to present an overview of the recent advances in the field by some of the leading researchers and educators. The book intends to educate and provide the readers with an introduction to some of the important areas in this multi-disciplinary domain. This broad overview is useful for students, engineers, and scientists who are interested in learning about the latest advances in this important field.

This book addresses a selection of important subjects in multi-dimensional imaging describing fundamentals, approaches, techniques, new developments, applications, and a relevant bibliography. It consists of 17 chapters and is divided into four parts that deal with multi-dimensional digital holographic techniques, multi-dimensional biomedical imaging and microscopy, multi-dimensional imaging and display, and spectral and polarimetric imaging. The chapters are written by some of the most prominent researchers and educators in the field.

We wish to thank the authors for their outstanding contributions, and the Wiley editors and staff for their support and assistance.

This book is dedicated to the memory of our departed friend, Dr Fumio Okano.

Bahram Javidi, Storrs, Connecticut, USA
Enrique Tajahuerce, Castelló, Spain
Pedro Andrés, Valencia, Spain

Acknowledgments

We are grateful to the authors, whom we have known for many years as friends and colleagues, for their outstanding contributions to this book. Special thanks go to John Wiley & Sons Editor, Ms Alex King, for her support and encouragement of this book from the initial stages to the end. We thank John Wiley & Sons production team Tom Carter and Genna E. Manaog, as well as Lynette Woodward and Sangeetha Parthasarathy, for their assistance in finalizing this book.

Be with those who help your being.
Rumi

Part One

Multi-Dimensional Digital Holographic Techniques

1

Parallel Phase-Shifting Digital Holography

Yasuhiro Awatsuji
Division of Electronics, Kyoto Institute of Technology, Japan

1.1 Chapter Overview

Parallel phase-shifting digital holography is a technique capable of not only instantaneously measuring the three-dimensional (3D) field but also motion picture measurement of time evolution in the 3D field. The recording and reconstruction flow of this technique are described. The technique has been experimentally demonstrated by a parallel phase-shifting digital holography system using a normal-speed camera, which lead to a high-speed camera being constructed and used so that 3D motion and phase motion picture capture were demonstrated at the rate of up to 262 500 frames per second (fps). As an ultrafast phase imaging technique, a parallel phase-shifting digital holography system using a femtosecond pulsed laser has been experimentally demonstrated. A portable parallel phase-shifting digital holography system will also be introduced here. Finally some function-extended parallel phase-shifting digital holography will be mentioned for the purpose of motion picture-measurement of 3D and color, 3D and spectral characteristics, 3D and polarization characteristics, and 3D motion picture microscopy.

1.2 Introduction

Holography is a technique for recording and reconstructing perfect wavefronts of objects [1]. The technique actively investigates not only three-dimensional (3D) displays but also 3D measurement of objects. In this technique, the complex amplitude distribution of an object is recorded as a form of an interference fringe image. The complex amplitude distribution consists of amplitude and phase distributions of objects, and can provide a 3D image. In conventional holography, a high-resolution photosensitive material, called the *holographic plate*, is used to record the interference fringe image. The medium in which the interference fringe image is recorded is the *hologram*.

Multi-dimensional Imaging, First Edition. Edited by Bahram Javidi, Enrique Tajahuerce and Pedro Andrés.
© 2014 John Wiley & Sons, Ltd. Published 2014 by John Wiley & Sons, Ltd.

Recently, there has been a great deal of progress in image sensors such as charge-coupled devices (CCDs) and complementary metal-oxide semiconductor (CMOS) image sensors, and such devices have been used in holography in place of holographic plates. Holography using image sensors is called *digital holography* [2,3]. Digital holography has the following attractive features: it does not require a wet and chemical process for developing; quantitative evaluation is easy for 3D images of objects; and focused images of 3D objects at the desired depth can be instantaneously recorded without a mechanical focusing process. Also, this technique can quantitatively provide phase distribution of an object. Thus, digital holography can serve as a quantitative 3D and phase-imaging video camera. The technique is used in many fields such as shape and deformation measurement, particle measurement, microscopy, endoscopy, object recognition, information security, and so on.

Since the pixel size and pixel pitch of image sensors are too large to record fine interference fringes that would be recorded on a photographic plate, in-line digital holography is frequently applied. In in-line digital holography, the object and reference waves almost orthogonally irradiate the image sensor. Indeed, in-line digital holography allows instantaneous measurement of the object wave in principle, but the reconstructed image is degraded because the undesired images are superimposed on the desired object wave. To obtain just the object wave, phase-shifting digital holography has been proposed [4].

Although phase-shifting digital holography can only derive the complex amplitude of an object wave at an arbitrary depth, it needs multiple holograms to reconstruct the object wave free of undesired images. The multiple holograms are sequentially recorded by using reference waves with different phase retardations. Indeed phase-shifting digital holography allows reconstruction of a clear object wave, but is useless for instantaneous measurement of moving objects. To achieve a phase-shifting method that can perform instantaneous measurement, parallel phase-shifting digital holography has been investigated [5–27]. The technique uses an ingenious arrangement of image sensor pixels and a phase-shifting array device.

In this chapter, the basic concept and processing flow of parallel phase-shifting digital holography are explained. Three parallel phase-shifting digital holography experimental systems and their results are described [23–26]. Also, a portable system based on parallel phase-shifting digital holography is introduced [27]. Finally, some function-extended parallel phase-shifting digital holography techniques are mentioned [28–35].

1.3   Digital Holography and Phase-Shifting Digital Holography

Digital holography is a technique for recording the interference fringe image by an image sensor and reconstructing the complex amplitude distribution of an object by computer [2,3]. A schematic of a system setup of digital holography is shown in Fig. 1.1. Generally, a laser is used as the optical source. A laser beam is divided into two beams. One beam illuminates the object and the beam scattered from the object is called the *object wave*. The object wave irradiates the image sensor. The other beam illuminates the image sensor directly and this beam is called a *reference wave*. An interference fringe image is generated by the object and reference waves and captured with the image sensor. The captured interference image is called a *digital hologram*. The complex amplitude distribution of the object is numerically reconstructed from the digital hologram by computer. Therefore, one instantaneous 3D image of an object can be reconstructed from a single hologram. By sequential capturing of holograms with a camera, a 3D motion picture image of the object can be recorded.

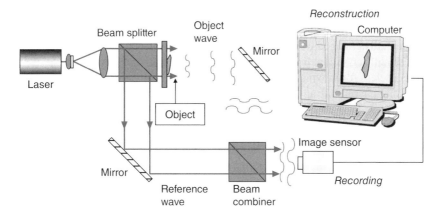

Figure 1.1 Schematic diagram of digital holography

To reconstruct the image in digital holography, a diffraction integral is generally applied to the hologram recorded with the image sensor. Although the use of only the diffraction integral is the simplest calculation scheme used to reconstruct the image and allows instantaneous measurement, the reconstructed image is degraded because the undesired images, which are the non-diffraction wave and the conjugate image, are superimposed on the desired object wave, which forms the image of the object. To extract just the object wave, phase-shifting digital holography has been proposed [4].

Figure 1.2 shows the optical setup schematic for phase-shifting digital holography [4]. More than two holograms are sequentially recorded using reference waves with different phase

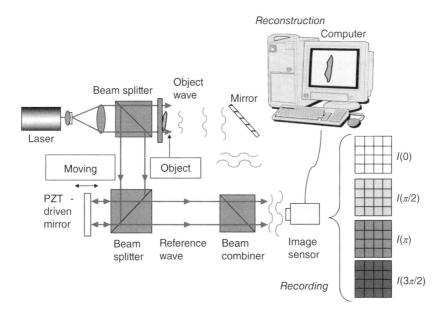

Figure 1.2 Schematic diagram of phase-shifting digital holography

retardations. A method of four-step phase-shifting of the reference wave, such as 0, $\pi/2$, π, and $3\pi/2$, is frequently adopted for phase-shifting digital holography. Usually, the retardation is sequentially changed by using a piezoelectric-transducer (PZT) mirror or wave plates. Indeed phase-shifting digital holography can only derive the complex amplitude of an object wave and is useless for moving objects. To obtain a clear reconstructed 3D image of moving objects, parallel phase-shifting digital holography has been proposed [5–27].

1.4 Parallel Phase-Shifting Digital Holography

The essence of parallel phase-shifting digital holography [5–27] is a single-shot technique for implementing phase-shifting digital holography. The single-shot technique uses a single image sensor and space-division multiplexing of holograms. Figure 1.3 shows a schematic diagram of the principle of parallel phase-shifting digital holography. Multiple holograms needed for phase-shifting digital holography are stuffed into a single hologram by using space-division multiplexing of the holograms pixel by pixel. To implement the multiplexing of the holograms, several ideas have been proposed. A micro phase-retarder array such as the micro glass-plate array is inserted in the reference wave path and imaged onto the image sensor [5]. High light efficiency is achieved by this arrangement, but precise alignment of the optical system for imaging of the micro phase-retarder array onto the image sensor pixel by pixel is needed. To make alignment easy, a spatial light modulator (SLM) consisting of a liquid crystal is used in the micro phase-retarder array [17]. Also a micro polarization-element array was proposed to achieve the multiplexing of the holograms. In this arrangement, a micro polarization-element array is attached to the image sensor [6,13,16,23–27]. The directions of the transmission axes of the micro polarizer array are alternately changed pixel by pixel. 2 × 2 configuration

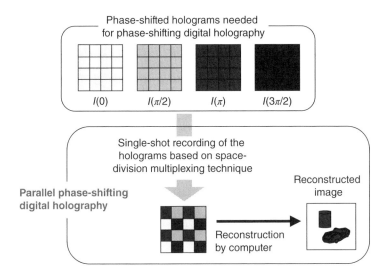

Figure 1.3 Schematic diagram of principle of parallel phase-shifting digital holography

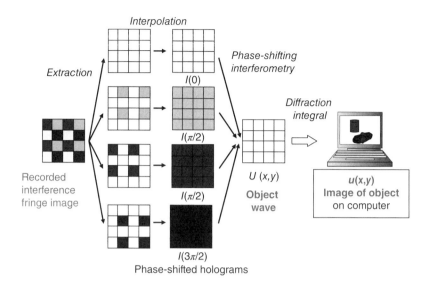

Figure 1.4 Schematic diagram of the principle of flow for image reconstruction in parallel phase-shifting digital holography

[5–7,11] and 2 × 1 configuration [10,11,13,14] of the unit of micro polarizer array have been reported for parallel four-step and parallel two-step phase-shifting digital holography, respectively. Light efficiency of this arrangement is lower than that using a micro phase-retarder array, but alignment of the optical element in the parallel phase-shifting digital holography system is quite easy.

Figure 1.4 shows a schematic diagram of a flow for image reconstruction in parallel phase-shifting digital holography. This figure shows one example of the implementation of parallel four-step-phase-shifting digital holography [5–7,11] and it uses four phase shifts. The pixels containing the same phase shift are extracted from the recorded single hologram. For each phase-shift, the extracted pixels are relocated in another 2D image at the same addresses at which they were located before being extracted. The values of blanked pixels in the 2D image are interpolated using the neighboring pixels. By this relocation and interpolation, multiple holograms, $I(0)$, $I(\pi/2)$, $I(\pi)$, $I(3\pi/2)$, are obtained. If the amplitude and phase distributions of the object are not drastic, Eq. (1.1), which is a calculation of the complex amplitude used in conventional sequential phase-shifting digital holography, gives almost the same distribution as the true complex amplitude distribution of the object wave on the image sensor plane $u(x, y)$.

$$u(x, y) = \frac{\{I(0) - I(\pi)\} + i\{I(\pi/2) - I(3\pi/2)\}}{4}. \tag{1.1}$$

The complex amplitude of distribution of the object wave at where the object was positioned in the recording step $U(X, Y)$, can be reconstructed by the diffraction integral of the derived complex amplitude. Fresnel transformation is one of the candidates for the diffraction integral as follows.

$$U(X, Y) = \int\limits_{-\infty}^{\infty} \int\limits_{-\infty}^{\infty} u(x, y) \exp\left[\frac{2\pi i}{\lambda} \left\{ Z + \frac{(X-x)^2 + (Y-y)^2}{2Z} \right\} \right] dxdy. \qquad (1.2)$$

Here, λ and i are the wavelength of the laser beam and imaginary unit, respectively. Z is the distance between the image sensor and the plane on which the complex amplitude is calculated.

Two-step phase-shifting digital holography can also be applied to parallel phase-shifting digital holography [10,11,13,14]. Equation (1.3), which is a calculation of the complex amplitude used in the case of $-\pi/2$ in the phase shift of Meng's two-step phase-shifting interferometry [36], gives the complex amplitude of the object wave at the image sensor plane $u(x, y)$.

$$u(x, y) = \frac{1}{2\sqrt{I_r}} \left[\{I(0) - a(x, y)\} - i \left\{ I\left(-\frac{\pi}{2}\right) - a(x, y) \right\} \right]. \qquad (1.3)$$

Here, $a(x, y)$ is defined as follows.

$$a(x, y) = \frac{v - \sqrt{v^2 - 2w}}{2}, \qquad (1.4)$$

$$v = I(0) + I\left(-\frac{\pi}{2}\right) + 2I_r, \qquad (1.5)$$

$$w = I(0)^2 + I\left(-\frac{\pi}{2}\right)^2 + 4I_r^2. \qquad (1.6)$$

Here, I_r is the intensity distribution of the reference wave. We know the $I(0)$, $I(-\pi/2)$, and I_r, so the complex amplitude distribution of the object wave can be reconstructed by the diffraction integral of $u(x, y)$. I_r can be measured before or after the recording of the holograms. The space bandwidth product of parallel two-step phase-shifting digital holography is twice that of parallel four-step phase-shifting digital holography [15].

1.5   Experimental Demonstration of Parallel Phase-Shifting Digital Holography

An experimental parallel phase-shifting digital holography system was constructed for the first time in study [23]. The system was based on parallel two-step phase-shifting digital holography. Figure 1.5 shows a schematic diagram and a photograph of the system. This system consisted of an interferometer and a polarization-imaging camera. Perpendicularly-polarized light is emitted from the laser and split into two beams by a beam splitter. One beam illuminates the objects. The scattered light from the objects passes through the polarizer and is changed to perpendicularly-polarized light. It then arrives at the image sensor of the original polarization-imaging camera and forms the object wave. The other beam passes through the quarter wave plate and then arrives at the image sensor. This wave is the *reference wave*.

A Nd:YVO$_4$ laser operated at 532 nm was used as the optical source. The developed polarization-imaging camera consists of a normal-speed camera and a micro-polarizer array, and can detect orthogonal two-linear polarizations, both at 2×1 pixels, implementing a 90° phase shift of the reference wave. The photographs of the camera and image sensor with a

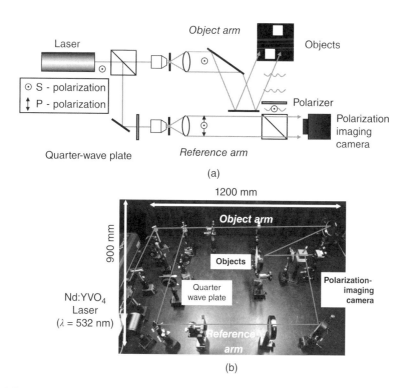

(a)

(b)

Figure 1.5 Parallel phase-shifting digital holography system using a normal-speed polarization-imaging camera. (a) Schematic diagram, (b) photograph

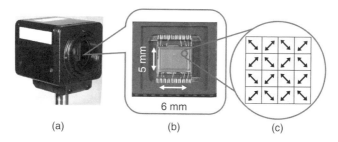

(a)   (b)   (c)

Figure 1.6 Polarization-imaging camera originally developed for a parallel two-step phase-shifting digital holography system. (a) Overview, (b) image sensor with a micro-polarizer array, (c) schematic diagram of the configuration of the transmission axis of the micro-polarizer array

micro-polarizer array are shown in Figs. 1.6(a) and (b). Figure 1.6(c) shows the transmission axis of each pixel. The number of pixels and pixel pitch of the image sensor are 1164(H) × 874(V) and 4.65 × 4.65 µm, respectively.

Parallel two-step phase-shifting digital holography was experimentally demonstrated by the constructed system. Figure 1.7 shows the objects: an origami crane and a die, located 470 and 600 mm away from the image sensor plane, respectively. Figures 1.8(a) and (b)

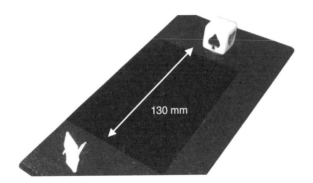

Figure 1.7 Objects used in the parallel phase-shifting digital holography system experiment using a normal-speed polarization-imaging camera. An origami crane and a die were located 470 and 600 mm away from the image sensor

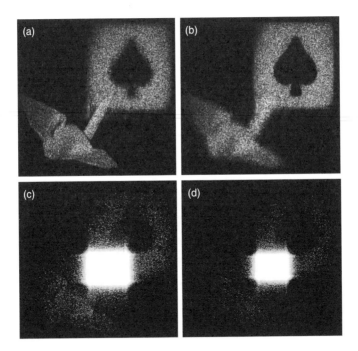

Figure 1.8 Reconstructed images. Images (a) and (b) are those reconstructed by the parallel phase-shifting digital holography system at the positions 470 and 600 mm away, respectively. Images (c) and (d) were reconstructed by the diffraction integral alone at the positions 470 and 600 mm away, respectively

show the images reconstructed by the parallel phase-shifting digital system at those positions. The in-focus origami crane was clearly reconstructed; the spot of the die was defocused and blurred in Fig. 1.8(a), *vice versa* in Fig. 1.8(b). Thus, the 3D imaging capability of the constructed system was experimentally confirmed. Focused images of the objects were also reconstructed from the same hologram by conventional in-line digital holography that does not use phase-shifting digital holography but the diffraction integral alone for comparison, and is shown in Figs. 1.8(c) and (d). The focused images reconstructed by just the diffraction integral are degraded because of superposition of the zeroth-order diffraction and the conjugate images. Thus, it has been experimentally demonstrated that the zeroth-order diffraction and the conjugate images were successfully removed from the reconstructed image by the constructed parallel two-step phase-shifting digital holography system.

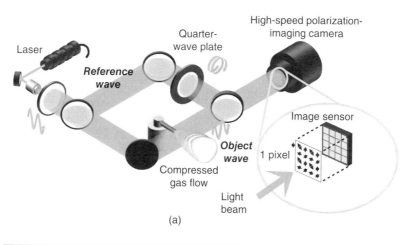

(a)

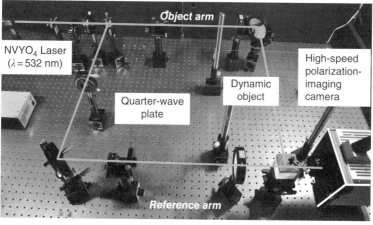

(b)

Figure 1.9 Parallel phase-shifting digital holography system using a high-speed polarization-imaging camera. (a) Schematic diagram, (b) photograph

1.6   High-Speed Parallel Phase-Shifting Digital Holography System

To demonstrate the high-speed 3D imaging capability of parallel phase-shifting digital holography, a high-speed parallel phase-shifting digital holography system was constructed [24,25]. A schematic diagram and a photograph of the system are shown in Fig. 1.9. This system consisted of a Mach–Zehnder interferometer and a high-speed polarization-imaging camera. Dynamic objects or fast phenomena are set in the path of the reference wave. A Photoron FASTCAM-SA5-P was used as the high-speed polarization-imaging camera. The pixel pitch of the camera was $20\,\mu m$. In general, the available number of pixels in a high-speed camera is approximately inversely proportional to the frame rate. As typical of the high-speed polarization-imaging camera used, 1024×1024 pixels, 512×512 pixels, 128×128 pixels, and 64×64 pixels were available at the rate of 7000 frames per second (fps), 15 000 fps, 150 000 fps, and 300 000 fps, respectively. A Nd:YVO$_4$ laser operated at 532 nm was used as the optical source.

To demonstrate the imaging capability of the dynamic phase of the constructed system, compressed gas flow sprayed from a nozzle was set for a dynamic object shown in Fig. 1.10. The nozzle was positioned at 19 cm away from the high-speed polarization-imaging camera. The inner diameter of the nozzle was 1 mm. Holograms at a rate of 20 000 fps were captured when the number of the pixels in the holograms was 512×512. Figure 1.11 shows the phase images reconstructed from the recorded holograms. The pixel values in the phase images were normalized in the range of 0–255. The pixel value of 255 represents a phase of 2π. The images in Fig. 1.11 were obtained at $t = 0$, 10, 15, 20, 65, 80, 85, 90, 95, and 100 ms (a–j, respectively). The heads of the two nozzles are positioned in the right and left parts in each image.

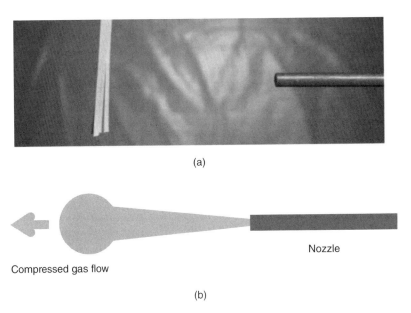

(a)

Nozzle

Compressed gas flow

(b)

Figure 1.10   Object used in the parallel phase-shifting digital holography system experiment using a high-speed polarization-imaging camera. Compressed gas flow sprayed from a nozzle. (a) Photograph, (b) schematic

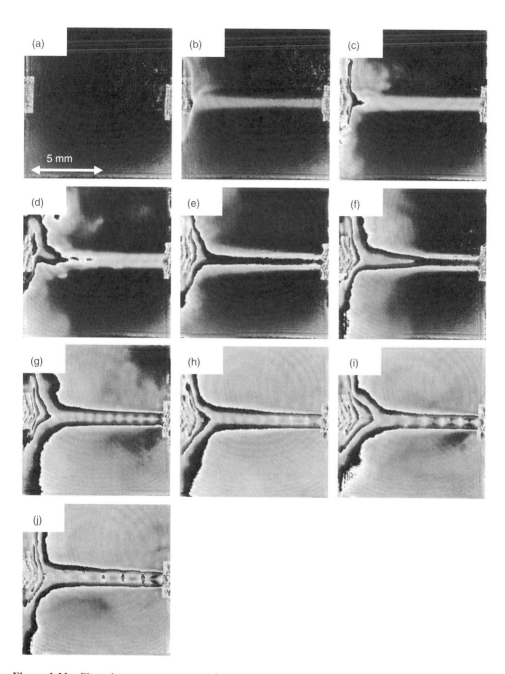

Figure 1.11 Phase images reconstructed from the recorded holograms at a frame rate of 20 000 fps. $t = 0, 10, 15, 20, 65, 80, 85, 90, 95$, and 100 ms (a–j respectively)

The abrupt transitions from white to black in each image were caused by phase wrapping from 2π to 0. Because the zeroth-order diffraction image and the conjugate image were eliminated by the use of phase-shifting digital holography, a clear and high-speed phase motion picture was obtained. First, the phase gradually increased as the flow rate of the compressed gas increased. Next, the phase increased from the opposite side of the sprayed nozzle, as shown in Fig. 1.11(f). After that, the phase of the background was changed by the gas, which was reflected by the left nozzle. Also, interesting spatially-periodic phase distributions appeared in the flow.

Figure 1.12 shows the phase images reconstructed from the holograms recorded at 180 000 fps. First, a mass of the compressed gas was sprayed from the nozzle and the gas flow was laminar. Next, the flow changed to turbulent as shown in Fig. 1.12(h). After that, the turbulent flow met the laminar one as shown in Fig. 1.12(i). Then, vortex-like phase distributions were observed in Fig. 1.12(j). Thus, a high-speed phase picture of a dynamic object was experimentally demonstrated by high-speed parallel parallel-phase shifting digital holography.

1.7 Single-Shot Femtosecond-Pulsed Parallel Phase-Shifting Digital Holography System

To demonstrate the ultrafast 3D imaging capability of parallel phase-shifting digital holography, a system was constructed [26]. A schematic diagram and a photograph are shown in Fig. 1.13 (Plate 1). The system contained a Mach–Zehnder interferometer and a polarization-imaging camera. The ultrafast phenomenon was generated in the path of the reference wave. The polarization-imaging camera was almost the same as that used in the first demonstration system of parallel phase-shifting digital holography, but this one was optimized for a light wave with a 800 nm wavelength. A mode-locked Ti:sapphire laser with a regeneration amplifier (Solstice, Spectra-Physics Inc.) was used as the optical source generating a single-shot femtosecond light pulse. The center wavelength and duration of the light pulse were 800 nm and 96 fs, respectively.

Two fine electrodes made of stainless steel were set facing each other in the reference wave path. The diameter of the two electrodes and the distance between them were 1.2 and 1.8 mm, respectively. The electrodes were positioned 31 cm away from the camera, 10 kV was applied to the electrodes and a spark discharge was induced between them. The spark discharge in the air at 1 atm was recorded. Figure 1.14 shows a photograph of the spark discharge. Figures 1.15(a) and (b) (also Plate 2) show the reconstructed phase images with and without the phase-shifting method, respectively. The reconstructed images were represented by pseudocolors of 256 gradations, and the relations between pixel (or phase) values and colors are shown by the color bar in Fig. 1.15 (Plate 2). The phase distribution between the two electrodes changed by the spark discharge is clearly reconstructed in Fig. 1.15(a). On the other hand, the image in Fig. 1.15(b) is significantly degraded and the phase changes are unclear because the zeroth-order diffraction image and the conjugate image were superimposed on the desired image of the object. Therefore, the effectiveness of the parallel phase-shifting digital holography system using a single femtosecond light pulse is confirmed. Thus, an ultrafast phase image of a dynamic object has been experimentally demonstrated by parallel parallel-phase shifting digital holography using a single-shot femtosecond-pulsed laser.

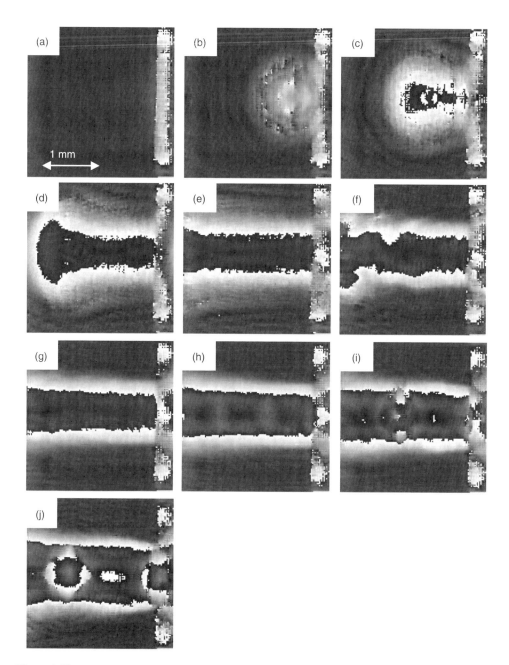

Figure 1.12 Phase images reconstructed from the recorded holograms at a frame rate of 180 000 fps. $t = 0, 3.2, 4.0, 4.8, 5.6, 24, 67, 87, 95,$ and 120 ms (a–j respectively)

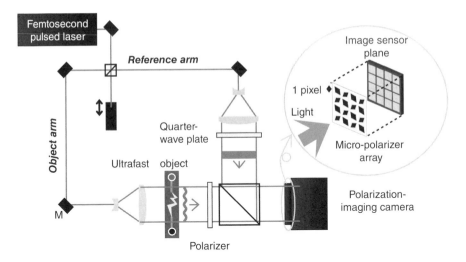

Figure 1.13 (Plate 1)   Schematic diagram of parallel phase-shifting digital holography system using a femtosecond pulsed laser. *See plate section for the color version*

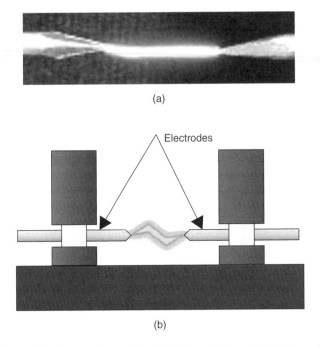

Figure 1.14   Object used in the experiment of parallel phase-shifting digital holography system using a high-speed polarization-imaging camera. Spark discharge between two fine electrodes. (a) Photograph, (b) schematic

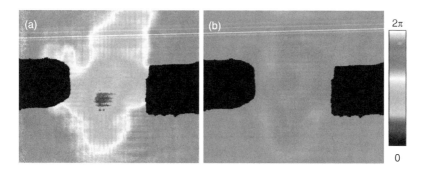

Figure 1.15 (Plate 2) Reconstructed images. (a) By parallel phase-shifting digital holography, (b) by a diffraction integral alone. *See plate section for the color version*

1.8 Portable Parallel Phase-Shifting Digital Holography System

For practical use of a parallel phase-shifting digital holography system, it is necessary to make the system portable by miniaturization. To demonstrate miniaturization, a portable parallel phase-shifting digital holography system was constructed [27]. Figure 1.16 shows a schematic diagram and a photograph of the constructed portable system.

A diode-pumped solid-state (DPSS) laser, which was developed for laser pointers, operated at 532 nm was used as the optical source because the laser is compact and the image sensor used is sensitive to green light. The camera in the portable system was the same as that used in the first demonstration system of parallel phase-shifting digital holography [23]. Optical components were bolted on to an aluminum board of a thickness of 3 mm in consideration of strength and weight. The size of the board was 400 × 200 mm. A black acrylic board 5 mm in thickness was used for the chassis. The size and weight of the constructed system were 450 (L) × 250 (W) × 200 mm (H) and 7 kg, respectively.

To verify the validity of the portable system, single-shot phase-shifting digital holography was experimentally demonstrated. Figure 1.17 shows the photograph of the objects. A bead of 5 mm diameter and a die of 7 mm at the sides were located 200 and 240 mm away from the image sensor, respectively. Figures 1.18(a) and (b) show the image reconstructed at the position from 200 mm and that from 240 mm by the constructed system, respectively. For comparison, the reconstructed image from a single hologram by the diffraction integral alone is shown in Fig. 1.18(c). As seen in Fig. 1.18, the zeroth-order diffraction wave and the conjugate image, which were not removed by the conventional in-line digital holography using diffraction integral alone, were successfully removed by the portable system. When the dot of the die was in focus the edge of the hole of the bead was not, and *vice versa*. Thus, the capability of single-shot phase-shifting interferometry has been experimentally verified by this portable system.

1.9 Functional Extension of Parallel Phase-Shifting Digital Holography

Several functional extensions have been proposed in parallel phase-shifting digital holography [28–35].

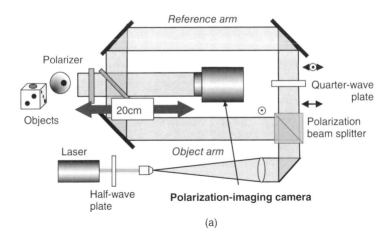

(a)

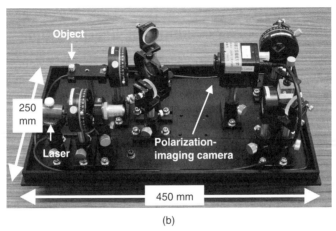

(b)

Figure 1.16 Schematic diagram of the portable parallel phase-shifting digital holography system. (a) Schematic, (b) photograph

1.9.1   Parallel Phase-Shifting Digital Holography Using Multiple Wavelengths

When three laser beams of red, green, and blue light (corresponding to the three primary colors) are simultaneously used in parallel phase-shifting digital holography, 3D structure and color motion picture measurement for dynamic objects are possible [28,29]. Also parallel phase-shifting digital holography using visible and near infrared laser beams has been reported for 3D motion picture measurement of visible and invisible information about a dynamic object. The simultaneous measurement of visible and invisible 3D information is valuable for simultaneous measurement of both inside and out of a living specimen [30]. If multiple wavelength laser beams are used in parallel phase-shifting digital holography, multi-spectral and 3D motion picture measurement are possible. This multiple-spectral and 3D motion picture enables us to measure the dynamics of 3D structures and analyze the chemical function of a dynamic object simultaneously.

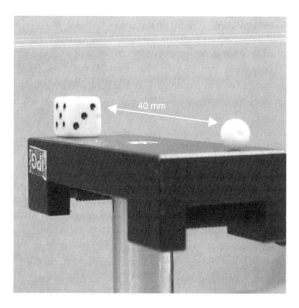

Figure 1.17   Photograph of the objects used in the experiment of the portable parallel phase-shifting digital holography system; A 5 mm diameter bead and a 7 mm sided die were located at the positions 200 and 240 mm away from the image sensor, respectively

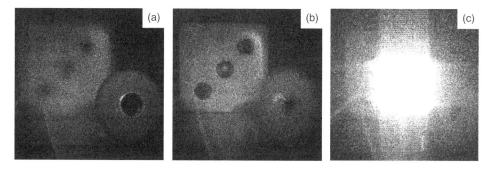

Figure 1.18   Reconstructed images. Images (a) and (b) were reconstructed by the portable parallel phase-shifting digital holography system. Images (a) and (b) were focused at the positions 200 and 240 mm away from the image sensor, respectively. Image (c) was reconstructed by diffraction integral alone and was focused at the position 200 mm away from the image sensor

1.9.2   *Parallel Phase-Shifting Digital Holography Using Multiple Polarized Light*

When two linearly-polarized laser beams, which are orthogonal to each other and of the same wavelength, are used in parallel phase-shifting digital holography, two phase images corresponding with the linear polarizations can be acquired [31]. By using the two phase-image and phase-unwrapping techniques, the measurement range can be extended several thousand times in depth compared to that using only single polarization and a single wavelength of light.

By using linearly-polarized laser beams orthogonal to each other and of the same wavelength (used in an optical system of parallel phase-shifting digital holography), two parallel phase-shifting digital holography processes are simultaneously carried out [32]. This technique enables us to simultaneously measure 3D structure and 3D distribution of polarization characteristics of a dynamic object, and contributes to the simultaneous measurement of 3D structure and mechanical properties such as stress distribution and analysis of chemical composition of that object.

1.9.3   Parallel Phase-Shifting Digital Holographic Microscope

For 3D motion picture measurement of dynamic and micro objects, parallel phase-shifting digital holographic microscopy has been proposed [33–35]. A parallel phase-shifting digital holographic microscope was constructed by introducing optical microscope objectives into a parallel digital holography system, and was experimentally demonstrated by using the same polarization-imaging camera used in the first demonstration system [33]. A parallel phase-shifting digital holographic color microscope was also proposed [34]. The color microscope simultaneously uses three lasers emitting red, green, and blue light. After this, a high-speed parallel phase-shifting digital holographic microscope was reported. The high-speed microscope consisted of an interferometer and the high-speed polarization-imaging camera used in the high-speed parallel phase-shifting digital holography system [35]. A 3D motion picture recording of a living specimen was achieved at up to 150 000 fps by the high-speed microscope.

1.10   Prospects and Conclusion

Parallel phase-shifting digital holography has been introduced as a technique for 3D motion picture measurement. This technique is not only capable of high-precision 3D motion picture measurement but also one of robust interferometry against disturbance. A phase motion picture of up to 262 500 fps and a 96-fs temporal resolution phase-image were achieved by parallel phase-shifting digital holography systems. The maximum recording speed of a 3D motion picture and the temporal resolution of the 3D image are determined by the frame rate of the high-speed camera and the pulse duration of the laser beam used in parallel phase-shifting digital holography system, respectively. This technique will contribute to many fields requiring 3D measurement of dynamic objects such as fluidics, particle measurement, stress measurement, displacement and deformation measurement, biological microscopes, flow cytometry, evaluation of micro-machines, mechanical characterization of materials, production inspections, and others.

Acknowledgments

The author thanks Professor Emeritus Toshihiro Kubota, Professor Shogo Ura, Mr Kenzo Nishio, Mr Peng Xia, and Mr Motofumi Fujii at Kyoto Institute of Technology, Dr Tatsuki

Tahara at Kansai University, Dr Takashi Kakue at Chiba University, and Professor Osamu Matoba for their technical support and helpful discussions.

This study was partly supported by the Industrial Technology Research Grant Program from New Energy and Industrial Technology Development Organization (NEDO) of Japan and by the Funding Program for Next Generation World-Leading Researchers GR064 from the Japan Society for the Promotion of Science (JSPS).

References

[1] Gabor, D., (1948) 'A new microscopic principle'. *Nature (London)* **161**, 777–778.

[2] Goodman, J. W. and Lawrence, R. W. (1967) 'Digital image formation from electronically detected holograms'. *Appl. Phys. Lett.* **11**, 77–79.

[3] Kronrod, M. A., Merzlyakov, N. S., and Yaroslavskii, L. P. (1972) 'Reconstruction of a hologram with a computer'. *Sov. Phys. Tech. Phys.* **17**, 333–334.

[4] Yamaguchi, I. and Zhang, T. (1997) 'Phase-shifting digital holography'. *Opt. Lett.* **22**, 1268–1270.

[5] Sasada, M., Awatsuji, Y., and Kubota, T. (2004) 'Parallel quasi-phase-shifting digital holography that can achieve instantaneous measurement'. *Technical Digest of the 2004 ICO International Conference: Optics and Photonics in Technology Frontier (International Commission for Optics, 2004)*, 187–188.

[6] Sasada, M., Fujii, A., Awatsuji, Y., and Kubota, T. (2004) 'Parallel quasi-phase-shifting digital holography implemented by simple optical set up and effective use of image-sensor pixels'. *Technical Digest of the 2004 ICO International Conference: Optics and Photonics in Technology Frontier (International Commission for Optics, 2004)* 357–358.

[7] Awatsuji, Y., Sasada, M., and Kubota, T. (2004) 'Parallel quasi-phase-shifting digital holography'. *Appl. Phys. Lett.* **85**, 1069–1071.

[8] Awatsuji, Y., Sasada, M., Fujii, A., and Kubota, T. (2006) 'Scheme to improve the reconstructed image in parallel quasi-phase-shifting digital holography'. *Appl. Opt.* **45**, 968–974.

[9] Awatsuji, Y., Fujii, A., Kubota, T., and Matoba, O. (2006) 'Parallel three-step phase-shifting digital holography'. *Appl. Opt.* **45**, 2995–3002.

[10] Awatsuji, Y., Tahara, T., Kaneko, A., Koyama, T., Nishio, K., Ura, S., *et al.* (2008) 'Parallel two-step phase-shifting digital holography'. *Appl. Opt.* **47**, D183D189.

[11] Kakue, T., Moritani, Y., Ito, K., Shimozato, Y., Awatsuji,Y., Nishio, K., *et al.* (2010) 'Image quality improvement of parallel four-step phase-shifting digital holography by using the algorithm of parallel two-step phase-shifting digital holography'. *Opt. Express* **18**, 9555–9560.

[12] Tahara, T., Shimozato, Y., Kakue, T., Fujii, M., Xia, P., Awatsuji, Y., *et al.* (2012) 'Comparative evaluation of the image-reconstruction algorithms of single-shot phase-shifting digital holography' *J. Electron. Imaging* **21**, 013021.

[13] Tahara, T., Awatsuji, Y., Kaneko, A., Koyama, T., Nishio, K., Ura, S., *et al.* (2010) 'Parallel two-step phase-shifting digital holography using polarization'. *Opt. Rev.* **17**, 108–113.

[14] Tahara, T., Ito, K., Kakue, T., Fujii, M., Shimozato, Y., Awatsuji, Y., *et al.* (2011) 'Compensation algorithm for the phase-shift error of polarization-based parallel two-step phase-shifting digital holography'. *Appl. Opt.* **50**, B31–B37.

[15] Tahara, T., Awatsuji, Y., Nishio, K., Ura, S., Kubota, T., and Matoba, O. (2010) 'Comparative analysis and quantitative evaluation of the field of view and the viewing zone of single-shot phase-shifting digital holography using space-division multiplexing'. *Opt. Rev.* **17**, 519–524.

[16] Xia, P., Tahara, T., Fujii, M. Kakue, T., Awatsuji,Y., Nishio, K., *et al*. (2011) 'Removing the residual zeroth-order diffraction wave in polarization-based parallel phase-shifting digital holography system'. *Appl. Phys. Express* **4**, 072501.

[17] Lin, M., Nitta, K., Matoba, O., and Awatsuji, Y. (2012) 'Parallel phase-shifting digital holography with adaptive function using phase-mode spatial light modulator'. *Appl. Opt.* **51**, 2633–2637.

[18] Tahara, T., Shimozato, Y., Xia, P., Ito, Y., Awatsuji, Y., Nishio, K., *et al*. (2012) 'Algorithm for reconstructing wide space bandwidth information in parallel two-step phase-shifting digital holography'. *Opt. Express* **20**, 19806–19814.

[19] Miao, L., Nitta, K., Matoba, O., and Awatsuji, Y. (2013) 'Assessment of weak light condition in parallel four-step phase-shifting digital holography'. *Appl. Opt.* **52**, A131–A135.

[20] Xia, P., Shimozato, Y., Tahara, T., Kakue, T., Awatsuji, Y., Nishio, K., *et al*. (2013) 'Image reconstruction algorithm for recovering high-frequency information in parallel phase-shifting digital holography' *Appl. Opt.* **52**, A210–A215.

[21] Tahara, T., Shimozato, Y., Xia, P. Ito, Y., Kakue, T., Awatsuji, Y., *et al*. (2013) 'Removal of residual images in parallel phase-shifting digital holography'. *Opt. Rev.* **20**, 7–12.

[22] Xia, P., Tahara, T., Kakue, T., Awatsuji, Y., Nishio, K., Ura, S., *et al*. (2013) 'Performance comparison of bilinear interpolation, bicubic interpolation, and B-spline interpolation in parallel phase-shifting digital holography'. *Opt. Rev.* **20**, in press.

[23] Tahara, T., Ito, K., Fujii, M., Kakue, T., Shimozato, T., Awatsuji, Y., *et al*. (2010) 'Experimental demonstration of parallel two-step phase-shifting digital holography'. *Opt. Express* **18**, 18975–18980.

[24] Kakue, T., Yonesaka, R., Tahara, T., Awatsuji, Y., Nishio, K., Ura, S., *et al*. (2011) 'High-speed phase imaging by parallel phase-shifting digital holography'. *Opt. Lett.* **36**, 4131–4133.

[25] Kakue, T., Fujii, M., Shimozato, Y., Tahara, T., Awatsuji, Y., Ura, S., *et al*. (2011) '262500-frames-per-second phase-shifting digital holography' *2011 OSA Topical Meeting and Exhibit, Digital Holography and Three-Dimensional Imaging (DH) Technical Digest*, DWC25.

[26] Kakue, T., Itoh, S., Xia, P., Tahara, T., Awatsuji, Y., Nishio, K., *et al*. (2012) 'Single-shot femtosecond-pulsed phase-shifting digital holography'. *Opt. Express* **20**, 20286–20291.

[27] Fujii, M., Kakue, T., Ito, K., Tahara, T., Shimozato, T., *et al*. (2011) 'Construction of a portable parallel phase-shifting digital holography system' *Opt. Eng.* **50**, 091304.

[28] Awatsuji, Y., Koyama, T., Kaneko, A., Fujii, A., Nishio, K., Ura, S., and Kubota, T. (2007) 'Single-shot phase-shifting color digital holography' *2007 IEEE LEOS Annual Meeting (LEOS 2007) Conference Proceedings*, 84–85.

[29] Kakue, T., Tahara, T., Ito, K., Shimozato, Y., Awatsuji, Y., Nishio, K., *et al*. (2009) 'Parallel phase-shifting color digital holography using two phase shifts'. *Appl. Opt.* **48**, H244–H250.

[30] Kakue, T., Ito, K., Tahara, T., Awatsuji, Y., Nishio, K., Ura, S., *et al*. (2010) 'Parallel phase-shifting digital holography capable of simultaneously capturing visible and invisible three-dimensional information'. *J. Display Technol.* **6**, 472–478.

[31] Tahara, T., Maeda, A., Awatsuji, Y., Nishio, K., Ura, S., Kubota, T., and Matoba, O. (2012) 'Parallel phase-shifting dual-illumination phase unwrapping'. *Opt. Rev.* **19**, 366–370.

[32] Tahara, T., Awatsuji, Y., Shimozato, Y., Kakue, T., Nishio, K., Ura, S., *et al*. (2011) 'Single-shot polarization-imaging digital holography based on simultaneous phase-shifting interferometry'. *Opt. Lett.* **36**, 3254–3256.

[33] Tahara, T., Ito, K., Kakue, T., Fujii, M., Shimozato, Y., Awatsuji, Y., *et al*. (2010) 'Parallel phase-shifting digital holographic microscopy'. *Biomed. Opt. Express* **1**, 610–616.

[34] Tahara, T., Kakue, T., Awatsuji, Y., Nishio, K., Ura, S., Kubota, T., and Matoba, O. (2010) 'Parallel phase-shifting color digital holographic microscopy'. *3D Res.* **1**, 04–05.

[35] Tahara, T., Yonesaka, R., Yamamoto, S., Kakue, T., Xia, P. Awatsuji, Y., *et al.* (2012) 'High-speed three-dimensional microscope for dynamically moving biological objects based on parallel phase-shifting digital holographic microscopy'. *IEEE J. Sel. Topics Quantum Electron.* **18**, 1387–1393.

[36] Meng, X. F., Cai, L. Z., Xu, X. F., Yang, X. L., Shen, X. X., Dong, G. Y., and Wang, Y. R. (2006) 'Two-step phase-shifting interferometry and its application in image encryption'. *Opt. Lett.* **31**, 1414–1416.

2

Imaging and Display of Human Size Scenes by Long Wavelength Digital Holography

Massimiliano Locatelli[1], Eugenio Pugliese[1], Melania Paturzo[2], Vittorio Bianco[2], Andrea Finizio[2], Anna Pelagotti[1], Pasquale Poggi[1], Lisa Miccio[2], Riccardo Meucci[1] and Pietro Ferraro[2]

[1]*CNR, Istituto Nazionale di Ottica, Largo E. Fermi, Italy*
[2]*CNR, Istituto Nazionale di Ottica, Sezione di Napoli, Italy*

2.1 Introduction

This chapter is devoted to the description of *Digital Holography* (DH) [1–4] in the infrared region. As we will see, this portion of the electromagnetic spectrum offers interesting and useful opportunities for imaging purposes from various points of view. *Infrared Radiation Digital Holography* (IRDH) is an almost unexplored research field. Recently, new breakthrough potentialities of IRDH have been demonstrated, allowing us to overcome some of the most binding conditions of DH at visible wavelengths. Hence, a completely new class of products can be provided, taking advantage of the longer wavelength employed, thus involving out-of-lab applications to be exploited in the field of safety. The next sections will be devoted to showing such capabilities.

2.2 Digital Holography Principles

A hologram is the recorded interference pattern between two waves emitted by a coherent source. The wavefront impinging on the target of interest, which is scattered toward the recording device, is commonly referred to as the *object beam*, whereas the wave directly reaching the detector is the *reference beam*. Holography is a two-step process, the first step aims to record

Multi-dimensional Imaging, First Edition. Edited by Bahram Javidi, Enrique Tajahuerce and Pedro Andrés.
© 2014 John Wiley & Sons, Ltd. Published 2014 by John Wiley & Sons, Ltd.

the hologram, and a second step whose purpose is to reconstruct the object wavefront or, as is usually said, to reconstruct the hologram [1].

In principle, any portion of the recording device can collect the information coming from each portion of the object, regardless of its size. In this configuration it is possible to use any portion, however small, of the hologram to reconstruct the entire wavefront, even though, as we will see, the smaller the portion of the hologram, the lower the final reconstruction resolution will be. In a typical holographic configuration the object beam and the reference beam travel different paths but, in order to obtain interference fringes, they must be coherent with each other and therefore, they are usually extracted from a single laser beam. For the same reason the corresponding optical paths cannot differ more than the laser coherence length. Furthermore, as in every interferometry experiment, in order to get good interference fringe visibility, the two beams should have comparable intensities and the differences between the two optical paths should be kept stable, within a fraction of the wavelength, during the time required to perform hologram recording. Finally, in order to obtain an undistorted reconstructed object wavefront, the reference beam amplitude has to be uniform across the recording device surface; strictly speaking every homogeneous reference wave could therefore be used. Usually, however, only plane waves or large curvature spherical waves are employed.

From the point of view of theory and purpose, DH is directly derived from analog holography but differs from it for the hologram recording medium and for the wavefront reconstruction method [2–10]. In DH the recording medium is a digital device, typically a CCD or a CMOS device. The interference pattern, containing the information about the wavefront to be studied, is sampled and digitized by the electronics of the device and stored in the memory of a computer as a matrix of numbers [11–19]. Hence, DH takes advantage of the joint capabilities of coherent imaging and numerical processing allowing us to obtain new products, including both qualitative imaging and quantitative phase microscopy. In this framework the capability of DH at visible wavelengths to see clearly through turbid fluids has been recently demonstrated (e.g., blood), paving the way for deep exploitation of this technique in industrial and biomedical research fields [20–22].

Current electronic recording devices, however, have a much smaller spatial resolution with respect to old photographic plates: the resolution attainable with a classic support can reach a maximum of about 7000 $cycles$/mm (where cycle means a pair of white/black fringes) while a CCD with square pixels of lateral dimensions greater than 5 μm (a typical values in CCD cameras) can solve a maximum of 100 $cycles$/mm. This limitation is particularly restrictive when working in off-axis holography where the maximum recordable number of $cycles$/mm directly imposes a limit on the angle θ between the reference beam and the normal to the sensor. If we refer, in particular, to speckle holography in optimal recording conditions, the recording device should be able to solve entirely the interference pattern resulting from the superposition of the reference wave with the waves spread by all the different points constituting the object under investigation. The maximum value of the angle θ between the reference and the object beam in off-axis configurations is thus related to the lateral dimension of the object, to the lateral dimension of the detector, and to the distance between them. If, for simplicity, we refer to the simplified bi-dimensional configuration sketched in Fig. 2.1, we can quantify these relations by means of simple geometrical considerations.

If we want all the object point sources to contribute to the interference pattern across the entire detector, it is necessary that the interference fringes between the reference beam and the highest spatial frequency component of the object wave can be recorded by the detector; if

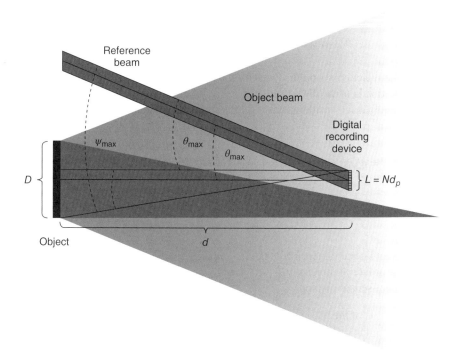

Figure 2.1 Maximum θ angle value according to the sampling condition. *Source:* M. Locatelli, E. Pugliese, M. Paturzo, V. Bianco, A. Finizio, A. Pelagotti, P. Poggi, L. Miccio, R. Meucci, and P. Ferraro 2013. Reproduced with permission from the Optical Society

we indicate with φ the angle formed by the highest spatial frequency originating in the object with respect to the detector axis, we observe [3], across the sensor, a sinusoidal interference fringe pattern between this highest spatial frequency and the reference beam with period

$$P = \frac{\lambda}{\sin\theta + \sin\varphi} = \frac{\lambda}{2\sin\left(\dfrac{\theta+\varphi}{2}\right)\cos\left(\dfrac{\theta-\varphi}{2}\right)}, \qquad (2.1)$$

which, for $\theta \cong \varphi$ and $\theta + \varphi = \psi$, becomes

$$P \cong \frac{\lambda}{2\sin\left(\dfrac{\psi}{2}\right)}, \qquad (2.2)$$

where λ is the employed wavelength. The maximum angle, ψ_{max}, allowed by Whittaker–Shannon sampling theorem is reached when the value of the fringe period reaches its minimum admitted value, P_{min}, equal to two times the detector pixel's lateral dimension d_p

$$P_{min} = \frac{\lambda}{2\sin\dfrac{\psi_{max}}{2}} = 2d_p \qquad (2.3)$$

the maximum angle ψ_{max} thus results as

$$\psi_{max} = 2\sin^{-1}\left(\frac{\lambda}{4d_p}\right) \overset{small\ angle}{\approx} \frac{\lambda}{2d_p}. \tag{2.4}$$

Considered the typical pixel pitch values for most used wavelengths, it follows that the small angle approximation is usually well satisfied.

With simple geometric considerations it is thus simple to calculate the maximum value θ_{max}, that is

$$\theta_{max} = \psi_{max} - \varphi = 2\sin^{-1}\left(\frac{\lambda}{4d_p}\right) - \tan^{-1}\left(\frac{\frac{D}{2} + \frac{L}{2}}{d}\right) \overset{small\ angle}{\approx} \frac{\lambda}{2d_p}$$

$$- \frac{D+L}{2d} \overset{D \gg L}{\approx} \frac{\lambda}{2d_p} - \frac{D}{2d}, \tag{2.5}$$

where
D = is the object lateral dimension
L = is the sensor lateral dimension
d = is the object-sensor distance.

The reconstruction of the wavefront is performed numerically by means of appropriate algorithms reproducing the diffraction process operated on the reconstruction beam by the optical transmittance constituting the hologram; it is thus possible to derive the desired information on the wavefront under investigation, both in amplitude and phase, and to reconstruct the wavefront in a digital version too [21–23].

To obtain numerically the analytical expression of such a wavefront we can exploit the Rayleigh–Sommerfeld formula [24] that, by placing the obliquity factor equal to 1 [3], can be written in this case as

$$\mathcal{E}(x_R, y_R) = \frac{1}{i\lambda} \iint_{-\infty}^{+\infty} \mathcal{R}(x_H, y_H) H(x_H, y_H) \frac{e^{i\frac{2\pi}{\lambda}\rho}}{\rho} dx_H dy_H, \tag{2.6}$$

where (Fig. 2.2)

$\mathcal{E}(x_R, y_R)$ is the complex wavefront in the reconstruction plane (X_R, Y_R)
$H(x_H, y_H)$ is the intensity of the interferogram in the hologram plane (X_H, Y_H)
$\mathcal{R}(x_H, y_H)$ is the wavefront reconstruction beam in the hologram plane
$\rho = \sqrt{d^2 + (x_R - x_H)^2 + (y_R - y_H)^2}$ is the distance between the generic point of the recording plane and the generic point of the reconstruction plane
d is the distance between the recording plane and the reconstruction plane.

From $\mathcal{E}(x_R, y_R)$ it is possible to extract the intensity value $I(x_R, y_R)$ and the phase value $\phi(x_R, y_R)$ of the object wavefront in the reconstruction plane

$$I(x_R, y_R) = \mathcal{E}(x_R, y_R)\, \mathcal{E}^*(x_R, y_R) \tag{2.7}$$

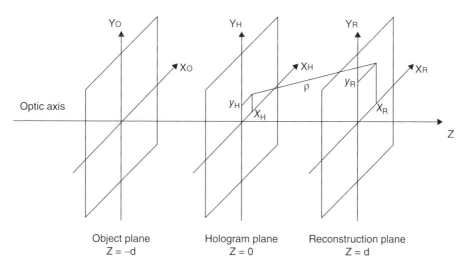

Figure 2.2   Object plane, hologram plane, and reconstruction plane. *Source:* Pelagotti A., Locatelli M., Geltrude A., Poggi P., Meucci R., Paturzo M., Miccio L., Ferraro P. 2010. Reproduced with permission from Springer

and

$$\phi(x_R, y_R) = \tan^{-1}\left(\frac{Im\left\{\mathcal{E}\left(x_R, y_R\right)\right\}}{Re\{\mathcal{E}(x_R, y_R)\}}\right). \tag{2.8}$$

A common way to operate this transformation, that is, to simulate the wavefront propagation from the recording plane to the reconstruction plane, is to follow the so-called *Fresnel method*.

2.2.1   Fresnel Method

For values of x_R, y_R, x_H, y_H small compared to the distance d between the reconstruction plane and the hologram plane, the expression of ρ can be approximated with its Taylor development

$$\rho \approx d + \frac{\left(x_R - x_H\right)^2}{2d} + \frac{\left(y_R - y_H\right)^2}{2d} - \frac{\left[\left(x_R - x_H\right)^2 + \left(y_R - y_H\right)^2\right]^2}{8d^3} + \cdots \tag{2.9}$$

The fourth term of this expression can be neglected if it is small compared to the wavelength [6], that is, if

$$\frac{\left[\left(x_R - x_H\right)^2 + \left(y_R - y_H\right)^2\right]^2}{8d^3} \ll \lambda \quad \rightarrow \quad d \gg \sqrt[3]{\frac{\left[\left(x_R - x_H\right)^2 + \left(y_R - y_H\right)^2\right]^2}{8\lambda}}. \tag{2.10}$$

Therefore, if we use the expression of the development up to the first order for the numerator (most critical factor) and up to the zero order term for the denominator (less critical factor) [4], we obtain

$$
\mathcal{E}\left(x_R, y_R\right) = \frac{e^{i\frac{2\pi}{\lambda}d}}{i\lambda d} e^{\frac{i\pi}{\lambda d}\left(x_R^2 + y_R^2\right)} \iint_{-\infty}^{+\infty} \mathcal{R}\left(x_H, y_H\right) H\left(x_H, y_H\right)
$$
$$
e^{\frac{i\pi}{\lambda d}\left(x_H^2 + y_H^2\right)} e^{\frac{i2\pi}{\lambda d}\left(-x_H x_R - y_H y_R\right)} dx_H dy_H. \tag{2.11}
$$

This equation is called the Fresnel approximation of the Rayleigh–Sommerfeld integral or *Fresnel transform*. If we now define the variables

$$
\mu = \frac{x_R}{\lambda d}, \qquad v = \frac{y_R}{\lambda d} \tag{2.12}
$$

the previous integral becomes

$$
\mathcal{E}\left(\mu, v\right) = \frac{e^{i\frac{2\pi}{\lambda}d}}{i\lambda d} e^{i\pi\lambda d\left(\mu^2 + v^2\right)} \iint_{-\infty}^{+\infty} \mathcal{R}\left(x_H, y_H\right) H\left(x_H, y_H\right)
$$
$$
e^{\frac{i\pi}{\lambda d}\left(x_H^2 + y_H^2\right)} e^{-i2\pi\left(x_H \mu + y_H v\right)} dx_H dy_H. \tag{2.13}
$$

The expression, in this way, has thus assumed, unless the multiplication factor out of the integral not depending on the variables x_H, y_H, the appearance of a two-dimensional Fourier transform and we can therefore write

$$
\mathcal{E}\left(\mu, v\right) = \frac{e^{i\frac{2\pi}{\lambda}d}}{i\lambda d} e^{i\pi\lambda d\left(\mu^2 + v^2\right)} \mathcal{F}\left\{ \mathcal{R}\left(x_H, y_H\right) H\left(x_H, y_H\right) e^{\frac{i\pi}{\lambda d}\left(x_H^2 + y_H^2\right)} \right\}, \tag{2.14}
$$

where $\mathcal{F}$ denotes the Fourier transform.

It should be recalled, at this point, that in digital holography, hologram recording is performed in the digital domain and $H(x_H, y_H)$ is therefore a discretized function; if we assume that the sensor is composed of a rectangular array of M by N pixels with spacing, along the axes X_H and Y_H, respectively equal to Δx_H and Δy_H, our hologram appears to be an array of numbers $H(k\Delta x_H, l\Delta y_H) = H(k, l)$ and the previous integrals are therefore to be transformed into discrete summations or, equivalently, the continuous Fourier transform has to be replaced by a discrete Fourier transform; the wavefront on the reconstruction plane is then, in turn, a discrete function $\mathcal{E}(m\Delta\mu, n\Delta v) = \mathcal{E}(m, n)$ of the discrete variables $m\Delta\mu, n\Delta v$; taking into account that the maximum spatial frequency is determined by the sampling range in the spatial domain and namely [3]

$$
M\Delta\mu = \frac{1}{\Delta x_H} \quad , \quad N\Delta v = \frac{1}{\Delta y_H}. \tag{2.15}
$$

It is possible to write

$$\mathcal{E}(m,n) = \frac{e^{i\frac{2\pi}{\lambda}d}}{i\lambda d} e^{i\pi\lambda d\left[\frac{m^2}{M^2\Delta x_H^2} + \frac{n^2}{N^2\Delta y_H^2}\right]} DF\left\{\mathcal{R}(k,l) H(k,l) e^{\frac{i\pi}{\lambda d}\left[(k\Delta x_H)^2 + (l\Delta y_H)^2\right]}\right\}, \quad (2.16)$$

where DF denotes the discrete Fourier transform.

Finally, it can be observed that, according to the Fourier transform relationship,

$$\Delta x_R = \frac{\lambda d}{M\Delta x_H}, \qquad \Delta y_R = \frac{\lambda d}{N\Delta y_H} \qquad (2.17)$$

and this means that the reconstructed wavefront in the plane (X_R, Y_R), is represented by a matrix consisting of $M \times N$ elements, each of which is called reconstruction pixel, with dimensions $\Delta x_R, \Delta y_R$. Suppose we have a square detector ($M = N$), with square pixels ($\Delta x_H = \Delta y_H = d_p$), we have

$$\Delta x_R = \Delta y_R = d_{pr} = \frac{\lambda d}{N d_p}. \qquad (2.18)$$

From these expressions it follows that the reconstruction resolution increases when the distance d decreases; unfortunately this distance cannot be reduced indefinitely because there will be a minimum value $d_{\min}$ for which the maximum angle $\theta_{\max}$ admitted by the sampling theorem coincides with the minimum value of the angle $\theta_{\min}$ required to have the diffraction orders separation; in this condition, sketched in Fig. 2.3, we thus have the maximum resolution for a certain object dimension; working at the minimum distance we thus have

$$\theta = \theta_{\max} = \theta_{\min} = \varphi = \frac{\psi_{\max}}{2} \qquad (2.19)$$

and, consequently

$$\frac{\lambda}{4d_p} = \frac{D+L}{2d_{\min}} \quad \rightarrow \quad d_{\min} = \frac{2d_p(D+L)}{\lambda} \overset{D\gg L}{\approx} \frac{2d_p D}{\lambda}. \qquad (2.20)$$

If we work at the minimum distance the reconstruction pixel pitch becomes

$$d_{pr} = \frac{2(D+L)}{N} = \frac{2D}{N} + 2d_p \overset{D\gg L}{\approx} \frac{2D}{N}. \qquad (2.21)$$

This means that, at the minimum distance allowed by the sampling theorem, the reconstruction pixel pitch does not depend on the wavelength used to irradiate the object and cannot be smaller than twice the detector pixel pitch. Furthermore, if a large sized object, with respect to the detector lateral dimension, is investigated the reconstruction resolution increases with the number of elements constituting the detector.

2.2.2   Advantages of Digital Holography

DH is characterized by a series of interesting features that make it more appealing with respect to its classical counterpart and to most standard imaging techniques. The key feature that differentiates DH from standard imaging techniques is the possibility to reconstruct both the

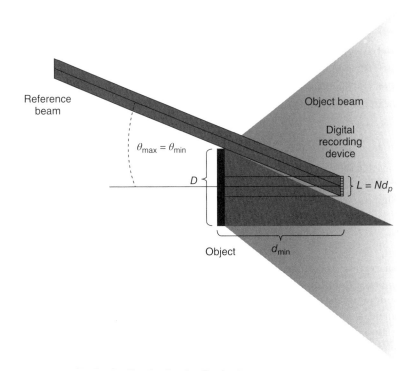

Figure 2.3 Minimum distance configuration. *Source:* Pelagotti A., Paturzo M., Geltrude A., Locatelli M., Meucci R., Poggi P., and Ferraro P., 2010. Reproduced with permission from Springer

amplitude images and the phase-contrast maps of the sample under investigation. Hence, DH allows us to perform quantitative phase microscopy of transparent samples with a label-free approach. The phase information, of great interest in a wide variety of applications, is obtained with a minimally invasive technique since the remarkable sensitivity, which characterizes modern recording digital devices allows us, if necessary, to use very low power laser radiation during the hologram recording process. Another interesting capability of DH is the possibility to perform, with a single acquisition, the reconstruction of the object image at different distances; this means that tunable focusing on particulars belonging to different planes that make up the scene can be obtained numerically during the reconstruction process starting from a single recorded hologram. Furthermore, as in classical holography, each portion of a hologram carries the information of the entire sample and can be used to retrieve it; this means that even if part of the sensor is covered during the recording process it is still possible to recover the information about the entire scene under investigation. The most important advantage of DH with respect to analog holography is that the digital hologram recording process is by far faster and easier than its analog counterpart: no delicate and time consuming photographic recording and development processes are required. Moreover, the times required for numerical reconstruction are, today, with modern computers, in the order of a second and so they do not represent a limitation in most applications.

2.3   Infrared Digital Holography

The idea behind holography is so general and rich in application that it is of interest in all regions of the electromagnetic spectrum. Obviously, the different wavelengths require different recording devices and the inhomogeneous technological development of the various acquisition devices has inevitably benefited certain regions of the spectrum with respect to others. In DH, for example, infrared (IR) sources were penalized, compared to visible sources, due to the critical issues related to the detection of this kind of radiation. However, increasing interest toward standard IR imaging techniques, firstly related to their military, industrial, and thermal efficiency applications has boosted the production of thermographic cameras and the availability of such IR cameras, based on the latest sensors generation, represents a significant stimulus to the development of DH techniques also in this region of the electromagnetic spectrum. This renewed interest could lead to innovative applications in security screening, night vision, and biological science, as it can extend holographic 3D imaging capabilities from visible light to the IR spectrum, up to the terahertz frequencies.

The advantages using long wavelength radiation to perform DH are related to some intrinsic characteristics of this technique. First of all, as we have seen in the previous paragraph, the radiation wavelength does not enter into the expression for the reconstruction pixel pitch when operating at the minimum distance admitted by the sampling theorem; the reason for this unexpected property resides in the fact that, according to the general expression of the reconstruction resolution, the longer the wavelength in use the larger the reconstruction pixel size should be (the worse the resolution) but, if we look at the expression of the maximum angle between object beam and reference beam, we see that a longer wavelength allows us to work at lower object-camera distances thus compensating for what was lost in terms of resolution. The reconstruction pixel pitch is indeed rather heavily dependent on the number and size of the elements constituting the recording device. In the past, infrared detectors were characterized by a number of sensitive elements much lower than the detectors in the visible range and, at the same time, the size of the sensing elements was much larger than those of typical CCDs. This deficiency in IR detection devices, combined with the ease of use of visible radiation, in the past prevented a significant development of DH techniques at these wavelengths. Today a typical high-level commercial uncooled mid-IR detector can have 1024×768 elements with a $20\,\mu m$ lateral dimension, a typical high-level commercial visible camera can have 2048×2048 elements with $6.5\,\mu m$ lateral dimension. With the expected progress of IR detection technologies, it is reasonable to assume that these values are going to be soon overcome so that the resolution gaps between visible detectors and IR detectors are intended to get smaller and smaller. Considering these typical sensors' technical values it is simple to evaluate advantages and disadvantages of *Infrared Digital Holography* (IRDH) with respect to visible DH: inserting the appropriate values in the equations for the minimum distance and for the reconstruction pixel pitch, for a fixed object dimension, D, it follows that IR radiation is the best choice if large objects are of interest, even if a slightly lower resolution is obtained. An object with 1-m lateral dimension, for example, must be at least 24 m away from a visible detector (when working at $0.632\,\mu m$) while it could stay at about 3 m if CO_2 radiation is used and even less than 0.5 m when working in the terahertz region at $100\,\mu m$. The reconstruction resolution instead maintains the same order in all cases.

There is also, however, an intrinsic advantage in using long IR radiation in DH, and more generally, in every interferometric experiment; that is, the inherent lower sensitivity to vibrations of long wavelength radiation. In fact, in the case of optical path variations due to seismic noise or any other vibration source, the longer the wavelength is that is used to create the interference pattern, the lower the phase variation between the reference and object beams. This benefit makes the stability of the object and of the measuring apparatus in general a less critical factor during hologram acquisition and is thus strongly indicated for DH, particularly in the investigation of large objects. IR radiation thus makes it possible to work in less restrictive conditions, giving much more versatility to the technique and opening the way to many possible out-of-laboratory applications. By means of long wavelength radiation it is possible to create video holograms (meaning holographic video of slowly changing dynamical scenes) without using a pulsed laser and the short acquisition times necessary when working with visible radiation. When analysis of very large sized samples is needed, CO_2 laser radiation is strongly recommended since the high output power that CO_2 lasers can reach is required to efficiently irradiate large object surfaces. Furthermore, the very high coherence length typical of CO_2 lasers represents an important advantage in every interferometric application. All these features make CO_2 lasers highly recommended in DH applications aiming to investigate large sized objects in out-of-laboratory applications.

Among the reasons that may encourage the use of IR in DH, there is also the transparency of various materials at certain wavelengths in the IR spectrum; this property could be exploited to study internal structures of such materials at different depths, to investigate heterogeneous materials or in security and safety applications. As we will see mid-IR radiation transmits well through smoke, while far-IR is well known for its capability of passing through plastic materials, clothes, paper, wood, and many other materials. Unfortunately water is strongly absorbing both at $10.6\,\mu m$ and in the terahertz region and, consequently, the exploitation of these wavelengths for biological purposes is restricted to thin sample investigation.

2.4   Latest Achievements in IRDH

In this section we will show some of the more recent and promising results obtained in the field of IRDH.

The very first results in this field were obtained in 2003 [25] employing a holographic setup in transmission configuration, using a CO_2 laser at $10.6\,\mu m$ with $190\,mW$ power and a pyroelectric camera, Pyrocam III by Spiricon, with 124×124 LiTaO3 elements of pixel size $85\,\mu m \times 85\,\mu m$ and center to center spacing of $100\,\mu m \times 100\,\mu m$. In this first experiment the wavefront transmitted through a small drilled metallic plate was recorded and numerically reconstructed both in amplitude and phase.

The first speckle IRDH results on large size samples were obtained in 2010 [2627] using a larger size and more sensitive detector, the microbolometer detector Miricle 307 K with 640×480 ASi pixels at $25\,\mu m$ pixel pitch; in order to guarantee a more efficient object irradiation a $100W$ CO_2 laser source and the experimental setup shown in Fig. 2.4 were employed. This is a standard holographic setup and all the experiments shown in the following sections are based on this basic configuration.

In this setup the CO_2 laser beam was first divided by a ZnSe beam splitter (BS1), which reflected 80% of the impinging radiation and transmitted the remaining 20%. The transmitted part, which constituted the reference beam, was reflected by means of a plane mirror (M1) toward a ZnSe variable attenuator (VA); the reference beam was then redirected by means of another two plane mirrors (M2, M3) toward the thermocamera but, before impinging on the

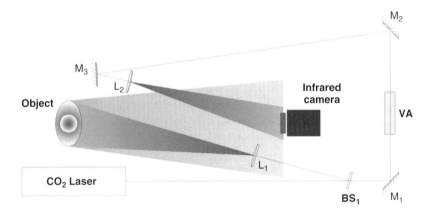

Figure 2.4 M1, M2, and M3 are plane mirrors; BS1 is a ZnSe 80/20 beam splitter; L1 and L2 are ZnSe 1.5 inch focal length spherical lenses. *Source:* Pelagotti A., Paturzo M., Locatelli M., Geltrude A., Meucci R., Finizio A., Ferraro P. 2012. Reproduced with permission from the Optical Society

detector, it encountered a ZnSe converging lens of 1.5 inch focal length (L2), which enlarged it in order to reach the thermocamera with enough low intensity and an almost planar wavefront. The reflected part of the fundamental beam indeed constituted the object beam; before impinging on the sample it passed through a ZnSe converging lens of 1.5 inch focal length (L1), which enlarged the beam so as to irradiate a fairly large surface area of the object depending on its distance from the sample. The interference pattern across the sensor was optimized in order to increase the fringe visibility, acting on the reference beam intensity by means of the variable attenuator. The hologram could be collected and digitally stored in a computer both in the form of a single image and in the form of a video if a dynamical scene was of interest. Thanks to the low sensitivity to vibration of the whole system the anti-vibration modality of the table was not activated during the experiments. Since the detector was only sensitive to IR radiation, artificial light and sunlight were not disruptive.

In Fig. 2.5 the hologram of a small bronze statue (about 10 cm high) reproducing Emperor Augustus, recorded with this configuration and its amplitude reconstruction, are shown. A second and larger size statuette (34 cm high) reproducing Benvenuto Cellini's Perseus sculpture was successfully investigated with the same configuration; in Fig. 2.6 a photo of the statue, the relative hologram, and the reconstructed amplitude wavefront are shown.

2.4.1 Super Resolution by Means of Synthetic Aperture

Since IR detector elements are larger and less numerous than typical visible detector elements, these first results lacked resolution compared to typical visible holograms. Since the hologram dimension (that is the detector surface) enters at the denominator in the reconstruction pixel pitch expression, and since the detector surface is, in general, significantly smaller than the interferometric pattern surface created in air by the object beam and the reference beam, it is possible to synthetically increase the numerical aperture of the system by means of an automatic technique capable of recording several shifted but partially superimposable holograms and stitching them together [28]. In order to collect the various holograms, the

Figure 2.5   (From left to right) small statue of Augustus; IR hologram of Augustus; numerically reconstructed hologram of Augustus. *Source:* Paturzo M., Pelagotti A., Finizio A., Miccio L., Locatelli M., Geltrude A., Poggi P., Meucci R., Ferraro P. 2010. Reproduced with permission from the Optical Society

Figure 2.6   (From left to right) statue of Perseus; IR hologram of Perseus; numerically reconstructed hologram of Perseus

basic setup described in the previous section was employed but the thermocamera was fixed at two motorized translational stages; by means of a remotely controlled routine, the sensor was moved horizontally and vertically along a serpentine path on the hologram plane and acquired various equally spaced, minimally overlapping, portions of the large interferometric pattern. The recording session had to be completed in a relatively short time so as to minimize changes in the interferential pattern, even if the task was facilitated by the low vibration sensitivity of long wavelength radiation. The numerous holograms recorded were then stitched together by means of an automated algorithm, obtaining a digital hologram with a synthetic but larger numerical aperture. The stitching procedure, called *registration*, is a standard image processing technique for determining the geometrical transformation that aligns points in one picture with corresponding points in another picture. Often, registration is performed manually

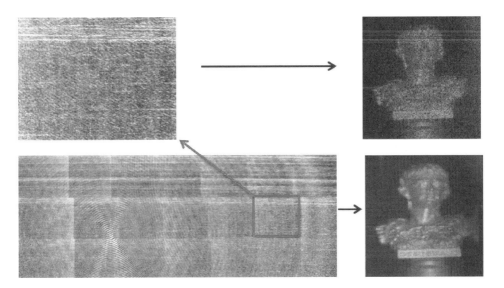

Figure 2.7 Augustus hologram stitching. (From left to right) Single hologram and single hologram amplitude reconstruction: 7 × 3 stitched holograms and stitched hologram amplitude reconstruction

or in a semi-automatic way by a user iteratively setting the parameters of the geometrical transformation. However, this approach is time consuming and can give subjective results. On the other sides most totally automated algorithms are not always applicable to aligning and stitching speckle holograms, which, due to their apparently random and very fine pattern, and to the inevitable inherent differences between images' structures and tones, are hard to match and experimentally result in noisy images. In this experiment a specific automated method based on the maximization of the mutual information (MMI) [29] was employed; this method was capable of taking into account any possible shift, rotation, and scaling among the different holograms, and showed excellent results compared to standard methods. With this technique it was thus possible to improve the reconstruction resolution of the previously investigated sample, the Augustus statue. In particular, a synthetic hologram composed of 7 × 3 standard holograms was employed to obtain the synthetic hologram and the corresponding super-resolved amplitude reconstruction of the Augustus image. The remarkable result obtained with this technique is shown in Fig. 2.7.

Again, the high power of the CO_2 laser employed in the experiment proved an advantageous parameter helping to obtain a larger interferential pattern with respect to the standard visible pattern, and therefore allowing us to acquire a larger number of single holograms for stitching together. The reconstructed image resolution, however, cannot increase indefinitely because when the synthetic aperture hologram reaches a certain dimension the reconstructed wavefront curvature, corresponding to different perspectives, cannot be properly represented with a 2D numerical reconstruction.

Noteworthy, the enhanced resolution makes it possible to exploit more efficiently the capability, peculiar to DH, to focus on different object planes: thanks to the increased numerical aperture of the synthetic hologram, the depth of focus is decreased and it becomes possible to

Figure 2.8   Stitched Augustus hologram amplitude reconstruction with the Augustus inscription in focus (on the left) and with Augustus' head in focus (on the right)

read the inscription on the base of the Augustus statue in the reconstructed image (left image of Fig. 2.8), or to focus on Augustus' head (right image of Fig. 2.8), simply by changing the numerical reconstruction distance.

2.4.2   Human Size Holograms

As we have already mentioned, by means of long wavelength DH, it is possible to obtain advantages in terms of a larger field of view and lower sensitivity to seismic noise. Furthermore, using a high power source like a CO_2 laser, it is possible to irradiate quite uniformly very large surfaces. With long wavelength radiation it becomes possible to record holograms in critical conditions including illuminated environments and not ground isolated setups. Moreover, at this wavelength, holographic videos of moving objects, although at limited speed, can be easily recorded and reconstructed. All these features are fundamental if holograms of large targets are to be recorded. However, one of the peculiar difficulties in recording a digital hologram of a large object is effectively irradiating the entire surface of the sample. In order to obtain the optimal homogeneous irradiation of the target, different configurations were tested to record human size holograms [30] and, what is even more interesting, to record live person holograms [31]. In order to compare the results obtained with each configuration, the object, a plastic mannequin 1.90 cm high (Fig. 2.9 (Plate 3)), was always maintained in the same position with respect to the thermocamera. In particular, the distance between the object and the thermocamera was the minimum distance for recording holograms of that size according to the equations discussed earlier.

In order to obtain a larger beam size across the sample, a high focusing lens to enlarge the object beam was employed in the basic holographic setup. This configuration, however, showed important disadvantages: since the peripheral part of the Gaussian object beam is characterized by a lower intensity, the object surface was not uniformly irradiated and this resulted in a poor and inhomogeneous reconstruction quality of the outer part of the sample. Furthermore, the object beam dimension was limited by the focusing power of the lens and, consequentially, only a portion of the sample could be fruitfully irradiated. In these conditions we could not take complete advantage of the increased field of view offered by the longer wavelength because the irradiated area was smaller than the maximum

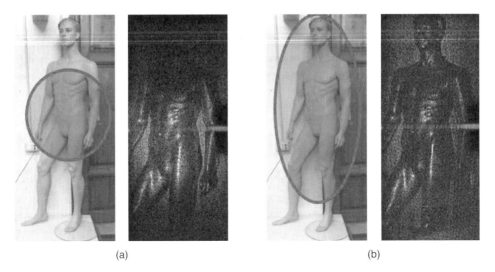

(a)   (b)

Figure 2.9 (Plate 3)   (a) Spherical lens configuration. Mannequin image with irradiated area in red; mannequin hologram amplitude reconstruction. (b) Cylindrical lens configuration. Mannequin image with irradiated area in red; mannequin hologram amplitude reconstruction. *See plate section for the color version*

recordable size (Fig. 2.9a). When the analysis of targets extending in a preferential direction, like the human-size mannequin, is needed a cylindrical lens can be fruitfully employed to obtain a stretched beam (Fig. 2.9b). In another configuration, based on a completely different approach, the object beam, expanded by the usual spherical lens was moved along the sample, changing its propagation direction by means of a plane mirror driven by two motion control devices. Using this configuration it was thus possible to uniformly irradiate an overall surface of more than $4\,\mathrm{m}^2$, an area larger than the allowed field of view.

In this case, a video hologram during which the object beam entirely scanned the mannequin surface was recorded in about half a minute. When faster scanning acquisitions are realized, the fringe visibility decreases progressively and the reconstruction resolution gradually reduces; this problem could be solved by means of a higher frame rate and a shorter exposure time. Such an acquisition procedure can be exploited in different ways: the video hologram can be reconstructed as a video that can be reproduced at normal velocity showing a slow scan of the sample, or it can be decimated and played at incremented speed so that the amplitude image of the whole object is perceived all at once; a third possibility consists of extracting the most significant frames and obtaining the whole image by their superposition as shown in Fig. 2.10 (Plate 4).

The natural following step was to test the possibility of acquiring live human holograms [31]; using the first configuration a human half bust was clearly recorded and reconstructed as shown in Fig. 2.11. Clearly, when operating with human beings, resolution gets worse because of body micro movements, and because of the lower reflectivity of human skin and clothes compared to the plastic surface of the mannequin. It is, however, important to point out that such a result would be very difficult, if not impossible, to reach with a CW visible laser in unisolated conditions.

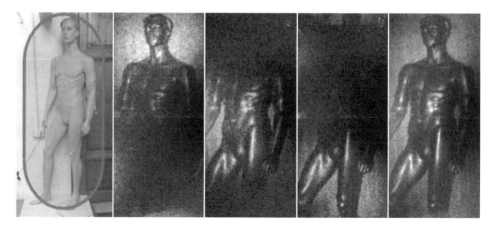

Figure 2.10 (Plate 4)   (From left to right) Mannequin image with irradiated area in red; mannequin hologram amplitude reconstruction at different scanning time and superposition of the most significant frames. *See plate section for the color version*

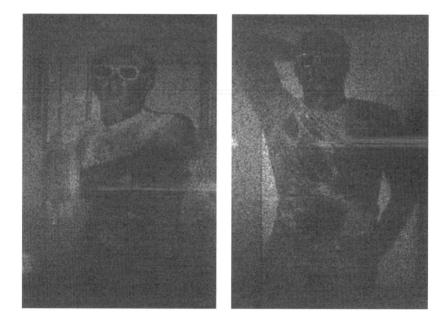

Figure 2.11   Amplitude reconstructions of a human half bust

2.4.3   *Visible Analog Reconstruction of IR Digital Holograms*

Hologram reconstruction in DH is usually implemented by means of numerical algorithms but nothing precludes to "write", somehow, the hologram on an appropriate medium in order to perform an analog reconstruction. Spatial light modulators (SLM) (arrays of pixels where each pixel modulates the phase and the amplitude of the light transmitted through or reflected from it) are the preferred devices to accomplish this particular task [32,33]. However, in general, the SLM pixel pitch differs from the pixel pitch of the employed recording device and, furthermore, an optical reconstruction makes sense only if visible light is used, different from the case we are dealing with. The so-called *imaging equations* [3,4] relate the coordinate of an object point with that of the corresponding point in the reconstructed holographic image, in most general conditions (i.e., for a rescaled hologram and for different recording/reconstruction wavelengths). Generally, when a 3D hologram reconstruction with a different wavelength is performed, the resulting image is affected by aberrations (e.g., spherical aberration, coma, astigmatism, field curvature, and distortion). The expression of these various aberrations can be deduced from the imaging equations and it can be demonstrated [34] that if the detector-object distance is equal to the distance between the detector and the reference beam origin (i.e., in the so-called lensless Fourier holography configuration), aberrations are minimized.

In an early work [35], holograms with $10.6\,\mu m$ laser source of Perseus statuette were recorded in the basic holographic arrangement but in a lensless Fourier holography configuration. In particular, 120 holograms were acquired while rotating, each time, 3° of the figurine around itself. In the reconstruction process, a diode pumped solid state laser emitting at $0.532\,\mu m$ was used. The laser beam was optically manipulated so to obtain a converging beam impinging on a reflective liquid crystal on silicon (LCoS) phase only SLM (PLUTO by Holoeye with 1920×1080 pixels, $8\,\mu m$ pixel pitch, and $60\,Hz$ frame rate). The acquired holograms were played back in sequence with regular frequency. With such an arrangement it was possible to obtain an analog reconstructed movie of the rotating figurine on a screen at a fixed distance from the SLM. An image of the optically reconstructed wavefront, acquired by means of a standard CCD camera, is compared with a standard IR numerical reconstruction image in Fig. 2.12.

The experimental reconstructed image position and its magnification with respect to the real object were in good agreement with the predicted theoretical values and no significant aberration was observed.

In a later work [36], a holographic digital video display of a 3D ghostlike image of multi-view recorded holograms of Perseus, floating in space, was obtained. To achieve this result a holographic video display system built from nine LCoS phase-only SLMs (Holoeye HEO-1080P with 1920×1080 pixels, $8\,\mu m$ pixel pitch, and $60\,Hz$ frame rate) forming a circular configuration, was used (Fig. 2.13). Removal of the gaps between the SLMs was provided by a beam splitter to tile them side by side [37] and to achieve a virtual alignment with a continuous increased field of view. In order to position the reconstructed 3D image slightly above the display setup and to avoid blocking of the observer's vision by the display components, the SLMs were also tilted up at a small angle. Negligible reduction in the quality of the reconstructions for a tilted illumination of up to 20° has been shown by experiments [36]. All SLMs were illuminated with a single astigmatic expanding wave by means of a cone mirror as shown in Fig. 2.13. With the help of this configuration, observers can see the 3D ghost-like image floating in space and can move and rotate it around (Fig. 2.14). These two works demonstrate the possibility of obtaining direct real-time 3D vision of IR-recorded digital holograms and, considering the previously underlined capability of IRDH to investigate large size samples, emphasize the potentialities of IRDH as a valuable candidate in the research field of real 3D TV and for possible virtual museum applications.

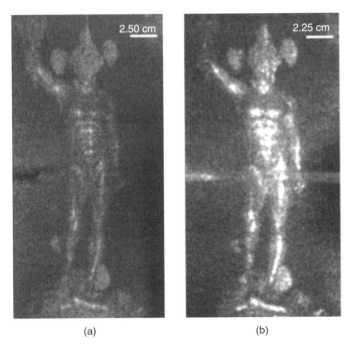

(a)   (b)

Figure 2.12 (a) Numerical reconstruction of Perseus statuette hologram. (b) SLM optical reconstruction of Perseus statuette hologram performed at a visible wavelength and collected by a monochromatic CCD

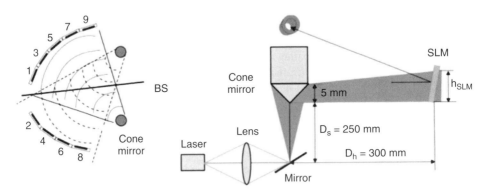

Figure 2.13 Circular holographic display. *Left*, arrangement of the nine phase-only SLMs, denoted as 1 ... 9. *Right*, illumination of a single SLM

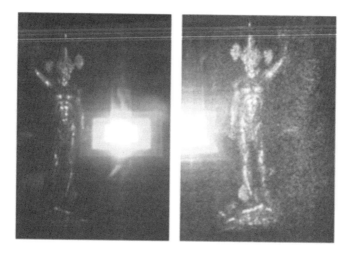

Figure 2.14 *Left*, single SLM optical reconstruction at 0.532 μm of the hologram captured at 10.6 μm and projected on a screen. *Right*, ghostlike multiple SLMs optical reconstruction of the same hologram

2.4.4 Smoke and Flames Hidden Object Holograms

In this section we report the latest results in the field of IRDH at 10.6 μm; in particular we illustrate one more useful peculiarity of this technique, which is the possibility to detect objects, or even animate beings, through smoke and, what is even more worthy of note, through a wall of flames. This possibility is a key and challenging target for possible industrial applications and, remarkably, for its possible safety purposes. Visible radiation is strongly affected by smoke or fog, and vision can be completely impaired in smoky environment. On the contrary, the last generation of uncooled IR microbolometer detectors, commercially available for imaging in the thermal infrared, allow passive or active (i.e., with IR laser illumination) clear vision through such scatterers since the IR electromagnetic radiation is just slightly scattered by fog droplets and smoke particles. This property of IR radiation is well known and in fact many fire departments already use these technologies for exploring fire scenarios. Unfortunately, in presence of flames, even these detectors cannot be of great help since the electromagnetic radiation emitted by flames can severely saturate them, as well as standard CCD or CMOS, occluding the scene behind the flames.

As previously discussed, DH at long IR wavelengths has considerable advantages with respect to visible DH, that make it very flexible and useful for recording real world large scenes. Another important characteristic to remark on is the possibility to use only a portion of the acquired hologram to reconstruct the entire scene, even if at the expense of resolution. Finally, DH usually employs a lensless setup to obtain out-of-focus acquisitions and to numerically recover the object wavefield at the desired focus plane in the reconstruction step. All these features, combined with the possibility to exploit IRDH to see behind smoke and flames, provide the unique possibility to perform real-time dynamic detection of moving people in fire scenes, independent of the chemical nature of the burning materials involved and from their emission spectrum. In order to demonstrate these unique imaging potentialities,

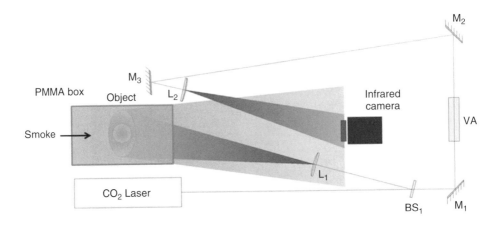

Figure 2.15   Imaging through smoke: employed setup

two kinds of experiments were conducted where the capability of seeing through smoke and flames by IRDH were respectively tested.

In an early series of experiments [31] the capability of IRDH to work efficiently in smoky environments was tested. In order to investigate this condition the basic holographic setup was used but the object under investigation was immersed in a thick blanket of smoke (Fig. 2.15). As a test object the same small Augustus statuette was used but, in order to obtain enough high smoke density around it, it was inserted inside a sealed Polymethyl methacrylate box. In one face of the box two windows in the IR range were fixed: an input AR/AR ZnSe input window through which the laser beam could reach the object and a germanium output window through which the light diffused by the object could reach the thermocamera.

By means of a lateral aperture placed in the box, smoke obtained by burning incense in a small furnace was injected inside the box and a holographic video of the interferometric pattern with the increasing smoke density was recorded. In order to have a quantitative measurement of the smoke concentration inside the box at any moment, the intensity attenuation of a 15 mW laser diode radiation travelling 6 cm inside the box was evaluated. In Fig. 2.16 the visible images of the empty box and after inletting smoke are shown; evidently at high smoke density the vision is completely impaired because of the severe scattering of the visible radiation; at the maximum smoke density the current in the photodiode was reduced by two orders of magnitude and it was impossible to see the statuette, even from the short side of the box.

In Fig. 2.16 the images obtained with the thermocamera, working in normal mode with its IR objective in the two conditions, are also shown. As expected, with a standard thermographic image the object immersed in the smoke is still clearly visible since IR light at such long wavelengths is only slightly scattered by smoke particles. Finally, the reconstructed IR holographic images in the two conditions, shown in Fig. 2.16, demonstrate that clear vision through smoke can be obtained by means of IRDH too.

It is important to note here that the random movements of the scattering particles (smoke and dust particles), which represents a noise source in the standard thermographic image, on the contrary may contribute to getting clearer vision in the holographic reconstruction. Indeed, it has been shown that, if multiple acquisitions are reconstructed and opportunely averaged, better quality images with less noise can be obtained [31].

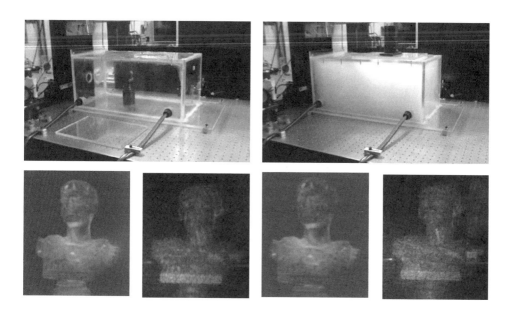

Figure 2.16 (a) Augustus statuette in the box without smoke. (b) Augustus statuette in the box filled with smoke. (c) Box without smoke: thermographic imaging of Augustus on the left and hologram amplitude reconstruction on the right. (d) Box with smoke: thermographic imaging of Augustus on the left and hologram amplitude reconstruction on the right

In a second series of experiments [31] the capability of IRDH to detect an object hidden behind a curtain of flames was tested. In these tests the basic holographic setup with flames inserted between the thermocamera and the object (Fig. 2.17) was used.

The tests were performed employing portable mini stoves to create the wall of flames and a live human half bust as a target. A large portion of the object was covered by flames, impairing vision by means of a common CCD camera. Noteworthy in this case, a clear thermographic vision of the whole scene with no blind areas was impaired as well. On the contrary, the holographic recorded image did not show such a problem, allowing visibility through flames without significant resolution loss. This capability can be explained by the fundamental intrinsic features of holography. First of all, since no objective is required during hologram recording, the IR radiation energy emitted by the flames is not focused on the detector but is distributed over its whole surface; this means that no image of the flames is formed on the detector and, consequently, no pixel saturation effect is obtained. In other words, since out of focus images are recorded in DH, the typical saturation effect of the IR camera elements observed in standard imaging configurations is avoided and the sensor is not blinded by the flame emission. If the radiation energy intercepted by the sensor was so high to disturb the detector somehow, it would be, however, always possible to use a narrow band filter around 10.6 to completely remove its contribution. A second important aspect of the question, strictly related to the interferometric nature of DH technique, is that the flames' radiation is not coherent with the radiation used to get the interferogram and consequently does not affect the interferometric pattern in any way. Furthermore, thanks to the ability of holography to reconstruct

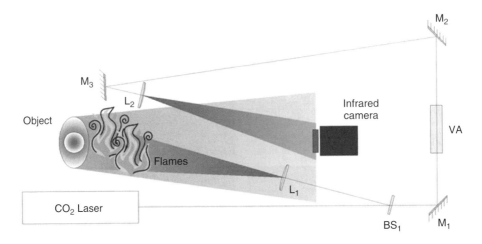

Figure 2.17 Imaging through flames: employed setup

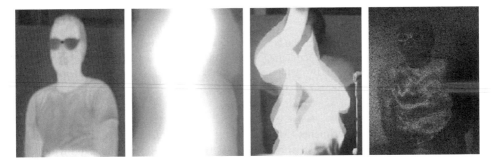

Figure 2.18 Imaging of a live human being seen through flames. From left to right; thermographic image without flames, thermographic image with flames, visible image with flames, holographic reconstructed image with flames

the object wavefront entirely from a smaller portion of the hologram, this technique allows enhanced vision even if some macro-particles, often present in real fire-scenarios, obstruct direct imaging. The advantages of digital holographic image system with respect to standard thermographic and visible recording systems are clearly noticeable in the comparison of the visible, thermal, and holographic reconstructed images in Fig. 2.18.

2.5 Conclusion

As we have seen, IRDH at 10.6 μm has paved the way for large sized object investigation, especially for non-destructive testing and, thanks to its ability to see through smoke, dust particles and flames. It could be employed in the future, to large kinds of applications. Mid-IRDH could also play a very important role in research activity into 3D displays: holographic television, commonly regarded as the holy grail of holography, is one of the most promising and challenging developments for the future display market. Only holography can provide all depth information necessary to obtain the reconstruction of natural-looking 3D scenes; IRDH, thanks to its capacity to record whole wavefront information of large size samples in daylight and not isolated conditions, and thanks to the possible developments introduced by the progresses of spatial light modulators, may play an important role in this challenge.

References

[1] Gábor D., A new microscopic principle, *Nature* **161**, pp. 777–778 (1948).

[2] Goodman J. W. and Lawrence R. W., Digital image formation from electronically detected holograms, *Applied Physics Letters* **11**, pp. 77–79 (1967).

[3] Schnars U. and Jüptner W. P. O., *Digital Holography, Digital Hologram Recording, Numerical Reconstruction and Related Techniques*, [1st edn], Springer (2005).

[4] Goodman J. W., *Introduction to Fourier Optics*, [2nd edn], McGraw-Hill (1996).

[5] Grilli S., Ferraro P., De Nicola S., Finizio A., Pierattini G., and Meucci R., Whole optical wavefields reconstruction by digital holography, *Optics Express* **9**, pp. 294–302 (2001).

[6] De Nicola S., Ferraro P., Finizio A., and Pierattini G., Correct-image reconstruction in the presence of severe anamorphism by means of digital holography, *Optics Letters* **26**, pp. 974–976 (2001).

[7] De Nicola S., Ferraro P., Finizio A., De Natale P., Grilli S., Pierattini G., and A Mach-Zender interferometric system for measuring the refractive indices of uniaxial crystals, *Optics Communications* **202**, pp. 9–15 (2002).

[8] Ferraro P., De Nicola S., Finizio A., Coppola G., Grilli S., Magro C., and Pierattini G., Compensation of the inherent wave front curvature in digital holographic coherent microscopy for quantitative phase-contrast imaging, *Applied Optics* **42**, pp. 1938–1946 (2003).

[9] De Nicola S., Ferraro P., Finizio A., Grilli S., and Pierattini G., Experimental demonstration of the longitudinal image shift in digital holography, *Optical Engineering* **42**, pp. 1625–1630 (2003).

[10] Grilli S., Ferraro P., Paturzo M., Alfieri D., De Natale P., De Angelis M., *et al.*, In-situ visualization, monitoring and analysis of electric field domain reversal process in ferroelectric crystals by digital holography, *Optics Express* **12**, pp. 1832–1834 (2004).

[11] Ferraro P., De Nicola S., Coppola G., Finizio A., Alfieri D., and Pierattini G., Controlling image size as a function of distance and wavelength in Fresnel-transform reconstruction of digital holograms, *Optics Letters* **29**, pp. 854–856 (2004).

[12] De Angelis M., De Nicola S., Finizio A., Pierattini G., Ferraro P., Grilli S., and Paturzo M., Evaluation of the internal field in lithium niobate ferroelectric domains by an interferometric method, *Applied Physics Letters* **85**, pp. 2785–2787 (2004).

[13] Paturzo M., Alfieri G., Grilli S., Ferraro P., De Natale P., De Angelis M., *et al.*, Investigation of electric internal field in congruent LiNbO3 by electro-optic effect, *Applied Physics Letters* **85**, pp. 5652–5654 (2004).

[14] De Nicola S., Finizio A., Pierattini G., Alfieri D., Grilli S., Sansone L., and Ferraro P., Recovering correct phase information in multiwavelength digital holographic microscopy by compensation for chromatic aberrations, *Optics Letters* **30**, pp. 2706–2708 (2005).

[15] Ferraro P., Grilli S., Miccio L., Alfieri D., De Nicola S., Finizio A., and Javidi B., Full color 3-D imaging by digital holography and removal of chromatic aberrations, *Journal of Display Technology* **4**, pp. 97–100 (2008).

[16] Merola F., Miccio L., Paturzo M., De Nicola S., and Ferraro P., Full characterization of the photorefractive bright soliton formation process using a digital holographic technique, *Measurement Science & Technology* **20**, 045301 (2009).

[17] Miccio L., Finizio A., Puglisi R., Balduzzi D., Galli A., and Ferraro P., Dynamic DIC by digital holography microscopy for enhancing phase-contrast visualization, *Biomedical Optics Express* **2**, pp. 331–344 (2011).

[18] Paturzo M., Finizio A., and Ferraro P., Simultaneous multiplane imaging in digital holographic microscopy, *Journal of Display Technology* **7**, pp. 24–28 (2011).

[19] Memmolo P., Finizio A., Paturzo M., Miccio L., and Ferraro P., Twin-beams digital holography for 3D tracking and quantitative phase-contrast microscopy in microfluidics, *Optics Express* **19**, pp. 25833–25842 (2011).

[20] Paturzo M., Finizio A., Memmolo P., Puglisi R., Balduzzi D., Galli A., and Ferraro P., Microscopy imaging and quantitative phase contrast mapping in turbid microfluidic channels by digital holography, *Lab Chip* **12**, pp. 3073–3076 (2012).

[21] Bianco V., Paturzo M., Finizio A., Balduzzi D., Puglisi R., Galli A., and Ferraro P., Clear coherent imaging in turbid microfluidics by multiple holographic acquisitions, *Optics Letters* **37**, pp. 4212–4214 (2012).

[22] Bianco V., Paturzo M., Finizio A., Ferraro P., and Memmolo P., seeing through turbid fluids: a new perspective in microfluidics, *Optics and Photonics News* **23**(12), pp. 33–33 (2012).

[23] Bianco V., Paturzo M., Memmolo P., Finizio A., Ferraro P., and Javidi B., Random resampling masks: a non-Bayesian one-shot strategy for noise reduction in digital holography, *Optics Letters* **38**(5), pp. 619–621 (2013).

[24] Born M. and Wolf E., *Principles of Optics, Electromagnetic Theory of Propagation, Interference and Diffraction of Light*, [7th edn (expanse)], Cambridge University Press (1999).

[25] Allaria E., Brugioni S., De Nicola S., Ferraro P., Grilli S., and Meucci R., Digital holography at 10.6 µm, *Optics Communications* **215**, pp. 257–262 (2003).

[26] Pelagotti A., Locatelli M., Geltrude A., Poggi P., Meucci R., Paturzo M., *et al.*, Reliability of 3D imaging by digital holography at long IR wavelength, *Journal of Display Technology*, **6**(10), pp. 465–471 (2010).

[27] Pelagotti A., Paturzo M., Geltrude A., Locatelli M., Meucci R., Poggi P., and Ferraro P., Digital holography for 3D imaging and display in the IR range: challenges and opportunities, *3D Research*, **1**(4/06), pp. 1–10 (2010).

[28] Pelagotti A., Paturzo M., Locatelli M., Geltrude A., Meucci R., Finizio A., and Ferraro P., An automatic method for assembling a large synthetic aperture digital hologram, *Optics Express* **20**(5), pp. 4830–4839 (2012).

[29] Maes F., Vandermeulen D., and Suetens P., Medical image registration using mutual information, *Proc IEEE* **91**(10), 1699–1722 (2003).

[30] Geltrude A., Locatelli M., Poggi P., Pelagotti A., Paturzo M., and Meucci R., Infrared digital holography for large object investigation *Proc. SPIE* 8082–8012 (2011).

[31] Locatelli M., Pugliese E., Paturzo M., Bianco V., Finizio A., Pelagotti A., *et al.*, Imaging live humans through smoke and flames using far-infrared digital holography, *Optics Express* **21**(5), pp. 5379–5390 (2013).

[32] Onural L., Yaraş F., and Kang H., Digital holographic three-dimensional video displays, *Proc. IEEE* **99** 576 (2011).

[33] Yaraş F., Kang H., and Onural L., State of the art in holographic displays: A survey, *Journal of Display Technology* **6**(10), 443–454 (2010).

[34] Meier R. W., Magnification and third-order aberrations in holography, *Journal of the Optical Society of America* **55**(8), 987–992 (1965).

[35] Paturzo M., Pelagotti A., Finizio A., Miccio L., Locatelli M., Geltrude A., *et al.*, Optical reconstruction of digital holograms recorded at 10.6 μm: route for 3D imaging at long infrared wavelengths, *Optics Letters* **35**(12), pp. 2112–2114, (2010).

[36] Stoykova E., Yaraş F., Kang H., Onural L., Geltrude A., Locatelli M., *et al.*, Visible reconstruction by a circular holographic display from digital holograms recorded under infrared illumination, *Optics Letters* **37**(15), pp. 3120–3122 (2012).

[37] Yaraş F., Kang H., and Onural L., Circular holographic video display system, *Optics Express* **19**(10), 9147–9156 (2011).

3

Digital Hologram Processing in On-Axis Holography

Corinne Fournier, Loïc Denis, Mozhdeh Seifi and Thierry Fournel
Laboratoire Hubert Curien, Saint Etienne University, France

3.1 Introduction

The quantitative three-dimensional reconstruction and tracking of micro or nano objects spread in a volume is of great interest in many fields of science, such as in biomedical fields (e.g., tracking of markers), fluid mechanics (e.g., the study of turbulence or evaporation phenomena), and chemical engineering (e.g., the study of reactive multiphase flow), among many other applications. The development of accurate and high-speed 3-D imaging systems is crucial in these fields. Several imaging techniques have been investigated during the last 20 years, such as 3-D Particle Tracking Velocimetry with four cameras (Virant and Dracos 1997) or extended Laser Doppler Anemometry (Volk *et al.* 2008). Three-dimensional tracking has been performed with single-molecule fluorescent microscopy using nanometer-sized fluorescent markers based on astigmatism optics (Huang *et al.* 2008), double-helix PSF (Pavani *et al.* 2009), or multi-plane detection (Pavani *et al.* 2009). Each of these techniques has its own advantages and limitations but none of these techniques can yet compete in accuracy with digital holography (DH) to reconstruct 3-D trajectories and size of high speed moving objects.

DH is a non invasive 3-D metrological tool that is suitable for fast moving object reconstruction and sizing. It has proved to be efficient in many fields. Some recent examples include: Verpillat *et al.* (2011); Chareyron *et al.* (2012); El Mallahi *et al.* (2013); Lamadie *et al.* (2012); Moon *et al.* (2013) and Seifi *et al.* (2013b).

Two setups are commonly used in DH: the on-axis setup and the off-axis setup. Although off-axis setups are well adapted to the reconstruction of object surface, the on-axis setups are more suited to accurate reconstruction of micro or nano objects in a volume. In contrast to off-axis holography, on-axis hologram exploits the whole frequency bandwidth of the sensor to encode the depth of objects with high accuracy. Furthermore because it does not involve

Multi-dimensional Imaging, First Edition. Edited by Bahram Javidi, Enrique Tajahuerce and Pedro Andrés.
© 2014 John Wiley & Sons, Ltd. Published 2014 by John Wiley & Sons, Ltd.

beam splitters, mirrors, and lenses, the in-line setup (i.e., Gabor setup) is less sensitive to vibrations. This imaging technique is also called "lensless imaging" (Faulkner and Rodenburg 2004; Repetto *et al.* 2004; Allier *et al.* 2010; Fienup 2010), because it involves no lens between the object and the sensor. The disadvantage of on-axis setups comes from the superimposition of a background to the hologram signal, thereby reducing the dynamic range of the signal of interest.

Over the past decade, numerous algorithms for the analysis of digital holograms have been proposed (several journal special issues were published on the subject: see Poon *et al.* 2006; Coupland and Lobera 2008, Kim *et al.* 2013). These reconstruction algorithms are mostly based on a common approach (hereafter denoted the classical approach): a digital reconstruction based on the simulation of hologram diffraction.

In contrast to this optical approach, signal processing tools commonly used in the image processing of other imaging modalities provide a rigorous way to process on-axis holograms leading to optimal image processing in certain cases. Rather than transforming the hologram, the aim is to find the reconstruction that best models the measured hologram. This "inverse problems" approach extracts more information from the hologram and is proved to solve two essential problems in digital holography: the improvement of accuracy of reconstruction and enlargement of the studied field beyond the physical limit of the sensor size (Soulez *et al.* 2007a,b). It also leads to almost unsupervised algorithms (only few tuning parameters are used). These approaches are sometimes referred to as *compressive sensing methods* (Brady *et al.* 2009, Lim *et al.* 2011, Rivenson *et al.* 2010). The drawback of these approaches is a computational load heavier than with classical techniques. The parameters that increase the processing time are the size of the reconstructed volume, the number of parameters to estimate (in the case of parametric reconstruction), and the model complexity. Accelerations have been recently proposed to reduce processing time.

The second section of the chapter defines the framework of hologram processing and introduces hologram-image formation models and the mathematical notations used in the following. In the third section, we briefly remind that, from a signal processing point of view, the light propagation operator classically used to reconstruct holograms does not invert hologram formation. Then, Section 3.4 gives a unified presentation of hologram processing methods based on inverse problems. We present in Section 3.5 an estimate of the parameter accuracy lower bound reachable using such algorithms. In Section 3.6, recent algorithms aimed at reducing the complexity of the reconstruction are presented.

3.2   Model of Hologram Image Formation

In this section, we remind the reader of the mathematical model of hologram formation (Goodman 1996) that will be used in the reconstruction methods described in the next sections. We also introduce the matrix notations, commonly used in inverse problem frameworks, allowing us to account easily for sampling and cropping of the signal for theoretical analysis of the problem. We consider an on-axis holography setup where n small objects are illuminated with a collimated laser beam. We assume that the Royer criterion is satisfied (i.e., that the surface of the projected objects on the sensor is less than 1% of the sensor area, see Royer

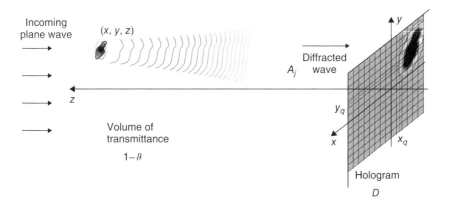

Figure 3.1   Illustration of in-line hologram formation model

1974). The digital camera records both the object wave–the wave that has propagated from the objects to the sensor–and the reference wave–the illuminating wave–(see Fig. 3.1). This diffraction phenomenon can also be modeled as interferences between the waves diffracted by each object aperture ϑ_j and the illuminating wave A_{ref} (assumed to be unaltered by the diffracting objects).

We consider small objects and that Fresnel's diffraction approximation is valid (Goodman 1996, p. 69), that is, for a propagating distance z, a width of object l and a wavelength λ the following condition is satisfied: $z^3 \gg \pi \, l^4/(64\lambda)$. For most experimental conditions Fresnel's approximation is valid (e.g., for a laser wavelength $\lambda = 532$ nm, and an object width $l \approx 100$ µm the minimum distance to the detector $z_{min} = 0.5$ mm). Under such conditions and for n particles of radii r_j and 3-D positions (x_j, y_j, z_j), the intensity measured by the detector at position (x, y) is given by (Seifi *et al.* 2013b):

$$I(x, y) = I^0_{ref}(x, y) - 2\sqrt{I^0_{ref}(x, y)} \sum_{j=1}^{n} \eta_j.\Re[(h_{z_j} * \vartheta_j)(x, y)] + \beta(x, y), \quad (3.1)$$

where I^0_{ref} stands for the intensity of the reference wave on the hologram plane (background image), the real factors η_j accounting for possible variations of incident energy seen by an object due to the non-uniformity of the reference wave, ϑ_j is the complex aperture of the jth object, and h_{z_j} is the impulse response function for free space propagation over a distance z_j (distance of the jth object to the hologram). β represents the sum of second-order terms of diffraction. Under Fresnel approximation, the impulse response is the so-called Fresnel function:

$$h_{z_j}(x, y) = \frac{1}{i\lambda z_j} \exp\left(\frac{i\pi(x^2 + y^2)}{\lambda z_j}\right). \quad (3.2)$$

Let us note that other kernels can be used depending on the experimental conditions (e.g., Rayleigh–Sommerfeld kernel: Goodman 1996).

For small objects and large distances between objects and sensor (i.e., $\pi l^2/(4\lambda z) \ll 1$) the second order terms of Eq. (3.1) are negligible. The model then simplifies to a linear model:

$$I(x,y) = I_{ref}^0(x,y) + \sqrt{I_{ref}^0(x,y)} \sum_{j=1}^{n} \alpha_j . m_j(x,y),$$

$$\text{with} \quad m_j(x,y) = -\Re(h_{z_j} * \vartheta_j)(x,y),$$

$$\alpha_j = 2 \eta_j. \tag{3.3}$$

The digitization of intensity I on an N-pixel camera leads to a digital hologram. To remove the terms in Eq. (3.3) that don't depend on the object patterns (e.g., the background I_{ref}^0), a background image is usually calculated either by taking an image of an empty volume, or recording a video of holograms and calculating the mean image of this video. To efficiently remove the effect of background, an element-wise subtraction of I_{ref}^0 followed by an element-wise division of the result by $\sqrt{I_{ref}^0}$ is performed on the digital hologram. The digital image obtained D is therefore modeled as a sum of diffraction patterns:

$$D(x,y) = \sum_{j=1}^{n} \alpha_j . m_j(x,y). \tag{3.4}$$

The digital hologram can be expressed in a vector form d of N grayvalues. Depending on the application, it may be related to the diffraction pattern of each object (FI, see illustration Fig. 3.2), or to the opacity distribution of the objects (FII, see illustration Fig. 3.3):

$$(\text{FI}) \qquad d = M\,\alpha + \epsilon \qquad \leftrightarrow \qquad \begin{bmatrix} D(x_1,y_1) \\ \vdots \\ D(x_N,y_N) \end{bmatrix} = \begin{bmatrix} \sum_j \alpha_j m_j(x_1,y_1) + \epsilon_1 \\ \vdots \\ \sum_j \alpha_j m_j(x_N,y_N) + \epsilon_N \end{bmatrix} \tag{3.5}$$

$$(\text{FII}) \qquad d = H\,\vartheta + \epsilon \qquad \leftrightarrow \qquad \begin{bmatrix} D(x_1,y_1) \\ \vdots \\ D(x_N,y_N) \end{bmatrix} = \begin{bmatrix} \sum_k \left[h_{z_k} * \vartheta_k \right](x_1,y_1) + \epsilon_1 \\ \vdots \\ \sum_k [h_{z_k} * \vartheta_k](x_N,y_N) + \epsilon_N \end{bmatrix} \tag{3.6}$$

Equations (3.5) and (3.6) are written in compact form using matrix notation. In words, Eq. (3.5) expresses the recorded hologram d as the sum of the diffraction patterns of each object ($M\,\alpha$), a perturbation term accounting for the different sources of noise and for our modeling approximations (ϵ). The term $M\,\alpha$ is the product between a $N \times n$ matrix (M) and a n elements vector (α). Matrix M may be thought of as a dictionary of the diffraction patterns of n objects (the j-th column of matrix M corresponds to the N graylevels of the diffraction pattern of the j-th object: $[m_j(x_1,y_1), \cdots, m_j(x_N,y_N)]^t$). Vector α defines the amplitude of each of the n diffraction patterns. Equation (3.5) thus corresponds to a discretization of Eq. (3.3).

Equation (3.6) expresses the hologram d as the sum of diffraction patterns $H\,\vartheta$ created by the opacity distribution ϑ and a noise term (ϵ). If the opacity distribution is defined over K planes of L pixels, ϑ is a vector of $K \cdot L$ elements corresponding to the stacking of all opacity values.

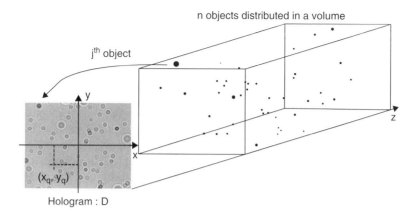

Figure 3.2 Illustration of parametrical objects image-hologram formation (FI)

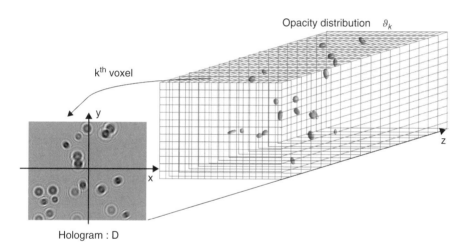

Figure 3.3 Illustration of the image-hologram formation computed from opacity distribution of the studied volume (FII)

H is then a $N^2 \times K \cdot L^2$ matrix corresponding to a (discrete) diffraction operator. Each column of H is a discretization of the impulse response kernel h, that is, the diffraction pattern on the hologram created by a point-like opaque object at a given 3-D location. $H\,\vartheta$ corresponds to the summation of the convolution of the opacity distribution in each plane z by the impulse response kernel of distance z.

Matrices M and H are written formally to clarify the proposed models and the derived reconstruction in the subsequent sections. It is worth noting that, in practice, they are neither stored nor explicitly multiplied to vectors α and ϑ. Due to (transversal) shift-invariance of models m_j and kernels h_j, the products $M\,\alpha$ and $H\,\vartheta$ can be computed using fast Fourier transforms (Denis *et al.* 2009b, Soulez *et al.* 2007b).

Pixel integration on the camera can be taken into account in matrices M and H by convolving the diffraction patterns m_j and diffraction kernels h_{z_k} (which form the matrix columns) with a 2-D rectangular function with the same area as the pixel's sensitive area.

3.3   DH Reconstruction Based on Back Propagation

Most of the methods for reconstructing digital holograms are based on the simulation of an optical reconstruction, followed by analysis of the 3-D reconstructed volume. In all-optical holography, after a hologram has been recorded and the holographic plate has been processed, the plate is re-illuminated with the reference wave. Hologram diffraction creates a virtual (i.e., defocused) and a real (i.e., focused) image. In digital holography, the holographic plate is replaced by a digital camera whose sensor size and resolution is worse by several orders of magnitude. The simulation of hologram diffraction, though straightforward to implement (and fast), leads to sub-optimal reconstructions with distortions due to boundary effects and the presence of the virtual (twin) image. In this section, we present hologram-diffraction based approaches and their limitations.

The classical 3-D reconstruction of digital holograms is performed in two steps. The first step is based on a numerical simulation of the optical reconstruction. A 3-D image volume V_{rec} is obtained by computing the diffracted field in planes located at increasing distances from the hologram (see Fig. 3.4). Different techniques to simulate diffraction have been proposed: Fresnel transform (Kreis 2005), fractional Fourier transform (Pellat-Finet 1994; Ozaktas *et al.* 1996), wavelets transform (Liebling *et al.* 2003). Using a convolution-based diffraction model, V_{rec} is given by:

$$V_{\mathrm{rec}}(x_p, y_q, z_r) = [D * h_{z_r}](x_p, y_q) \quad \leftrightarrow \quad v = H^t d. \tag{3.7}$$

Using Eq. (3.6), v can be expressed as:

$$v = H^t H \vartheta + \epsilon \tag{3.8}$$

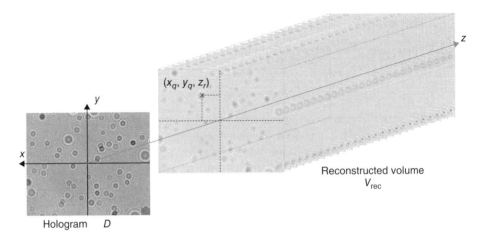

Figure 3.4   Illustration of classical reconstruction based on hologram diffraction. The z axis is magnified versus x and y axis. The red rectangle corresponds to the real size of the hologram

Unfortunately, hologram diffraction does not invert hologram recording: operator H^tH is far from the identity (i.e., the impulse response of the system "hologram recording" + "linear reconstruction" is a spatially variant halo).

The second step consists of localizing and sizing each object in the obtained 3-D image. The best focusing plane for each object has to be detected. Various criteria are suggested in the literature. Some are based on the local analysis of the sampled reconstructed volume. For example, Murata and Yasuda (2000) searched for the minimum gray level on the z-axis crossing the object center and Malek *et al.* (2004) computed the barycenter of the labeled object image after thresholding the 3-D reconstructed image. Pan and Meng (2003) used the imaginary part of the reconstructed field. Other approaches are based on an analysis of the object's 3-D image. Liebling and Unser (2004) used the criterion of the sparsity of wavelet's coefficients and Dubois *et al.* (2006) used the minimization of the integrated reconstructed amplitude. Hologram-diffraction based approaches suffer from various limitations:

- the lateral field of view is limited and, in practice, must be restricted to the center of the reconstructed images to reduce border effects;
- under-sampled holograms can lead to artifacts in the reconstructed volume (e.g., ghost images);
- twin-images of the objects superimposed on the real image can bias the localization and sizing of the objects;
- multiple intensity peaks can occur close to the actual in-focus depth location of each object (Fournier *et al.* 2004), leading to biased measurements when searching for in-focus plane using an intensity criterion;
- several tuning parameters depending on the experiment must be adjusted by the user.

In spite of these drawbacks, this approach is successfully used for many applications due to its short processing time and to satisfying accuracy in the center of the field of view. In the next section, we present signal processing approaches to reconstruct holograms that overcome the previously mentioned limits.

3.4 Hologram Reconstruction Formulated as an Inverse Problem

In Section 3.2, two linear models of the hologram formation were described. The equation FI (Eq. 3.5) models holograms of objects that have known diffraction patterns stored in a dictionary M. This is the case for simple shaped objects described by few parameters, with diffraction patterns, which are given by an analytical formula (e.g., radius and 3-D position for opaque spheres with a diffraction pattern model given by Tyler and Thompson 1976, or the Mie scattering formula, see Bohren and Huffman 2008). The non zero values of the vector α give the amplitude of the diffraction patterns that are present on the hologram. Equation FII (Eq. 3.6) models the diffraction of more complex objects (i.e., non parametric objects) that can be described by their opacity distribution ϑ sampled on a 3-D grid. The amplitude of the objects α or the opacity distribution ϑ can be estimated by inverting the hologram formation model, using a suitable regularization as typically done when dealing with ill-conditioned inverse problems.

Assuming, in our hologram models, that the noise ϵ is Gaussian and described by an inverse covariance matrix W, data are then distributed following a distribution of the form:

$$\text{(FI)} \quad p(d|\alpha) \propto \exp\ [-(M\ \alpha - d)^t W(M\ \alpha - d)], \tag{3.9}$$

$$\text{(FII)} \quad p(d|\vartheta) \propto \exp\ [-(H\ \vartheta - d)^t W(H\ \vartheta - d)]. \tag{3.10}$$

Noise is generally considered white, so that W is diagonal: $W = \text{diag}(w)$. Non uniform w can account for a signal-dependant variance. It can also be used to model missing data (e.g., $w_k = 0$ for pixels k that are outside the hologram support and $w_k = 1$ for pixels k that are inside the hologram support). Using this rigorous way to account for the limited size of the sensor permits an increase of the field-of-view size. Soulez et al. (2007b) showed that the field of view can be enlarged by a factor of 16. Chareyron et al. (2012) and Seifi et al. (2013b) showed that it is also possible to use such a binary mask to exclude, from the hologram analysis, some regions of the signal that cannot be explained with a simple mathematical model.

The negative log-likelihood $\mathcal{L}$ is given, up to an additive and a multiplicative constant, by:

$$\text{(FI)} \quad \mathcal{L}_I(d, \alpha) = -\log\ p(d|\alpha) = \|M\ \alpha - d\|_w^2, \tag{3.11}$$

$$\text{(FII)} \quad \mathcal{L}_{II}(d, \vartheta) = -\log\ p(d|\vartheta) = \|H\ \vartheta - d\|_w^2. \tag{3.12}$$

where $\|u\|_w^2$ is the weighted L_2 norm[1]. To get rid of a non perfect background removal that can leave a residual offset, we use zero mean data $(\bar{d})$ and zero mean diffraction model $(\bar{M}\ \alpha)$ on the hologram support. The neg-log-likelihood $\mathcal{L}$ is then given by:

$$\text{(FI)} \quad \mathcal{L}_I(d, \alpha) = \|\bar{M}\ \alpha - \bar{d}\|_w^2, \tag{3.13}$$

$$\text{(FII)} \quad \mathcal{L}_{II}(d, \vartheta) = \|\bar{H}\ \vartheta - \bar{d}\|_w^2, \tag{3.14}$$

with the zero-mean variables expressed with weighted scalar product[1]:

$$\bar{d} = d - \mathbf{1}\langle 1, d\rangle_w = \begin{bmatrix} d_1 - \sum_k w_k d_k / \sum_k w_k \\ \vdots \\ d_N - \sum_k w_k d_k / \sum_k w_k \end{bmatrix},$$

$$\bar{M} = [\bar{m}_1, \ldots, \bar{m}_n],$$

$$\forall j,\ \bar{m}_j = \mathbf{1}\langle 1, m_j\rangle_w - m_j = \begin{bmatrix} \sum_k w_k m_k / \sum_k w_k\ - m_1 \\ \vdots \\ \sum_k w_k m_k / \sum_k w_k\ - m_N \end{bmatrix},$$

----

[1]

weighted scalar product is defined as: $\langle u, v\rangle_w\ = \dfrac{\sum_k w_k u_k v_k}{\sum_k w_k}$

weighted L_2 norm is defined as: $\|u\|_w^2\ = \langle u, u\rangle_w\ = \dfrac{\sum_k w_k u_k^2}{\sum_k w_k}$

$$\overline{H} = [\overline{\mathbf{h}}_{z_1}, \ldots, \overline{\mathbf{h}}_{z_K}],$$

$$\forall k, \ \overline{\mathbf{h}}_k = \mathbf{1}\langle \mathbf{1}, \mathbf{h}_k \rangle_{\mathbf{w}} - \mathbf{h}_k = \begin{bmatrix} \sum_k w_k h_k / \sum_k w_k \ - h_1 \\ \vdots \\ \sum_k w_k h_k / \sum_k w_k \ - h_{N^2} \end{bmatrix}.$$

When the objects can be parameterized (e.g., disks), they are detected and localized using form (FI), as detailed in Section 3.4.1. More complex objects require the reconstruction of the opacity distribution using form (FII), see Section 3.4.2.

3.4.1 Reconstruction of Parametric Objects (FI)

The hologram model of objects that can be described with few parameters (e.g., 3-D location, shape, optical index,...), is also parametric. It can be used to create a dictionary M of diffraction patterns to model the hologram as a linear summation of the dictionary elements (form FI). Since the 3-D location of an object is continuous, the dictionary M should also be continuous (i.e., with infinite elements). The problem then amounts to finding the best match (least squares solution) between a linear summation of diffraction pattern models and the captured hologram. Several authors already suggest fitting models to the hologram leading to accurate and impressive results (Lee *et al.* 2007; Cheong *et al.* 2010; Fung *et al.* 2011; Cheong *et al.* 2011). However, they use a starting point for a fitting algorithm that is provided by the user or by a segmentation of the back propagated field that can be biased (e.g., for out-of-classical field objects) and requires tuning parameters.

Our approach, proposed in Soulez *et al.* (2007a,b), leads to an unsupervised algorithm and makes possible object reconstruction out of the classical field of view. It solves the problem iteratively, that is, objects are detected one after the other, aiming in each iteration to find the best fit between the model and the hologram. It consists of three steps:

- A *global detection step* (or a coarse estimation step), which finds the best-matching element in a discrete dictionary M (i.e., the diffraction pattern for a given 3-D location and shape). It is also called the *exhaustive search step*.
- A *local optimization step* (or a refinement step), which fits the selected diffraction pattern to the data for sub-pixel estimation.
- A *cleaning step*, which subtracts the detected pattern from the hologram to increase the signal-to-noise ratio of the remaining objects.

The procedure is then repeated on the residuals until no more object is detected. This approach to hologram reconstruction corresponds to the class of greedy algorithms (Denis *et al.* 2009a) known in signal processing as the *Matching Pursuit* (Mallat and Zhang 1993), or in radio-astronomy as the CLEAN algorithm (Högbom 1974).

3.4.1.1 Global object detection

In the first step, the best matching diffraction pattern of a sampled dictionary M is searched. The element that leads to the largest decrease of the neg-log-likelihood $\mathcal{L}_1$ is identified as the most probable (i.e., detected):

$$\underset{\substack{\alpha \geq 0 \\ \overline{m} \in \{\overline{m}_1, \dots, \overline{m}_n\}}}{\arg \min} \quad \|\alpha \, \overline{m} - \overline{d}\|_w^2. \tag{3.15}$$

By replacing α by its optimal value in Eq. (3.15), the diffraction pattern $\overline{m}^\dagger$ that minimizes $\mathcal{L}_1$ is also the one that maximizes the criterion $C(\overline{m})$ (Soulez *et al.* 2007b):

$$\overline{m}^\dagger = \underset{\overline{m} \in \{\overline{m}_1, \dots, \overline{m}_n\}}{\arg \max} \quad C(\overline{m}) \qquad \text{subject to} \qquad \langle \overline{m}, \overline{d} \rangle_w \geq 0, \tag{3.16}$$

$$\text{with} \quad C(\overline{m}) = \frac{\langle \overline{m}, \overline{d} \rangle_w^2}{\|\overline{m}\|_w^2}. \tag{3.17}$$

The detected object is the one whose diffraction pattern has the highest correlation with the data: $C(\overline{m})$ corresponds to the square of a weighted normalized correlation between a model

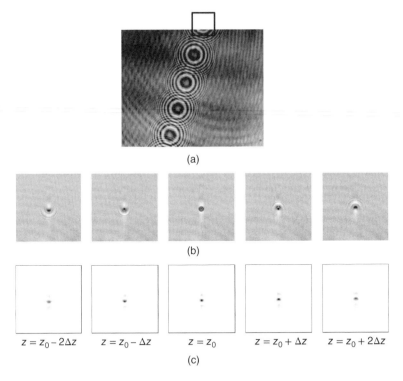

(a)

(b)

$z = z_0 - 2\Delta z$ $z = z_0 - \Delta z$ $z = z_0$ $z = z_0 + \Delta z$ $z = z_0 + 2\Delta z$

(c)

Figure 3.5 Illustration of classical reconstruction compared with the criterion map: (a) experimental in-line hologram of droplets (b) classical reconstruction based on hologram diffraction at different depths z. Artifacts appear during the numerical reconstruction due to the truncation of diffraction rings on the hologram boundary. (c) Criterion computation based on "Inverse Problems" approach at different depths z (see Eq. 3.17). For the sake of visualization, the contrast is inversed. The images represented in (b) and (c) correspond to the square area drawn on the hologram (a). $z_0 = 0.273$ m corresponds to the in-focus distance, $\Delta z = 6$ mm

and the hologram. Since the diffraction patterns are shift-invariant, Soulez *et al.* (2007b) show that the correlations in Eq. (3.16) can be computed using fast Fourier transforms.

Note that Gire *et al.* (2008) show that this global detection is less sensitive to ghost images compared with the classical reconstruction. Furthermore, "Border effects", which classically lead to measurement bias, are removed by taking into account the boundaries of the sensor by means of a binary mask w. Figure 3.5(c) shows the values of the criterion $C(\overline{m})$ on several consecutive reconstructed planes (for the sake of visualization, the contrast is inversed). Unlike in classical reconstruction, the maximum criterion value in these planes is on the in-focus plane.

3.4.1.2   Local optimization

The global detection step gives a rough estimation of the objects parameters. In the local optimization step these parameters are used as the first guess of an optimization algorithm to get sub-pixel accuracy.

Figures 3.6 and 3.7 illustrate an application of this algorithm to detect spherical droplets.

3.4.1.3   Cleaning

Once the local optimization step is finished, the accurate estimated parameters are used to simulate the diffraction pattern and remove it from the data.

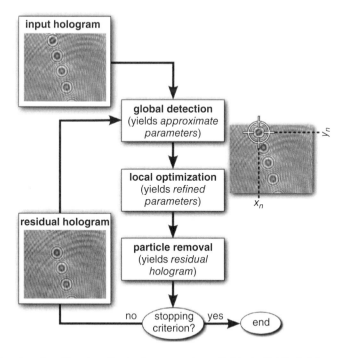

Figure 3.6 Iterative algorithm to estimate the parameters of objects distributed in a volume. *Source*: Soulez F., Denis L., Thiebaut E., Fournier C., and Goepfert C., 2007b. Reproduced with permission from the Optical Society

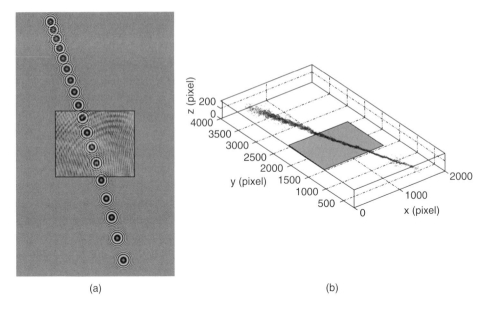

(a)   (b)

Figure 3.7   Illustration of droplets detection located out-of-field (from Soulez *et al.* 2007b): (a) super-imposition of one hologram of the series and the model of this hologram calculated from 16 detected particles (including 12 out-of-field); (b) represents the 3-D jet obtained by the detection of all particles located in a field equal to more than 16 times the hologram surface. The corresponding surface of the sensor is represented by a gray rectangle. The droplet detection is realized without significant bias, even for particles located far away from the sensor. *Source*: Soulez F, Denis L, Thiebaut E, Fournier C and Goepfert C 2007b. Reproduced with permission from the Optical Society

Repeating the detection, localization, and cleaning steps on the residual signal improves the signal-to-noise ratio of remaining objects with fainter signatures, in particular, particles distant from the camera center, and prevents from detecting the same particle multiple times. An illustration of cleaning is shown on Fig. 3.8 and on a video.

The algorithm stops when no more reliable particle can be detected ($\alpha < 0$). This algorithm is implemented in an online free Matlab toolbox called "HoloRec3D".[2]

This greedy algorithm was used by Soulez *et al.* (2007a,b) to 3-D reconstruct water droplets accurately, behaving as opaque spheres with diameters of about 100 µm. Grier's team used a simple model fitting algorithm and a Lorentz Mie Theory hologram formation model to reconstruct colloidal spherical particles and their optical index with diameters of about 1 µm (Lee *et al.* 2007; Cheong *et al.* 2010).

3.4.2   Reconstruction of 3-D Transmittance Distributions (FII)

When the objects are too complex to be parameterized by few parameters, or when the purpose is to reconstruct unknown objects, form (FII) is considered: an opacity distribution sampled on a 3-D grid is reconstructed from the hologram. Due to the ill-posed nature of this inversion

---

[2] http://labh-curien.univ-st-etienne.fr/wiki-reconstruction/index.php

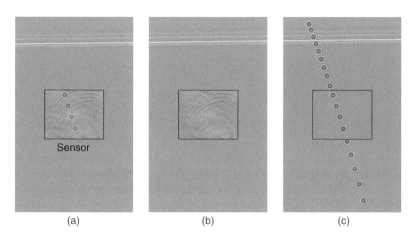

Figure 3.8 Illustration of the cleaning step: (a) experimental hologram, (b) cleaned hologram, and (c) simulated hologram using the estimated parameters

problem, it is mandatory to regularize it. The reconstructed 3-D distribution $\boldsymbol{\vartheta}$ is then given by the *Maximum A Posteriori* estimate (MAP):

$$\boldsymbol{\vartheta}^{(\mathrm{MAP})} = \arg \min_{\boldsymbol{\vartheta}} \quad \|\overline{\boldsymbol{H}}\boldsymbol{\vartheta} - \overline{\boldsymbol{d}}\|_w^2 + \beta \, \Phi_{\mathrm{reg}}(\boldsymbol{\vartheta}). \tag{3.18}$$

Several regularizations Φ_{reg} have been proposed to reconstruct holograms. When extended objects are considered, an edge-preserving smoothness prior, like total variation (the sum of the spatial gradient norm), is generally chosen (see Sotthivirat and Fessler 2004; Brady *et al.* 2009; Marim *et al.* 2010, 2011):

$$\boldsymbol{\vartheta}^{(\mathrm{MAP})} = \arg \min_{\boldsymbol{\vartheta}} \quad \|\overline{\boldsymbol{H}}\boldsymbol{\vartheta} - \overline{\boldsymbol{d}}\|_w^2 + \beta \mathrm{TV}(\boldsymbol{\vartheta}),$$

$$\text{with} \quad \mathrm{TV}(\boldsymbol{\vartheta}) = \sum_k \sqrt{(\boldsymbol{D_x}\,\boldsymbol{\vartheta})_k^2 + (\boldsymbol{D_y}\,\boldsymbol{\vartheta})_k^2},$$

where $\boldsymbol{D_x}$ and $\boldsymbol{D_y}$ are the finite difference operators along x and y (i.e., tranversal) axes.

Denis *et al.* (2009b) showed that enforcing a sparsity constraint through an ℓ^1 norm is sufficient to reconstruct holograms of diluted volumes:

$$\boldsymbol{\vartheta}^{(\mathrm{MAP})} = \arg \min_{\boldsymbol{\vartheta}} \quad \|\overline{\boldsymbol{H}}\boldsymbol{\vartheta} - \overline{\boldsymbol{d}}\|_w^2 + \beta \, \|\boldsymbol{\vartheta}\|_1, \quad \text{with} \quad \|\boldsymbol{\vartheta}\|_1 = \sum_k |\vartheta_k|. \tag{3.19}$$

A positivity constraint and spatially-variant regularization weights $\Phi_{\mathrm{reg}}(\boldsymbol{\vartheta}) = \sum_k \beta_k |\vartheta_k|$ improve the reconstruction and make it possible to extend the field of view, as illustrated in Fig. 3.9.

Note that the ℓ^1 norm minimization can also be applied to the object detection problem described in the previous section. Joint detection of all objects is more robust in the case of many objects than iterative detection of one object at a time. Intermediate procedures have been proposed in the compressed sensing literature (Needell and Tropp 2009) that detect several objects at a time, in a greedy fashion, and which can be adapted to include the local optimization step used to model a continuous dictionary.

Figure 3.9 Reconstruction of an experimental Gabor hologram of a glass reticle (from Denis *et al.* 2009b): hologram (*left*); classical linear reconstruction (*center*); MAP estimate with sparsity enducing prior to Eq. (3.19) and positivity constraint (*right*). Regularized reconstruction of holograms makes it possible to extend the field the view and suppresses twin-image artifacts. *Source*: Denis L, Lorenz D, Thiebaut E, Fournier C and Trede D 2009b. Reproduced with permission from the Optical Society

3.5 Estimation of Accuracy

The estimation and the improvement of accuracy are key issues in DH. (Jacquot *et al.* 2001; Stern and Javidi 2006; Garcia Sucerquia *et al.* 2006; Kelly *et al.* 2009). As the accuracy depends on several experimental parameters (e.g., sensor definition, fill factor, and recording distance) experimenters are in need of criteria to tune the experimental setup and to select the reconstruction algorithm that will provide the best achievable accuracy. The commonly used approach for accuracy estimation is to evaluate the Rayleigh resolution by estimating the width of the point spread function of the digital holographic system in the reconstructed planes. Fournier *et al.* (2010) suggested a methodology based on parametric estimation theory (see Kay 2008) to estimate the single point resolution defined in Dekker and den Bos (1997) (i.e., the standard deviation of the 3-D coordinates of a point source) in on-axis DH. This methodology can be applied to many DH configurations by adapting the hologram formation model, and possibly changing the noise model.

According to Cramér–Rao inequality, the covariance matrix of any unbiased estimator $\hat{\theta} = \{\hat{\theta}_i\}_{i=1:n_p}$ of the unknown vector parameter θ^* is bounded from below by the inverse of the so-called Fisher information matrix:

$$var(\hat{\theta}_i) \geq [\boldsymbol{I}^{-1}(\theta^*)]_{i,i}, \qquad (3.20)$$

where $\boldsymbol{I}(\theta^*)$ is the $n_p \times n_p$ Fisher information matrix.

The Fisher information matrix is defined from the gradients of the log-likelihood function $\log p(\boldsymbol{d}; \theta)$ (Kay 2008):

$$[\boldsymbol{I}(\theta)]_{i,j} \stackrel{\text{def}}{=} E\left[\frac{\partial \log p(\boldsymbol{d}; \theta)}{\partial \theta_i} \frac{\partial \log p(\boldsymbol{d}; \theta)}{\partial \theta_j}\right], \qquad (3.21)$$

where θ stands for the parameters vector of the object (e.g., (x, y, z, r) for a sphere located at (x, y, z) of radius r).

In the case of additive white Gaussian noise model (see Section 3.4.1), Fisher information matrix can be computed using gradients of the model $m(\theta)$ (Fournier *et al.* 2010):

$$[\boldsymbol{I}(\boldsymbol{\theta})]_{i,j} = \alpha^2 \left\langle \frac{\partial \boldsymbol{m}(\boldsymbol{\theta})}{\partial \theta_i}, \frac{\partial \boldsymbol{m}(\boldsymbol{\theta})}{\partial \theta_j} \right\rangle_w. \tag{3.22}$$

Note that w, as in the reconstruction algorithms, accounts for the finite sensor support size or for excluded data region of the analysis. The Cramér–Rao lower bound (CRLB) is asymptotically (for large samples) reached by maximum likelihood estimators. In digital holography, where the signal is distributed on the whole sensor, estimation is performed using a large set of independent identically distributed measurements (typically more than one million). The maximum likelihood estimator then approaches the CRLB. Note that if the optimization technique used for maximization of the likelihood fails to reach the global minimum, or if the noise level is too high, the resulting estimation error will exceed CRLB.

In a previous study (Fournier *et al.* 2010) about single point resolution estimation, we presented closed-form expressions of resolutions. It showed that:

- the CRLB predicted resolution behaves on optical axis as the classical Rayleigh resolution predicts;
- the CRLB give the resolution out of the optical axis and even out of the classical field of view;
- estimated parameters are correlated (an error on one parameter influences the estimation of the others).

Examples of standard deviation maps calculated using the described methodology are presented in Fig. 3.10 (Plate 5).

3.6 Fast Processing Algorithms

Reducing the processing time is one of the main issues in digital holography. One way to fix it is to use hardware device (e.g., Graphics Processing Units or multiprocessors: see Ahrenberg *et al.* 2009, Shimobaba *et al.* 2008, Page *et al.* 2008). A second way is to decrease the complexity of the algorithms. We worked on the latter issue tackling the time processing bottleneck of the global detection step in parametric object reconstruction (see 3.4.1). We present, in this section, two of our contributions aimed at reducing the complexity of this step while preserving the optimality of the signal processing approach.

3.6.1 Multiscale Algorithm for Reconstruction of Parametric Objects

Considering parametric objects described by n_θ parameters (e.g., 3-D position and intrinsic parameters such as radius, optical index etc.), the exhaustive-search step requires a time consuming exploration of the sampled parameter space of n_θ dimensions. Let us consider the

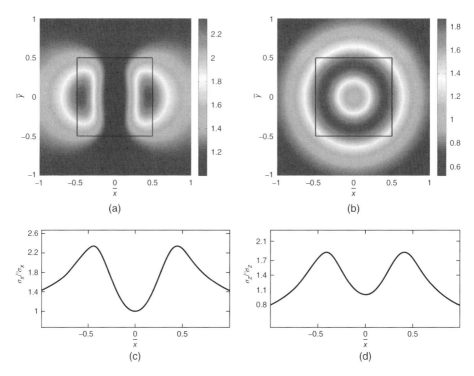

Figure 3.10 (Plate 5)   Single point resolution in a transversal plane (from Fournier *et al.* 2010): (a) *x*-resolution map normalized by the value of *x*-resolution on the optical axis; (b) normalized *z*-resolution map; (c) *x*-resolution for $\bar{y} = 0$; (d) *z*-resolution for $\bar{y} = 0$; for $z = 100$ mm, $\lambda = 0.532$ μm, $\Omega = 8.6.10^{-3}$ and $SNR = 10$. The squares in the center of figures (a) and (b) represent the sensor boundaries. *Source*: Fournier C, Denis L and Fournel T 2010. Reproduced with permission from the Optical Society. *See plate section for the color version*

simple case of spherical opaque objects with only four parameters (i.e., (x, y, z, r) 3-D coordinates, and radius). The search step is performed in 4-D space. To reach pixel-accuracy in (x, y) and sufficient accuracy in other parameters, hundreds of (z, r) pairs may need to be considered for each (x, y) location, leading to hundreds of millions or billions of quadruples (x, y, z, r) to be tested. Shift-invariance of the model can be exploited by using the Fast Fourier Transforms (FFT). The search is thus reduced for each exhaustive-search step to the computation of hundreds of convolutions to evaluate the generalized maximum correlation criterion given in 3.17 (the criterion requires 7 FFTs for each (z, r) pair (Soulez *et al.* 2007b). This is then repeated for each object unless multiple object detection is implemented (Needell and Tropp 2009). The main feature of the multiscale algorithm is to replace the computationally intensive exhaustive search by coarse-to-fine processing. The exhaustive search is carried out only on a down-sampled version of the hologram.

A sketch of the algorithm is illustrated in Fig. 3.11 (for more detail please refer to Seifi *et al.* 2013a). Since exhaustive search is a computational bottleneck, we build a multi-resolution

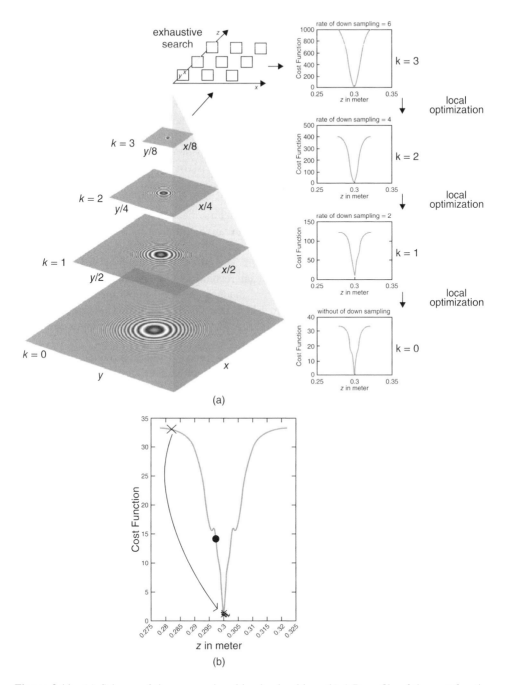

Figure 3.11 (a) Schema of the proposed multiscale algorithm, (b) 1-D profile of the cost function computed on the original hologram: the black crosses show the results of the estimation after each step of pyramidal multiscale algorithm on the profile of the cost function, the black circle shows an example of a coarse estimation from exhaustive search in the single-scale approach ($k = 0$). *Source*: Seifi M, Fournier C, Denis L, Chareyron D and Marie JL 2012. Reproduced with permission from the Optical Society

pyramid from the hologram and perform an exhaustive search only on the coarsest scale. Local optimization is then performed on increasingly fine scales, restarting numerical optimization each time from the parameters obtained at the previous (coarser) scale. The down-sampled hologram at level k is computed by low-pass filtering and downsampling the full-resolution hologram d by a linear filter. Using a coarse resolution hologram for the exhaustive search step not only reduces the number of (x, y) samples by a factor proportional to the layer number but also makes the log-likelihood smoother. Sampling of parameters z and r (i.e., depth and radius of a particle) can also be made coarser in this way. Figure 3.11(a) illustrates the widening of cost function when coarser resolution holograms are considered (for the sake of illustration a profile of the cost function along axis z is drawn). The risk of getting trapped in a local minimum is then much weaker. This fact relaxes sampling constraints that guarantee the estimation to be within reach of the global minimum.

The processing time gain depends on two main factors: the maximum downsampling period that can be used and the stopping criteria of the iterative optimization operations. The maximum downsampling period is chosen considering that a minimum number of fringes should remain on the coarsest downsampled hologram. The criterion for stopping the optimization process is given by the CRLB estimation for each resolution of the pyramid. Indeed, the optimization can be stopped when the parameters changes are equal to the theoretical standard deviations. We have validated our algorithm using a collection of simulated holograms and real holograms. The results indicate a factor of four increase in speed for a three-layer multiscale pyramid.

An other advantage of the proposed coarse-to-fine approach is that it provides an early estimation of parameters with additional accuracy after each refinement step. These coarse results can provide a quick feedback for huge stacks of holograms generated by high-speed cameras while off-line processes can refine the estimations using the finer scales.

3.6.2   Dictionary Size Reduction for Fast Global Detection

Direct matching of diffraction patterns on the in-line holograms dramatically improves the quality of reconstructed images, as discussed in Sections 3.4 and 3.5. Reconstruction methods dedicated to parametric objects (form FI described in Section 3.4.1) can be extended to a larger class of objects by considering the collection (or dictionary) of all objects of interest C. Matrix C is formed by collecting columns c_i, each representing a different object. If the dictionary is large enough (i.e., if C has many columns), any object of the considered class can be well approximated by its closest representative c_i.

Objects are not directly observed in holography, only their diffraction patterns are captured. Object recognition can then be performed in the framework of inverse problems by matching diffraction patterns. The dictionary of all possible diffraction patterns $\overline{M}$ introduced in Section 3.4.1 is obtained by considering for each object c_i diffraction patterns at various distances and all possible (x, y) translations. Let K be the dictionary of geometrically-centered diffraction patterns, that is, the collection of the diffraction patterns for all objects, at all considered depths, for objects centered on the optical axis. Dictionary K captures the variability of diffraction patterns of different objects with different recording distances. Due to object variability (the number of columns of dictionary C) and depth range, the dictionary of diffraction patterns K may be very large. Direct application of the greedy algorithm described in Section 3.4.1 would then lead to prohibitive computation time.

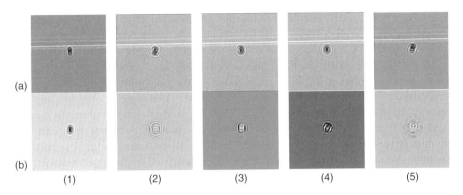

Figure 3.12 (a) Five random patterns of a 600-pattern dictionary and (b) the first five modes of the same dictionary. Patterns of the dictionary are calculated respectively for: (a-1) a "1" placed at 0.15 m from sensor, (a-2) a "2" placed at 0.1583 m, (a-3) a "3" placed at 0.1683 m, (a-4) a "6" placed at 0.1817 m, (a-5) a "7" placed at 0.19 m. The sensor is a 400×400 pixel camera with a fill-factor of 0.7 and a pixel size of 20 μm. The wavelength of the illuminating laser beam is 0.532 μm. The depth range of patterns in the dictionary is [0.15 m, 0.2 m]

Diffraction patterns in dictionary K exhibit various degrees of correlation and most of their variability can be captured in a low-dimensional sub-space:

$$K \approx \sum_{i=1}^{t} u_i \sigma_i v_i^{t} , \qquad (3.23)$$

where K is approximated by the best rank-t matrix as obtained by the singular value decomposition (SVD) considering the singular vectors u_i and v_i associated with the t largest singular values $\{\sigma_1, \dots, \sigma_t\}$. Within this approximation, diffraction pattern k_j is represented by the linear combination $\sum_i \beta_{i,j} u_i$, where coefficient $\beta_{i,j}$ is equal to $\sigma_i v_i(j)$. Vectors u_i represent so-called modes of the diffraction-patterns.

Using this approximation, the correlation terms in Eq. (3.17) are approximated as a linear combination of the correlation of each of the t modes with the data:

$$\langle k_j, \overline{d} \rangle_w \approx \sum_{i=1}^{t} \beta_{i,j} \langle u_i, \overline{d} \rangle_w . \qquad (3.24)$$

In Eq. (3.24), the scalar product $\langle u_i, \overline{d} \rangle_w$ does not depend on the considered diffraction-pattern k_j and can thus be computed once for all the diffraction-patterns. For more details, please refer to Seifi *et al.* (2013a).

3.7 Conclusion

Digital holography is very efficient for quantitative 3-D tracking and sizing of high speed objects spread in a volume. Classical reconstruction methods, based on back-propagation, have been used successfully since the 1980s to perform 3-D reconstructions. However, image processing techniques based on signal processing approaches and aiming to invert the image formation have been used increasingly in recent years to achieve accurate 3-D reconstruction. These approaches provide a rigorous way to process on-axis holograms. They lead, in certain cases, to optimal image processing so that their accuracy gets close to the Cramér–Rao lower bound. By processing the hologram directly, they also get rid of all sources of bias appearing when simulating the back-propagation of the hologram. In this framework, our team proposed two reconstruction algorithms dedicated to two types of objects: simple shape parametric objects and sparse object fields. These algorithms lead to accurate reconstruction and enlargement of the field of view. Let us note that the algorithm dedicated to parametric objects reconstruction is unsupervised and the one dedicated to the reconstruction of sparse field object opacity distribution requires only tuning of a single hyper-parameter. However, the processing time of inverse problems algorithms can be huge when the parameters to reconstruct are numerous. To tackle this drawback one can use dedicated hardware acceleration and/or decrease the complexity of the algorithms. Recently, we suggested two algorithms with reduced complexity: a coarse-to-fine algorithm based on a multiscale resolution pyramid and an algorithm aiming to reduce the diffraction-pattern dictionary (Fig. 3.12). We expect strong development and generalization of this new family of algorithms based on inverse approaches. Some current issues are the optic modeling of image-hologram formation for some specific objects and the reduction of the algorithm's complexity.

References

Ahrenberg L, Page AJ, Hennelly BM, McDonald JB and Naughton TJ 2009 Using commodity graphics hardware for real-time digital hologram view-reconstruction. *J. Display Technol* **5**, 111–119.

Allier CP, Hiernard G, Poher V and Dinten JM 2010 Bacteria detection with thin wetting film lensless imaging. *Biomedical Optics Express* **1**(3), 762–770.

Bohren CF and Huffman DR 2008 *Absorption and Scattering of Light by Small Particles*. Wiley-VCH Verlag, Weinheim.

Brady DJ, Choi K, Marks DL, Horisaki R and Lim S 2009 Compressive holography. *Optics Express* **17**(15), 13040–13049.

Chareyron D, Marie JL, Fournier C, Gire J, Grosjean N, Denis L 2012 Testing an in-line digital holography "inverse method" for the Lagrangian tracking of evaporating droplets in homogeneous nearly isotropic turbulence. *New Journal of Physics* **14**(4), 043039.

Cheong FC, Krishnatreya BJ and Grier DG 2010 Strategies for three-dimensional particle tracking with holographic video microscopy. *Optics Express* **18**(13), 13563–13573.

Cheong FC, Xiao K, Pine DJ and Grier DG 2011 Holographic characterization of individual colloidal spheres' porosities. *Soft Matter* **7**(15), 6816–6819.

Coupland J and Lobera J 2008 Special issue: Optical tomography and digital holography. *Measurement Science and Technology* **19**(7), 070101.

Dekker AJD and den Bos AV 1997 Resolution: a survey. *Journal of the Optical Society of America A* **14**(3), 547–557.

Denis L, Lorenz D and Trede D 2009a Greedy solution of ill-posed problems: error bounds and exact inversion. *Inverse Problems* **25**, 115017.

Denis L, Lorenz D, Thiébaut E, Fournier C and Trede D 2009b Inline hologram reconstruction with sparsity constraints. *Optics Letters* **34**(22), 3475–3477.

Dubois F, Schockaert C, Callens N and Yourassowsky C 2006 Focus plane detection criteria in digital holography microscopy by amplitude analysis. *Optics Express* **14**(13), 5895–5908.

El Mallahi A, Minetti C and Dubois F 2013 Automated three-dimensional detection and classification of living organisms using digital holographic microscopy with partial spatial coherent source: application to the monitoring of drinking water resources. *Applied Optics* **52**(1), A68–A80.

Faulkner HML and Rodenburg JM 2004 Movable aperture lensless transmission microscopy: a novel phase retrieval algorithm. *Physical Review Letters* **93**(2), 23903.

Fienup JR 2010 Coherent lensless imaging *Imaging Systems* Topical Meeting, Tucson, AZ, June 7–10.

Fournier C, Denis L and Fournel T 2010 On the single point resolution of on-axis digital holography. *Journal of the Optical Society of America A* **27**(8), 1856–1862.

Fournier C, Ducottet C and Fournel T 2004 Digital in-line holography: influence of the reconstruction function on the axial profile of a reconstructed particle image. *Measurement Science & Technology* **15**, 1–8.

Fung J, Martin KE, Perry RW, Kaz DM, McGorty R and Manoharan VN 2011 Measuring translational, rotational, and vibrational dynamics in colloids with digital holographic microscopy. *Optics Express* **19**(9), 8051–8065.

Garcia-Sucerquia J, Xu W, Jericho SK, Klages P, Jericho MH and Kreuzer HJ 2006 Digital in-line holographic microscopy. *Applied Optics* **45**(5), 836–850.

Gire J, Denis L, Fournier C, Thiébaut E, Soulez F and Ducottet C 2008 Digital holography of particles: benefits of the "inverse problem" approach. *Measurement Science and Technology* **19**, 074005.

Goodman JW 1996 *Introduction to Fourier Optics: Electrical and Computer Engineering.* McGraw-Hill.

Högbom JA 1974 Aperture synthesis with a non-regular distribution of interferometer baselines. *Astronomy and Astrophysics Supplement Series* **15**, 417.

Huang B, Wang W, Bates M and Zhuang X 2008 Three-dimensional super-resolution imaging by stochastic optical reconstruction microscopy. *Science* **319**(5864), 810–813.

Jacquot M, Sandoz P and Tribillon G 2001 High resolution digital holography. *Optics Communications* **190**(1–6), 87–94.

Kay SM 2008 *Fundamentals of Statistical signal Processing: Estimation Theory* 12th edn. Prentice Hall.

Kelly DP, Hennelly BM, Pandey N, Naughton TJ and Rhodes WT 2009 Resolution limits in practical digital holographic systems. *Optical Engineering* **48**, 095801–1,095801–13.

Kim MK, Hayasaki Y, Picart P and Rosen J 2013 Digital holography and 3D imaging: introduction to feature issue. *Appl. Opt.* **52**(1), DH1.

Kreis TM 2005 *Handbook of Holographic Interferometry, Optical and Digital Methods.* Wiley-VCH Verlag, Berlin.

Lamadie F, Bruel L and Himbert M 2012 Digital holographic measurement of liquid-liquid two-phase flows. *Optics and Lasers in Engineering* **50**, 1716–1725.

Lee SH, Roichman Y, Yi GR, Kim SH, Yang SM, van Blaaderen A, *et al.* 2007 Characterizing and tracking single colloidal particles with video holographic microscopy. *Optics Express* **15**(26), 18275–18282.

Liebling M and Unser M 2004 Autofocus for digital fresnel holograms by use of a Fresnelet-sparsity criterion. *Journal of the Optical Society of America A* **21**(12), 2424–2430.

Liebling M, Blu T and Unser M 2003 Fresnelets: new multiresolution wavelet bases for digital holography. *Image Processing, IEEE Transactions on* **12**(1), 29–43.

Lim S, Marks DL and Brady DJ 2011 Sampling and processing for compressive holography [invited]. *Applied Optics* **50**(34), H75–H86.

Malek M, Allano D, Coetmellec S, Ozkul C and Lebrun D 2004 Digital in-line holography for three-dimensional-two-components particle tracking velocimetry. *Measurement Science & Technology* **15**(4), 699–705.

Mallat SG and Zhang Z 1993 Matching pursuits with time-frequency dictionaries. *Signal Processing, IEEE Transactions on* **41**(12), 3397–3415.

Marim M, Angelini E, Olivo-Marin JC and Atlan M 2011 Off-axis compressed holographic microscopy in low-light conditions. *Optics Letters* **36**(1), 79–81.

Marim MM, Atlan M, Angelini E and Olivo-Marin JC 2010 Compressed sensing with off-axis frequency-shifting holography. *Optics Letters* **35**(6), 871–873.

Moon I, Anand A, Cruz M and Javidi B 2013 Identification of malaria infected red blood cells via digital shearing interferometry and statistical inference. *IEEE Photonics Journal* **5**(5) DOI: 10.1109/JPHOT.2013.2278522.

Murata S and Yasuda N 2000 Potential of digital holography in particle measurement. *Optics and Laser Technology* **32**(7–8), 567–574.

Needell D and Tropp JA 2009 CoSaMP: iterative signal recovery from incomplete and inaccurate samples. *Applied and Computational Harmonic Analysis* **26**(3), 301–321.

Ozaktas HM, Arikan O, Kutay MA and Bozdagt G 1996 Digital computation of the fractional Fourier transform. *Signal Processing, IEEE Transactions on* **44**(9), 2141–2150.

Page AJ, Ahrenberg L and Naughton TJ 2008 Low memory distributed reconstruction of large digital holograms. *Optics Express* **16**(3), 1990–1995.

Pan G and Meng H 2003 Digital holography of particle fields: reconstruction by use of complex amplitude. *Applied Optics* **42**, 827–833.

Pavani SRP, Thompson MA, Biteen JS, Lord SJ, Liu N, Twieg RJ, *et al.* 2009 Three-dimensional, single-molecule fluorescence imaging beyond the diffraction limit by using a double-helix point spread function. *Proceedings of the National Academy of Sciences* **106**(9), 2995–2999.

Pellat-Finet P 1994 Fresnel diffraction and the fractional-order Fourier transform. *Optics Letters* **19**(18), 1388–1390.

Poon TC, Yatagai T and Juptner W 2006 Digital holography–coherent optics of the 21st century: introduction. *Applied Optics* **45**(5), 821.

Repetto L, Piano E and Pontiggia C 2004 Lensless digital holographic microscope with light-emitting diode illumination. *Optics Letters* **29**(10), 1132–1134.

Rivenson Y, Stern A and Javidi B 2010 Compressive Fresnel holography. *Display Technology, Journal of* **6**(10), 506–509.

Royer H 1974 An application of high-speed microholography: the metrology of fogs. *Nouv. Rev. Opt* **5**, 87–93.

Seifi M, Denis L and Fournier C 2013a Fast and accurate 3D object recognition directly from digital holograms. *JOSA A* **30**(11), 2216–2224.

Seifi M, Fournier C, Denis L, Chareyron D and Marie JL 2012 Three-dimensional reconstruction of particle holograms: a fast and accurate multiscale approach. *JOSA A* **29**(9), 1808–1817.

Seifi M, Fournier C, Grosjean N, Meess L, Marié JL and Denis L 2013b Accurate 3D tracking and size measurementof evaporating droplets using an in-line digital holography and inverse problems reconstruction approach. *Optics Express* **21**(23), 27964–27980.

Shimobaba T, Ito T, Masuda N, Abe Y, Ichihashi Y, Nakayama H, *et al.* 2008 Numerical calculation library for diffraction integrals using the graphic processing unit: the GPU-based wave optics library. *J. Opt. A: Pure Appl. Opt* **10**(075308), 075308.

Sotthivirat S and Fessler JA 2004 Penalized likelihood image reconstruction for digital holography. *Journal of the Optical Society of America A* **21**(5), 737–750.

Soulez F, Denis L, Fournier C, Thiébaut E and Goepfert C 2007a Inverse problem approach for particle digital holography: accurate location based on local optimisation. *Journal of the Optical Society of America A* **24**(4), 1164–1171.

Soulez F, Denis L, Thiébaut E, Fournier C and Goepfert C 2007b Inverse problem approach in particle digital holography: out-of-field particle detection made possible. *Journal of the Optical Society of America A* **24**(12), 3708–3716.

Stern A and Javidi B 2006 Improved-resolution digital holography using the generalized sampling theorem for locally band-limited fields. *Journal of the Optical Society of America A* **23**(5), 1227–1235.

Tyler G and Thompson B 1976 Fraunhofer holography applied to particle size analysis a reassessment. *Journal of Modern Optics* **23**(9), 685–700.

Verpillat F, Joud F, Desbiolles P and Gross M 2011 Dark-field digital holographic microscopy for 3D-tracking of gold nanoparticles. *Optics Express* **19**(27), 26044–26055.

Virant M and Dracos T 1997 3D PTV and its application on Lagrangian motion. *Measurement Science and Technology* **8**(12), 1539.

Volk R, Mordant N, Verhille G and Pinton JF 2008 Laser doppler measurement of inertial particle and bubble accelerations in turbulence. *EPL (Europhysics Letters)* **81**(3), 34002.

4

Multi-dimensional Imaging by Compressive Digital Holography

Yair Rivenson[1], Adrian Stern[2], Joseph Rosen[1], and Bahram Javidi[3]

[1]*Department of Electrical and Computer Engineering, Ben-Gurion University of the Negev, Israel*
[2]*Department of Electro-Optics Engineering, Ben-Gurion University of the Negev, Israel*
[3]*Department of Electrical and Computer Engineering, University of Connecticut, USA*

4.1 Introduction

Unlike standard imaging, digital holography allows an indirect way to capture the complex field amplitude of a wavefront originating from an object. This provides three-dimensional (3D) information on the object recorded in a single two-dimensional (2D) recording device. Currently, digital holography is captured using a semiconductor based device and often reconstructed using numerical means on a computer. This type of holography is referred to as *Digital Holography* (DH) [1]. Digital holography is used in many areas including 3D imaging, digital holographic microscopy, aberration correction, holographic interferometry, and object surface and tomographic imaging.

During the last few years, holography was successfully combined with the rapidly growing signal acquisition-reconstruction scheme known as compressive sensing (CS) [2–5]. With introduction of CS, a theory that introduced a dramatic breakthrough in signal acquisition, implementations in optics were pursued. Shortly, research groups working on the implementation of the CS principle in optics realized that holography is a natural field for applications of CS principles [6–11]. The synergy between holography and compressive sensing has yielded new applications and also addressed classical holographic problems. Many compressive digital holographic sensing applications were demonstrated using different optical setups and having different goals. Amongst them are compressive Gabor holography [8], compressive Fresnel holography [9,11], off-axis frequency shifting holography [10], millimeter wave compressive holography [12], off-axis holography of diffuse objects [13], sectioning from optical scanning

Multi-dimensional Imaging, First Edition. Edited by Bahram Javidi, Enrique Tajahuerce and Pedro Andrés.
© 2014 John Wiley & Sons, Ltd. Published 2014 by John Wiley & Sons, Ltd.

holography [14], super-resolved wide field florescent microscopic holography [15,16], recovery of an object from holography in low illumination conditions [17], reduced scanning effort in incoherent multiple view-projection holography [18], video rate microscopic tomography [19], the use of compressive holography to see through partially occluding objects [20], single shot acquisition of spatial, spectral, and polarimetric information using a single exposure acquired hologram [21], nanometer accuracy object localization [22] and improved tomographic object reconstruction with multiple illumination angles [23,24]. The works in [25,26] provide an overview of the subject. The growing number of publications in the field indicates the importance of an obvious indication for the importance of the field.

This chapter reviews both the theoretical and the applicative aspects for using compressive sensing in digital holography. First, we equip the reader with the relevant background on compressive sensing. We then show how compressive sensing can be applied to DH, where the presentation is divided into three main parts, corresponding to different DH aspects that benefit from the CS theory:

1. Sensor design, particularly reducing the number of detector pixels, or baseline projections.
2. Reconstruction of an object from its truncated wavefront after encountering partially opaque obstacles.
3. Reconstruction of 3D object tomography from a single 2D projection.

4.2   Compressive Sensing Preliminaries

This section surveys briefly the theory of compressive sensing [2,3]. CS theory asserts that one can recover sparse signals and images from far fewer samples or measurements than traditional methods. Sparsity expresses the idea that the "information rate" of a continuous signal may be much smaller than its bandwidth, or in other words, that a discrete signal depends on a number of degrees of freedom that is much smaller than its (finite) length. More precisely, compressive sampling exploits the fact that many natural signals are sparse or compressible in the sense that they have concise representations when expressed in the proper basis or dictionary, Ψ. The sparsifying transform Ψ may be the Fourier or some wavelet basis, or a dictionary of waveforms tailored from *a-priori* information about the object.

The sensing mechanism requires correlating the signal with a small number of fixed waveforms that are incoherent (in the sense defined in Section 4.2.1) with the sparsifying basis Ψ. Signal reconstruction is performed numerically using appropriate algorithms. A block diagram of the CS process is shown in Fig. 4.1.

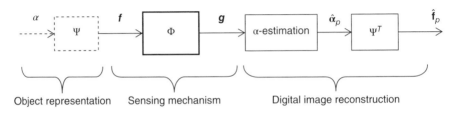

Figure 4.1   Imaging scheme of compressed sensing [27]. *Source*: A. Stern and B. Javidi 2007. Reproduced with permission from The Optical Society

The signal\image f consisting of N samples\pixels is sensed by taking a set, g, of M projections. We assume that f has a sparse representation in some known domain so that it can be composed by a transform Ψ and only S nonzero coefficients of a vector α, that is, $f = \Psi\alpha$, where only S ($S \ll N$) entries of α are nonzero. We refer to such an object as an S-sparse object. The transform Ψ can be for instance the Fourier or wavelet transforms that are commonly used in digital compression techniques. The sensing step is represented by the operator Φ in Fig. 4.1. Mathematically, Φ can be represented by an M by N matrix, hence the ith component of the measurement vector g is given by:

$$g_j = \langle f, \phi_i \rangle, \qquad i = 1, 2, \dots M, \tag{4.1}$$

where ϕ_i is the ith row of Φ, and $\langle \cdot, \cdot \rangle$ denotes the inner product. That is, we simply correlate the object we wish to acquire with the waveforms ϕ_i. For instance, if the sensing waveforms are Dirac deltas ("spikes"), then g is a vector of sampled values of f in the time or space domain. Particularly, if the sensing waveforms are indicator functions of pixels, then g is the image collected by sensors in an ideal digital camera. CS applies in the case that $M < N$, meaning that the sensed signal is undersampled in the conventional sense. This means that there are more variables than equations leading to an underdetermined system of equations. The reconstruction of f from the measurement g is a highly ill-posed problem. A classic approach for such data inversion would be to minimize the root mean square error between α and the estimated solution $\hat{\alpha}$:

$$\hat{\alpha} = \min \|\alpha\|_2 \ such\ that\ \mathbf{g} = \Omega\alpha, \tag{4.2}$$

where $\Omega_{M \times N} = \Phi\Psi$, is the operator combining the sensing and sparsity operators. This solution, however, does not take advantage of the sparsity of the image and therefore does not necessarily lead to the correct reconstruction. Given the *a-priori* information that the signal can be sparsely represented, an intuitively more appropriate approach would be to apply the ℓ_0-norm solution that satisfies the given constraints. Thus, $\hat{\alpha}$ is defined as:

$$\hat{\alpha} = \min \|\alpha\|_0 \ such\ that\ \mathbf{g} = \Omega\alpha, \tag{4.3}$$

where $\|\alpha\|_p = \left(\sum_{i=1}^{N} |\alpha_i| \right)^{1/p}$ is the ℓ_p norm of α. The ℓ_0-norm solution program (4.3) simply counts the number of nonzero terms in all possible α (that satisfy the acquisition model, $\mathbf{g} = \Omega\alpha$) and chooses the one with the fewest number of terms. Unfortunately, solving this problem essentially requires exhaustive searches over all subsets of columns of Ω, a procedure which is combinatorial in nature and has exponential complexity. This computational intractability has led researchers to develop alternatives to the ℓ_0-norm solution [2,3,28]. One of the approaches (frequently used in CS theory) is the minimum ℓ_1-norm solution according to which $\hat{\alpha}$ is found by:

$$\hat{\alpha} = \min \|\alpha\|_1 \ such\ that\ \mathbf{g} = \Omega\alpha. \tag{4.4}$$

Unlike the ℓ_0-norm, which counts the nonzero coordinates, the ℓ_1-norm is convex and thus can be recast as linear programming. A linear program is solved in polynomial time, while the ℓ_0-norm is solved in combinatorial time. In many cases, the minimum ℓ_1-norm solution is a good approximation of the minimum ℓ_0-norm solution [2,4,28,29].

Another minimization problem, which is often used in the context of compressive sensing, is formulated using total variation (TV) minimization [30] as follows:

$$\min_{f} TV(f) \quad such \quad that \quad \mathbf{g} = \mathbf{\Phi f}$$

$$with \ TV(f) = \sqrt{(f_{i+1,j} - f_{i,j})^2 + (f_{i,j+1} - f_{i,j})^2}. \tag{4.5}$$

This framework is extensively being used in compressive imaging applications, and only recently its performance guarantees were formulated [31].

The uniqueness of the solution of Eq. (4.4) and the equivalence between the ℓ_1 norm and the ℓ_0 norm solution holds if the number of compressive measurements M obeys certain conditions. These conditions can be derived from the signal sparsity and the coherence between the sensing and sparsifying operators and quantified using the *coherence parameter*. The coherence parameter expresses the idea that objects having a sparse representation in $\mathbf{\Psi}$ must be spread out, in the domain in which they are sensed, just as a spike in the time domain is spread out in the frequency domain. We shall distinguish between two different definitions of the coherence parameter, each applicable for different sensing system schemes.

4.2.1   The Coherence Parameter

4.2.1.1   Compressive Sensing by Uniformly Random Subsampling

In this measurement scheme we uniformly place our detectors at random in our measurement plane [4,29]. Mathematically, it is described as uniformly picking M out of N rows of $\mathbf{\Phi}$ at random, where $\mathbf{\Phi}$ is an $N{\times}N$ matrix describing the optical sensing operator in nominal sampling conditions. In this case, the appropriate coherence parameter definition is:

$$\mu_1 = \max_{i,j} |\langle \phi_i, \psi_j \rangle|, \tag{4.6}$$

where ϕ_i is a row vector of $\mathbf{\Phi}$, ψ_j is a column vector of $\mathbf{\Psi}$, and $\langle \cdot, \cdot \rangle$ denotes inner product. Thus, μ_1 measures the incoherence, or dissimilarity, between sensing and sparsifying operators. In the common case that $\mathbf{\Phi}$ and $\mathbf{\Psi}$ are orthonormal bases it can be shown that $1/\sqrt{N} \leq \mu_1 \leq 1$ [4,29]. According to the CS theory the signal can be reconstructed by taking M uniformly at random projections obeying [4,29]:

$$\frac{M}{N} \geq C\mu_1^2 S \log N, \tag{4.7}$$

where C is a small constant. It is clear from Eq. (4.7) that the smaller μ_1 is, the smaller the relative number of measurements required to allow for accurate reconstruction of the signal. The uniformly at random sampling scheme is useful when reducing the sensing effort is desired. For example, this may be the case when only a relatively small number of detectors are allowed due to the substantial cost of each detector.

4.2.1.2   Compressive Sensing by Structured Subsampling

This sampling scheme refers to the case when we cannot idealize (in the CS sense) the subsampling mechanism to be randomly uniform or that the sensing operator cannot be considered as an orthonormal basis (prior to its subsampling). In this case, the coherence parameter should be calculated as follows [5]:

$$\mu_2 = \max_{i \neq j} \frac{|\langle \omega_i, \omega_j \rangle|}{\|\omega_i\|_2 \|\omega_j\|_2}, \tag{4.8}$$

where $\langle \cdot, \cdot \rangle$ denotes inner product, ω_i is the column vector of $\mathbf{\Omega} = \mathbf{\Phi\Psi}$, $\mathbf{\Omega} \in \mathbb{C}^{M \times N}$, and $\|.\|_2$ is the ℓ_2-norm. It can be shown that $\sqrt{(N-M)/[M(N-1)]} \leq \mu_2 \leq 1$. Using this definition, an S-sparse signal reconstruction guarantee is given by [5]:

$$S \leq \frac{1}{2} \left\{ 1 + \frac{1}{\mu_2} \right\}. \tag{4.9}$$

As μ_2 gets smaller we can accurately reconstruct higher dimensional S-sparse signals. As the number of measurements $M \to N$, the coherence parameter $\mu_2 \to 0$, for large M. The structured subsampling case is more suitable for describing sensing mechanisms where we wish to extract more information from a measurement where its subsampling mechanism is imposed by the physical attributes of a given system, as we shall see in the rest of the chapter.

4.3   Conditions for Accurate Reconstruction of Compressive Digital Holographic Sensing

4.3.1   Compressive Sensing Reconstruction Performance for a Plane Wave Illuminated Object

In this subsection we discuss reconstruction of an object where the measurement is given by its Fresnel transform. Let us consider the 2D free-space propagation conditions in the Fresnel approximation. The input object, $f(x,y)$, is illuminated by a plane wave of wavelength λ and the complex values of a propagating wave are measured at a plane which lies at distance z away from the input plane (as illustrated in Fig. 4.2) such that:

$$g(x, y) = f(x, y) * \exp\left\{ \frac{j\pi}{\lambda z} \left(x^2 + y^2 \right) \right\}$$

$$= \exp\left\{ \frac{j\pi}{\lambda z} \left(x^2 + y^2 \right) \right\} \iint f(\xi, \eta) \exp\left\{ \frac{j\pi}{\lambda z} \left(\xi^2 + \eta^2 \right) \right\} \exp\left\{ \frac{-j2\pi}{\lambda z} \left(x\xi + y\eta \right) \right\} d\xi d\eta. \tag{4.10}$$

where "$*$" denotes convolution in Eq. (4.10). In order to capture $g(x,y)$ any of the well-known digital holographic recording techniques can be used [1]. The quadratic phase term

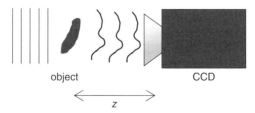

object   CCD

z

Figure 4.2   Fresnel propagation from an object illuminated by a plane wave [26]. *Source*: Y. Rivenson, A. Stern, and B. Javidi 2013. Reproduced with permission from The Optical Society

$\exp\{j\pi(x^2 + y^2)/(\lambda z)\}$ determines the behavior of the Fresnel integral. In the Fraunhofer approximation regime as $\lambda z \to \infty$, the Fresnel transform approaches the Fourier transform. Fourier transform is extensively used in CS literature as a preferred sensing operator [4,29,32] because it holds low coherence with the canonical (unit) and several wavelet expansions [33]. This provides the main motivation for applying CS in DH. However, as $\lambda z \to 0$ the captured field approaches the object's field distribution. In such a case, the coherence parameter between the object and hologram plane receives its maximal value, meaning that the compressive sensing ratio $M/N \to 1$; that is, the number of measurements needs to be exactly the same as the number of pixels representing the object. Thus, we see that the Fresnel sensing basis is dependent on the reconstruction distance z and the wavelength λ, therefore, the performance of compressive digital holographic sensing depends on these parameters as well.

In order to analyze the dependence of compressive digital holographic sensing on z and other optical system parameters, we need to also account for the fact that in DH the numerical version of the Fresnel wave propagation (Eq. 4.10) is used. To do so, we need to distinguish between *near* and *far* field numerical approximations [34]. The numerical near field approximation is given by:

$$g(p\Delta x_o, q\Delta x_o) = \mathcal{F}_{2D}^{-1} \exp\left\{ -j\pi\lambda z \left(\frac{m^2}{N\Delta x_0^2} + \frac{n^2}{N\Delta y_0^2} \right) \right\} \mathcal{F}_{2D}\{f(l\Delta x_0, k\Delta y_0)\}, \quad (4.11)$$

where Δx_0, Δy_0 are object and CCD resolution pixel size, with $0 \leq p, q, k, l \leq \sqrt{N} - 1$ and $\mathcal{F}_{2D}$ is the 2D Fourier transform. We assume that the size of the object and sensor size are $\sqrt{N}\Delta x_0 \times \sqrt{N}\Delta y_0$. The near field model is valid for the regime where $z \leq z_0 = \sqrt{N}\Delta x_0^2/\lambda$ [34]. For the working regime of $z \geq z_0 = \sqrt{N}\Delta x_0^2/\lambda$ the far field numerical approximation is given by the following:

$$g(p\Delta x_z, q\Delta y_z)$$
$$= \exp\left\{ \frac{j\pi}{\lambda z} \left(p^2\Delta x_z^2 + q^2\Delta y_z^2 \right) \right\} \mathcal{F}_{2D}\left[f\left(k\Delta x_0, l\Delta y_0 \right) \exp\left\{ \frac{j\pi}{\lambda z} \left(k^2\Delta x_0^2 + l^2\Delta y_0^2 \right) \right\} \right],$$
$$(4.12)$$

where $\Delta x_z = \lambda z/(\sqrt{N}\Delta x_0)$ is the output field's pixel size.

Let us consider the case where one wishes to design a sensing system that samples the object's diffraction field at some distance away from it using a small number of detectors. In this case the sensing system is best described using the randomly uniform subsampling scheme. If the object is sparse in the spatial domain, that is, $\Psi = \mathbf{I}$, it is shown in [11] that the coherence parameter for the near field numerical approximation is given by:

$$\mu_{1(near\ field)} = N[\Delta x_0^2/(\lambda z)]^2, \quad (4.13)$$

which means that the number of compressive measurements that are needed to accurately reconstruct the object is given by:

$$M \geq CN_F^2 \frac{S}{N} \log N, \quad (4.14)$$

where N_F denotes the recording device Fresnel number $N_F = N\frac{\Delta x_0 \Delta y_0}{4\lambda z}$ and C is a small constant factor [29]. Equation (4.14) determines that as the working distance gets larger, N_F decreases

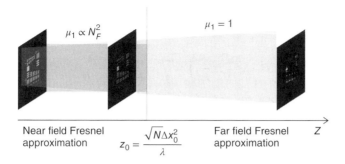

Figure 4.3 Illustration of numerical near and far field diffraction, and its relation with the coherence parameter, μ_1[26], *Source*: Y. Rivenson, A. Stern, and B. Javidi 2013. Reproduced with permission from The Optical Society

implying that fewer samples are required in order to reconstruct the signal accurately. It can be shown (see [11]) that when the near field approximation is not valid, the far field numerical approximation can be considered, and the coherence parameter becomes:

$$\mu_{1(near\ field)} = 1. \tag{4.15}$$

The number of required measurements is given by:

$$M \geq CS \log N, \tag{4.16}$$

which remains constant regardless of working distance. A physical intuition about these results is illustrated in Fig. 4.3 and explained as follows: It is known that the object's diffraction pattern spatial spread is inversely proportional to its Fresnel number. Thus, as we move away from the object plane (and the Fresnel number decreases) each sample contains information about a larger portion of the object. This implies that discarding some of the samples is possible since the missing information can be extracted from other samples, thus allowing reduction in the number of samples required to accurately reconstruct the object.

4.3.2 *Compressive Sensing Reconstruction Performance for a Spherical Wave Illuminated Object*

In many holographic applications, the object is illuminated by a spherical wavefront, especially in compact microcopy (lensless) systems. This illumination scheme is illustrated in Fig. 4.4.

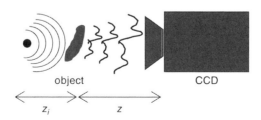

Figure 4.4 Fresnel propagation from an object illuminated by a diverging spherical wave

In this case, the calculation of the coherence parameter is slightly changed. Fresnel approximation for a 1D (for simplicity) diverging spherical wave in the free space is given by:

$$g(x) = \exp\left(j\pi\frac{x^2}{\lambda z_i}\right)f(x,y) * \exp\left(j\pi\frac{x^2}{\lambda z}\right) = \int \exp\left(j\pi\frac{\xi^2}{\lambda z_i}\right)f(\xi,\eta)\exp\left(j\pi\frac{(x-\xi)^2}{\lambda z}\right)d\xi$$

$$= \exp\left(j\pi\frac{x^2}{\lambda z}\right)\int f(\xi)\exp\left(j\pi\frac{\xi^2}{\lambda}\left(\frac{1}{z}+\frac{1}{z_i}\right)\right)\exp\left\{\frac{-j2\pi}{\lambda z}(x\xi)\right\}d\xi. \qquad (4.17)$$

For proper numerical representation, one can follow the same trail that leads to the determination of the near and far field numerical approximations for the plane wave case. Accordingly, the sampling criterion for the far field model is given by [34]:

$$\frac{\Delta x_0^2}{\lambda}\left(\frac{1}{z}+\frac{1}{z_i}\right) < \frac{1}{\sqrt{N}}, \qquad (4.18)$$

from which we can derive the far field limit of the Fresnel transform:

$$z > z_0 = \frac{\sqrt{N}\Delta x_0^2}{\lambda - \sqrt{N}\Delta x_0^2/z_i}. \qquad (4.19)$$

It is evident that the far field limit is dependent on the distance between the illumination source and the object and is higher compared with the plane wave case due to the $-\sqrt{N}\Delta x_0^2/z_1$ term in the denominator. Therefore, we can rewrite the coherence parameter, for the near field Fresnel approximation, when a spherical illumination is used as:

$$\mu_1 = \frac{\sqrt{N}\Delta x_0^2}{z(\lambda - \sqrt{N}\Delta x_0^2/z_i)}. \qquad (4.20)$$

As in the previous case, in the far field approximation $\mu_1 = 1$. In the limit of $\lambda z_i \to \infty$, Eq. (4.20) reduces to the coherence parameter, μ_1, for the case of planar field illumination:

$$\mu_1(\lambda z_i \to \infty) = \sqrt{N}\Delta x_0^2/(\lambda z) = \mu_{1(plane\ wave)}. \qquad (4.21)$$

Equation (4.21) shows us that the coherence factor obtained with the diverging spherical wave is higher than that with plane wave. Hence, the number of required samples with spherical wave illumination is higher. For the 2D case, it can be shown that [35]:

$$\mu^{2D} = \sqrt{N}\frac{\Delta x_0^2}{z(\lambda - \sqrt{N}\Delta x_0^2/z_i)}\sqrt{N}\frac{\Delta y_0^2}{z(\lambda - \sqrt{N}\Delta y_0^2/z_i)}, \qquad (4.22)$$

therefore, the required number of samples in the near field regime for the case where $\Delta x_0 = \Delta y_0$ is given by:

$$M \geq CN\left(\frac{\Delta x_0^2}{z_2\left(\lambda - \sqrt{N}\Delta x_0^2/z_1\right)}\right)^2 S\log N. \qquad (4.23)$$

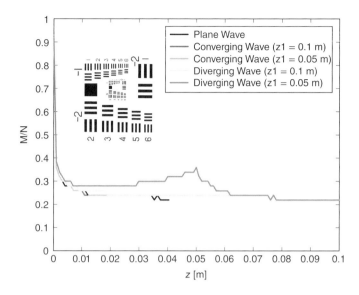

Figure 4.5 Compressive sensing ratio dependence on z for planar and spherical wave illumination. Simulation results obtained for the USAF 1951 resolution chart (*inset*)

Note that an illumination with a converging plane wave will give the opposite result, that is, the coherence parameter would be smaller for the same z distance of the object from the detector. The coherence parameter for the 1D converging wave is given by:

$$\mu_1 = \frac{\sqrt{N}\Delta x_0^2}{z(\lambda + \sqrt{N}\Delta x_0^2/z_i)}, \tag{4.24}$$

which means that for some applications it might be beneficial to use the converging wave illumination when the system number of pixels is the main issue rather than its axial compactness.

Figure 4.5 shows the compressive measurement ratio M/N as a function of imaging distance for different illumination schemes (planar wave, converging wave, and diverging wave). The results were obtained by simulating a compressive sensing Fresnel holography with undersampling the hologram field. The object used was a 1024×1024 pixel 1951 USAF resolution chart. The pixel spacing of the resolution chart is $\Delta x_0 = 5$ µm and the illuminating wavelength was $\lambda = 632.8$ nm. The detector was assumed to have 1024×1024 pixels. Fresnel samples were randomly chosen for different values of z. The algorithm ceased whenever it reached 32 dB reconstruction PSNR. From Fig. 4.5, it is evident that the compressive measurement ratio M/N was worse for the diverging wave illumination than for plane wave illumination. In this example converging wave illumination does not show any significant advantage over the planar wave illumination.

4.3.3   Reconstruction Performance for Non-Canonical Sparsifying Operators

When the object is not sparse in the spatial domain, that is, the sparsifying operator $\Psi \neq \mathbf{I}$, the number of necessary measurements, M, may differ according to the coherence parameter, μ_1,

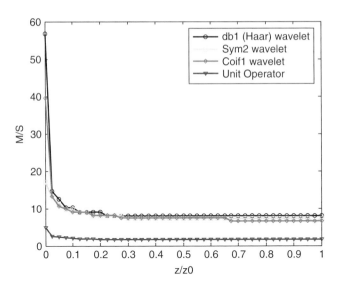

Figure 4.6 (Plate 6)   Simulation results showing the normalized compressive sampling ratio for different sparsifying bases [26]. *Source*: Y. Rivenson, A. Stern, and B. Javidi 2013. Reproduced with permission from The Optical Society. *See plate section for the color version*

between the Fresnel transform and the sparsifying operator. Unfortunately, deriving analytical results such as those in Eqs (4.13) and (4.15) for other sparsifying operators, can be extremely tedious. However, we shall demonstrate empirically that the trend of the analysis shown in [11], and predicted by Eqs. (4.13), (4.15) is valid for other popular sparsifying transforms. In Fig. 4.6 (Plate 6), we show simulation results for the normalized samples number ratio M/S obtained with various sparsifying bases: Haar, Coiflet, and Symlet wavelet expansions, compared with those obtained with the canonical representation $\Psi = I$. The object is a 1024 × 1024 pixel USAF 1951 resolution target. In order to reflect the dependence of the number of required samples, M, on z, the curves in Fig. 4.6 were normalized with respect to the sparsity level, S. This is because S depends on the wavelet type being used. For example, in this simulation the Haar expansion yielded $S/N = 0.017$, the Coiflets expansion yielded $S/N = 0.024$, the Symlet expansion yielded $S/N = 0.025$, while $S/N = 0.21$ for the canonical basis. It is witnessed from Fig. 4.6 that the ratio M/S, obtained with various wavelet sparsifying operators, Ψ, shares the same trend as predicted by Eq. (4.13) for $\Psi = I$ [11]; that is, it monotonically decreases with the working distance, z, until it approaches a constant asymptote, as predicted by Eq. (4.13). Figure 4.6 also shows that the lowest M/S ratio is achieved by the canonical sparsifying basis. This is expected from the relation between the Fresnel and Fourier transforms, and the fact that Fourier transform holds minimum coherence with the spikes' basis (see Section 4.2).

4.4   Applications of Compressive Digital Holographic Sensing

In this section, we consider holography based applications that take advantage of the CS paradigm. The applications presented here are versatile and include reconstruction of holograms captured with sparse pixel arrays, sparse camera positions for incoherent holography, reconstruction of an image through a partially occluding aperture, and object tomography from its recorded hologram. These applications, along with others, demonstrate why digital holography has based itself as a leading sensing modality for optical compressive sensing applications.

4.4.1 Compressive Fresnel Holography by Undersampling the Hologram Plane

In this subsection, the problem we wish to address is how to maintain high reconstruction accuracy of the scene while substantially reducing the number of detector pixels. These properties may be useful for reducing detector costs, scanning effort, reducing the captured data volume, or for extracting more information from a measurement. This may be of particular importance for holography in various spectral regimes (e.g., UV, IR, and THz), for incoherent holography or for improving imaging performance.

4.4.1.1 Improved Reconstruction using a Variable Density Subsampling Scheme

A more sophisticated approach can be applied when considering the subsampling scheme in light of the spatial distribution of the sparse coefficients obtained according to the optical setup, and not to just uniformly placing detectors at random across the sampling plane as prescribed by the conventional CS theory. In [9], it was shown that better reconstruction results are achieved when the sampling process puts more emphasis on sampling near the origin of the recorded Fresnel hologram, and less emphasis as we move away from the origin, according to some (non-uniform) probability density function [36].

In Fig. 4.7 we demonstrate this principle by applying a variable density subsampling to an off-axis Fresnel hologram. The object is a 2NIS coin recorded hologram with 100% of the pixels and is shown in Fig. 4.7(a). In Fig. 4.7(b) we see the reconstruction of the +1 order from standard Fresnel back propagation, from the full hologram. Subsampling using the proposed variable density of the full hologram is shown in Fig. 4.7(c), where only 6% of the hologram pixels were selected. This corresponds to only 6% of the measurements. In Fig. 4.7(d) the standard Fresnel back propagation from the subsampled hologram of Fig. 4.7(c) is shown. Due to the subsampling of data the reconstruction in Fig. 4.7(d) is much poorer than in Fig. 4.7(b). However, by applying the CS and reconstruction algorithm the image in Fig. 4.7(e) is obtained, which is almost similar to that obtained with the standard back propagation from full data.

4.4.1.2 Reducing the Scanning Effort in Multiple View Point Projection Incoherent Holography

Compressive digital holographic sensing was also demonstrated to dramatically improve the performance of multiple view projection (MVP) incoherent holography [37]. Multiple view projection holography is a method to obtain a digital hologram using a simple optical setup operating under spatially and temporarily "white" light illuminating conditions. The method requires only a conventional digital camera as a recording device. The MVP method

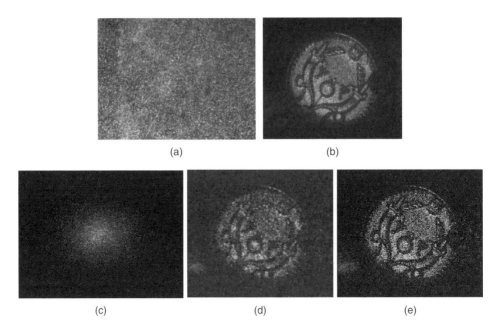

Figure 4.7   CDHS for a reflection, single shot, off-axis hologram. (a) Fresnel hologram of a 2NIS coin. (b) Back propagation reconstruction from the fully sampled hologram in (a). (c) 6% variable density random subsampling of (a). (d) Back propagation reconstruction from (c). Compressive sensing reconstruction in (e) yielding 31.2 dB PSNR (for the region of interest)

is basically divided into two steps as illustrated in Fig. 4.8. The first one is a scene acquisition step, in which multiple views of the scene are acquired by a camera translation. This step usually involves a tedious scanning effort, because it requires a separate camera exposure for each hologram pixel generation [37]. For instance, in order to record a hologram suitable for high definition display, $600 \times 800 \approx 2.88 \times 10^5$ projections should be acquired. The second step is referred to as the digital stage, where each view is digitally multiplied by a corresponding phase functions, and added afterwards. This process ultimately yields a digital Fourier or Fresnel hologram [37].

It was shown in [18] that by adopting the compressive digital holographic sensing approach to MVP holography, a significant reduction of the scanning effort in the acquisition step is possible. This is because (as discussed in Section 4.2), it requires about only $M = S\log N$ hologram pixels in order to accurately reconstruct the scene. Therefore, only $M = S\log N$ different views in the MVP method need to be acquired, instead of N views. Hence, an accurate reconstruction of the 3D scene can be obtained with only a fraction of the nominal number of exposures. The procedure works as follows:

1. Acquire $O(S\log N)$ projections of the 3D scene, instead of N projections required for the original MVP algorithm.

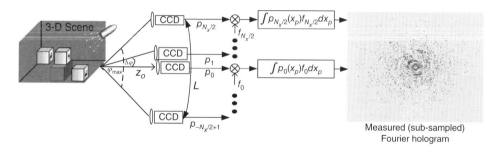

Figure 4.8   An illustration of compressive multiple view projection incoherent holography [18,37]. Using a CCD camera located at distance z_0 from a scene, $K \log N$ projections (denoted by p_i) are captured. Each acquired projection is digitally multiplied and summarized by a corresponding complex function to generate a subsampled hologram. *Source*: Y. Rivenson, A. Stern, and J. Rosen 2011 and N. T. Shaked, B. Katz, and J. Rosen 2009. Reproduced with permission from The Optical Society

2. Multiply each acquired projection by the corresponding phase function, as described previously. This way, a partial Fourier hologram is created, with only a fraction of the projection (coefficients).
3. Reconstruct the signal using the total variation (TV) minimization constraint (4.5). The 3D scene, is reconstructed with focusing on the different planes.

Figure 4.9 (Plate 7) demonstrates simulation results of compressive MVP holography. The simulation was carried as follows: a 3D scene was synthesized where the letters B,G,U were placed at different axial and transversal locations. Each acquired projection was multiplied with a phase function f_n, and then summed, as shown in Fig. 4.8. This way, we have generated a Fourier hologram. The hologram we created was of size 256×256 pixels, which corresponds to 256×256 projections. We than subsampled our resultant Fourier hologram according to the sampling scheme described in Fig. 4.8. For a real experiment, there is no need to acquire all the projections and then subsample; instead one can simply acquire a fraction of the projections needed to reconstruct the scene. The 100% samples were generated in this simulation in order to get a reference for our results. At the next stage a TV (4.5) or (ℓ_1) minimization (4.4) is performed. Figure 4.9 depicts simulation results for different number of compressive samples and different B,G,U letter displacements. No evident difference is found for compressive sensing based reconstruction from 6% of the samples (Figs. 4.9b and e) compared with reconstruction from the fully acquired hologram (Figs. 4.9a and d). The reconstruction results from the compressive sensing when using only 2.5% of the samples is quite satisfactory as well. Other, including real experimental results can be found in [18].

The proposed compressive digital holographic sensing approach for MVP holography holds a number of advantages: First, the acquisition effort is substantially reduced. For instance, if the projections are captured by a scanning process, then the total scanning time may be reduced in order of $15-20$ ($15-20$ times faster). Secondly, less data needs to be transmitted or stored in comparison with conventional MVP methods. The method does not require any additional hardware at the sensor level, and can remedy a temporal object's resolution limitations and bandwidth bottlenecks.

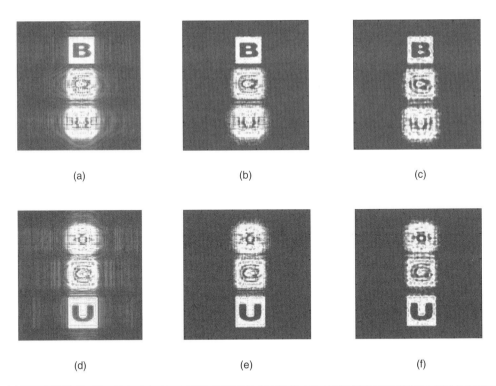

(a)   (b)   (c)

(d)   (e)   (f)

Figure 4.9 (Plate 7)   Reconstruction examples of the B (forward) and U (backward) planes. (a) Reconstruction of the B plane form 100% of the projections. (b) CS reconstruction of the B plane forms 6% of the projections. (c) CS reconstruction of the B plane forms 2.5% of the projections. (d) Reconstruction of the U plane forms 100% of the projections. (e) CS reconstruction of the U plane forms 6% of the projections. (f) CS reconstruction of the U plane forms 2.5% of the projections. *See plate section for the color version*

4.4.1.3   Other Applications Applying Compressive Fresnel Holography

The framework of compressive Fresnel holography can also be applied in order to extract super resolved information. This was demonstrated in work by Coskun *et al.* in [15], where the in-line holography framework was used to extract superresolved fluroscence beads and by Liu *et al.* [22], where a compressive digital holograhphy in-line setup was used in order to localize a moving object at an accuracy of (1/45)th of the detector's pixel size.

Another interesting effort is towards the reconstruction of multi-dimensional images as proposed in [21]. Owing CS, only a fraction of the pixels needs to be captured, therefore one can create a sensor in which a small number of pixels capture different information about the object, for example, polarization, color, and so on. Hence it is possible to reconstruct multi-dimensional data with a single shot and without the need of detection hardware for each dimension.

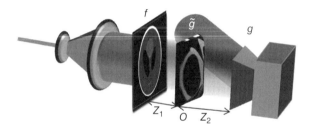

Figure 4.10 A schematic setup for the acquisition of a partially occluded object field. In the object and occluded planes, black represents totally opaque regions, the gray level represents turbid regions

4.4.2 Compressive Digital Holography for Reconstruction of an Object Set Behind an Opaque Medium

Here we show the application of compressive Fresnel holography reconstruction for an object that is set behind a partially opaque media. The sensing mechanism is based on the Fresnel transform of the object. The approach is briefly described in the following paragraphs.

4.4.2.1 Partially Occluded Object Recovery as a Compressive Sensing Problem

The schematic optical setup is shown in Fig. 4.10. Let us assume that the input object $f(x,y)$ is illuminated with a coherent plane wave of wavelength λ. The object's wavefront propagates a distance z_1 and hits a partially occluding plane, $o(x,y)$. The truncated and distorted wavefront propagates another distance, z_2, until it reaches the CCD sensor. This wavefront can be recorded using any of the various holographic techniques.

The occluded object wavefront can be described as follows:

$$g(x,y) = \left[f(x,y) * \exp\left\{ \frac{j\pi}{\lambda z_1} \left(x^2 + y^2 \right) \right\} \right] \times o(x,y) * \exp\left\{ \frac{j\pi}{\lambda z_2} \left(x^2 + y^2 \right) \right\}. \quad (4.25)$$

It is noted that essentially no information is added or discarded with the free space propagation from the occluding plane to the detector plane; therefore, the information loss is from $f(x,y)$ to $\widetilde{g}(x_{z1}, y_{z1})$. After discretization and applying the numerical far field Fresnel approximation given in Eq. (4.12), together with standard numerical Fresnel back propagation for distance $-z_2$, we obtain:

$$\widetilde{g}(p\Delta x_{z1}, q\Delta y_{z1}) = o(p\Delta x_{z1}, q\Delta y_{z1}) \times \exp\left\{ \frac{j\pi}{\lambda z} \left(p^2 \Delta x_{z1}^2 + q^2 \Delta y_{z1}^2 \right) \right\}$$

$$\times \mathcal{F}_{2D}\left[f\left(k\Delta x_0, l\Delta y_0 \right) \exp\left\{ \frac{j\pi}{\lambda z_1} \left(k^2 \Delta x_0^2 + l^2 \Delta y_0^2 \right) \right\} \right], \quad (4.26)$$

where $0 \leq p, q, k, l \leq N - 1$. The forward sensing model in Eq. (4.26) can be described as a subsampling or distortion (given by the fact that o is not a clear aperture) of the object's Fresnel wave propagation. In this sense it resembles the CS scheme. However, unlike conventional

CS, real world occluding environments are most likely to preserve some sort of structure and certainly cannot be modeled as randomly drawn samples from a uniform distribution.

4.4.2.2 Object Recovery Performance Guarantees

Examining Eq. (4.26) suggests that our ability to accurately reconstruct the object should be determined by μ_2 and Eq. (4.9). For the numerical far field approximation, it can be shown [20] that the coherence parameter μ_2 is given by:

$$\mu_2^{FF} = \max_{m \neq l} \frac{|\widehat{O}(m-l) \otimes \widehat{O}(m-l)|}{\|o\|_2^2}, \tag{4.27}$$

where $\otimes$ denotes the correlation operator, $\|.\|_2$ is the ℓ_2-norm operator and $\widehat{O}$ is the two dimensional Fourier transform of o, $\widehat{O} = \mathcal{F}_{2D}\{o\}$. The indices $0 \leq m, l \leq N - 1$ denote the sensing matrix columns. Equation (4.27) holds for the common case where $\Delta x_{z1} = \lambda z_1/(\sqrt{N}\Delta x_0)$ [1,34], while for the more general formulation the reader is referred to [20].

The result in Eq. (4.27) formulates the coherence parameter dependence on the structural properties of the occluding plane. The number of S-sparse signal elements that can be accurately recovered is inversely proportional to μ_2, according to Eq. (4.9).

4.4.2.3 Simulation – Reconstruction of an Object Through a Turbid, Partially Opaque Medium

In the following we illustrate the effectiveness of the method using a simulation. Real experimental results may be found in [20]. The proposed method is demonstrated using a MATLAB simulation. The schematic optical description is shown in Fig. 4.10.

The object in Fig. 4.11(a) is a 512×512 pixel phantom image. A hologram recording process is simulated, according to Eq. (4.26), where an occluding plane shown in Fig. 4.11(b) is composed from totally opaque regions denoted in black and non-opaque regions, which are composed from a complex random media. The non-opaque regions cover merely 28% from the field of view, thus almost three-quarters of the field is blocked. Noise is added at the detector, such that the hologram's SNR is 30 dB. Reconstruction from a noisy hologram of the occluded object using numerical back propagation is shown in Fig. 4.11(c). It can be seen that due to the occlusion most of the object's features are lost. In contrast, an almost perfect reconstruction is shown in Fig. 4.11(d) where the reconstruction is carried out by the proposed formulation of the problem as a compressive Fresnel holography problem.

Thus, we have shown that by using a compressive sensing formulation scheme for holographic imaging of objects located behind a complex partially occluding media an almost exact recovery is possible. This can be achieved using a single shot with an off-axis holography setup or a small number of acquisitions using a phase-shifting holography procedure. The results may be applicable to partially opaque, turbid, or non-linear media, where its physical properties are known in advance, or can be extracted during the object sensing process [20].

4.4.3 Reconstruction of 3D Tomograms from a 2D Hologram

Here we discuss the reconstruction of a 3D object tomography from its single recorded 2D hologram. Numerical reconstruction obtained by digitally focusing on different object depth

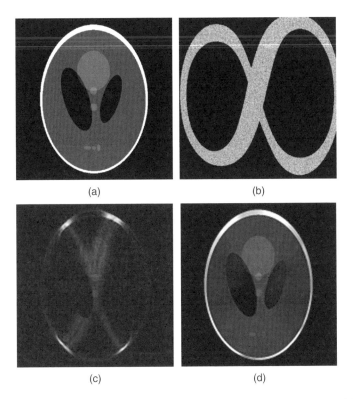

Figure 4.11 Simulation results for phantom reconstruction for an occluded phantom, with added detector noise (SNR of 30dB). (a) Original object (b) Occluding plane, which is composed from complex random areas (grayscale regions) and totally opaque (black regions). (c) Back propagation from the acquired hologram. (d) Reconstruction carried out using the proposed compressive Fresnel holography approach. The resultant reconstruction PSNR = 24.54 dB

planes may be distorted due to out-of-focus object points located in other object planes, as seen in Fig. 4.9. These disturbances are the result of an incomplete model of the system because the back propagation model of Eqs. (4.11)–(4.12) represents a 2D-2D model linking the hologram plane to a single depth plane, thus ignoring other object planes. Clearly, applying reconstruction techniques based on 2D-2D models for 3D-2D acquisition systems is subject to distortions at object points disobeying the model; that is, object points located in another depth planes.

4.4.3.1 Recasting the 3D Object Reconstruction from a Single 2D Hologram as a Compressive Sensing Problem

In order to avoid the out of focus distortions, a 3D-2D forward model relating all the $N_{object} = N_x \times N_y \times N_z$ voxels to $N_{holo} = N_x \times N_y$ hologram pixels should be used. We may formulate mathematically the (discrete) forward model relating the 3D object, $O(.)$, to its 2D wavefront recorded with the holography process, $U(.)$, as follows:

$$U(k\Delta x, l\Delta y) = \sum_{r=1}^{Nz} \mathcal{F}_{2D}^{-1}\{e^{-j\pi\lambda r\Delta z[(\Delta v_x m)^2 + (\Delta v_y n)^2]}e^{-j\frac{2\pi}{\lambda}r\Delta z} \times \mathcal{F}_{2D}[f(p\Delta x, q\Delta y; r\Delta z)]\}, \quad (4.28)$$

with $0 \leq p, m \leq N_x - 1$, $0 \leq q, n \leq N_y - 1$, and $1 \leq r \leq N_z - 1$. In Eq. (4.28) the 3D object space is partitioned in a grid with $N_x \times N_y \times N_z$ voxels, each of size $\Delta x \times \Delta y \times \Delta z$. The $\mathcal{F}_{2D}$ operator denotes the 2D discrete Fourier transform. The numerical model in Eq. (4.28) is based on the near field Fresnel approximation but other models may also be considered as discussed in Section 4.3.1. The spatial frequency variables are $\Delta v_x = 1/(N_x \Delta x)$ and $\Delta v_y = 1/(N_y \Delta y)$. This model assumes regular sampling intervals Δz between the different depth planes. The problem of reconstructing the 3D object from its 2D hologram is naturally ill-posed, since there are N_z times more variables than equations. In order to handle this problem we assume that the object is sparse, such that $S < N_{holo}$, therefore we may recast it as a compressive sensing problem. In case where the object in sparse in space domain, we solve the following minimization problem:

$$\min\{\|\mathbf{U} - \mathbf{\Phi O}^T\|_2^2 + \tau\|\mathbf{O}\|_1\}. \quad (4.29)$$

In Eq. (4.29) $\mathbf{\Phi}$ is the 3D-2D forward model from Eq. (4.28) written as a matrix-vector multiplication such that the vector $\mathbf{U}$ representing the impinging field is:

$$\mathbf{U} = \left[\mathcal{F}_{2D}^{-1}e^{-j\frac{2\pi}{\lambda}\Delta z}\mathbf{Q}_{\lambda^2\Delta z}\mathcal{F}_{2D}; \ldots ; \mathcal{F}_{2D}^{-1}e^{-j\frac{2\pi}{\lambda}N_z\Delta z}\mathbf{Q}_{\lambda^2 N_z\Delta z}\mathcal{F}_{2D}\right][\mathbf{o}_{\Delta z}; \ldots ; \mathbf{o}_{N_Z\Delta z}]^T = \mathbf{\Phi O}^T, \quad (4.30)$$

where the matrix $\mathbf{Q}_{\lambda^2 r\Delta z}$ is a diagonal matrix which accounts for the quadratic phase terms of Eq. (4.28) and $[\mathbf{o}_{\Delta z}; \ldots ; \mathbf{o}_{N_Z\Delta z}]^T$ is a lexicographical representation of the 3D object. In this case, the reconstruction guarantees for the accurate 3D object from its 2D holographic projection is given by [38]:

$$\mu_2 = \frac{\Delta x\Delta y}{\lambda\Delta z}. \quad (4.31)$$

Thus, by combining Eqs (4.31) and (4.9) we obtain the number of sparse object features that can be accurately reconstructed:

$$S \leq 0.5[1 + \lambda\Delta z/(\Delta x\Delta y)], \quad (4.32)$$

It can be shown that under certain conditions Eq. (4.32) implies axial super-resolution 3D object reconstruction. For more information the reader is referred to [38].

The most widely used tomographic 3D object reconstruction considers the solution of the following TV-norm minimization problem:

$$\min\{\|\mathbf{U} - \mathbf{\Phi O}^T\|_2^2 + \tau\|\mathbf{O}\|_{TV}\}, \quad (4.33)$$

which is the unconstraint formulation of Eq. (5), where $\|\| \|_{TV}$ is the 3D total variation operator given by:

$$\|\mathbf{O}\|_{TV} = \sum_l \sum_{i,j} \sqrt{(\mathbf{o}_{i+1,j,l} - \mathbf{o}_{i,j,l})^2 + (\mathbf{o}_{i,j+1,l} - \mathbf{o}_{i,j,l})^2}, \quad (4.34)$$

and τ is a regularization parameter that controls the ratio between the fidelity term and the sparsity level of the object.

This reconstruction approach using the 3D-2D forward model was successfully applied to reconstruction of a 3D object from a single recorded Gabor hologram [8], for reconstruction of a 3D object in the THz regime [12], tomographic reconstruction for diffuse objects from multiple off-axis exposures [13], for incoherent optical scanning holography [14], and also for video rate tomographic microscopy [19]. Recently it was shown that by changing the optical setup and recording the holographic projections of the object from different angles, using the same principles from classic tomography, it is possible to improve reconstruction accuracy [23,24] when combined with the compressive holography technique.

4.4.3.2 Improved Depth Resolution with the Multiple Projection Holography Technique

Here we discuss two other applications of this method. The first is using it for the MVP hologram reconstruction that we have presented in Subsection 4.4.1.2. Like other holography, object reconstruction using the 3D-2D reconstruction model also gives us the ability to improve the tomographic sectioning of the scene; that is, while still taking only a fraction of the projections as explained in Subsection 4.4.1.2. Demonstration of tomographic sectioning of a scene with white light illumination and standard cameras has been shown in [18].

Another benefit from using the compressive holography framework for the MVP generated hologram is described by the following: In multiple aperture systems, the axial resolution increases linearly with the extension of the system's baseline typically defined by its synthetic aperture. This means that the resolution of MVP is linearly proportional to the number of acquired views. However, as shown in [18], when using the compressive digital holographic sensing scheme the axial resolution increases at a higher rate than the required number of projections, by a ratio of $N/S \log N$. This stems from the rule of thumb, which predicts that as the amount of pixels in an image grows, its sparsity level increases at a slower rate. Therefore, the term N/S increases as the dimensionality of the problem increases, and in turn, the ratio of the axial resolution gain relative to the number of projections grows accordingly. Hence, it is possible to obtain the superior axial resolution benefits of the MVP method while reducing the scanning effort. In Fig. 4.12, we illustrate the benefits of CS applied to multiple view projection holography using the following numerical experiment: A synthetic object composed from the letters CDHS was generated where the letters DH were placed at one plane and the letters CS were placed in a more distant plane. A multiple view projection hologram was recorded using only 3275 projections (a representative projection is shown in Fig. 4.12a), which form just about 5% of the nominal number of the 256×256 projections required in the multiple view projection holography method. The obtained hologram is illustrated in Fig. 4.12(b). For contrast, reconstruction using standard numerical back propagation from the subsampled hologram is shown in Fig. 4.12(c). While the reconstruction in Fig. 4.12(c) suffers from severe out-of-plane crosstalk noise, the reconstruction in Fig. 4.12(b) exhibits clear depth slicing.

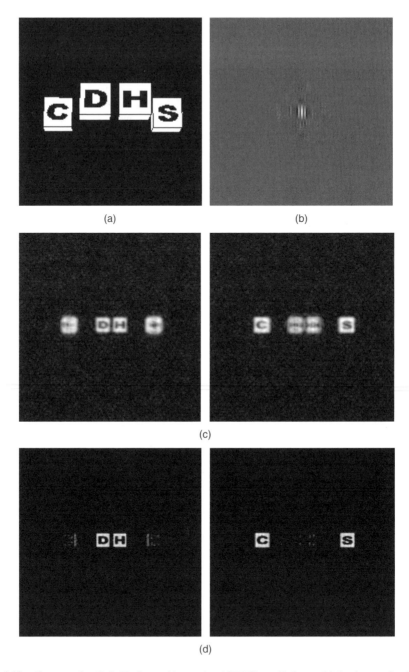

(a)   (b)

(c)

(d)

Figure 4.12 Compressive digital holographic sensing (CDHS) applied to multiple view projection incoherent holography. (a) One of the captured views of the scene. (b) Acquired, subsampled hologram, where only 5% of the nominal number views are acquired. (c) Standard numerical back propagation of two of the object planes from (b). (d) The two corresponding planes from (c) reconstructed using the compressive sensing approach. The depth sectioning is evident [26]. *Source*: Y. Rivenson, A. Stern, and B. Javidi 2013. Reproduced with permission from The Optical Society

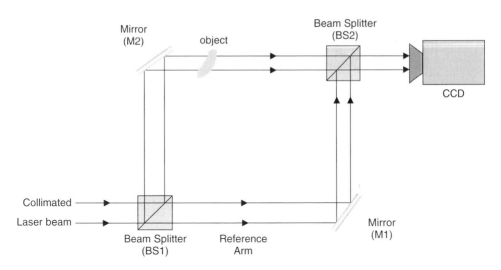

Figure 4.13 A schematic single exposure in line (SEOL) holography setup [26]. *Source*: Y. Rivenson, A. Stern, and B. Javidi 2013. Reproduced with permission from The Optical Society

4.4.3.3 Improved Depth Discrimination using the Single-Exposure In-Line Holography and Compressive Sensing

Another application that was recently demonstrated combined the 3D-2D compressive holography framework with the single exposure in line (SEOL) holography recording setup which was proposed in [39]. The SEOL digital holography was designed for capturing dynamic events in a 3D scene, a micro-organism, or its movement [39–43]. As illustrated in Fig. 4.13, SEOL digital holography utilizes a Mach–Zehnder interferometer setup to record the Fresnel diffraction field of the 3D object in a way similar to phase-shifting digital holography. However, in contrast to phase-shifting on-line digital holography techniques, SEOL digital holography uses only a single exposure. Prior works [39–43] that employed the SEOL holography setup have applied image processing, image recognition, and statistical inference techniques in order to perform tasks such as recognition, tracking, and visualization of micro-organisms. Although it was shown that most interference terms, that is, bias and twin image effects, can often be reduced or neglected in SEOL digital holographic microscopy [42], there are still reconstruction distortions due to out-of-focus object fields located in other object planes, which may impair the analysis tasks.

As explained in Section 4.4.3.1, using the compressive sensing approach with the 3D-2D forward operator leads to improved reconstruction results. This improvement is also apparent when the SEOL recording scheme is employed. Experimental results can be found in [43].

Furthermore, it can be shown that combining the SEOL and compressive holography framework can yield an almost perfect setup for practicing compressive holography. This is owing to three properties of the SEOL. First, the resolution of the system is the same as the in-line holography setup, which is at least twice as high when compared to achievable resolution from an off-axis recording setup. The second property is that for practical microscopic objects, the hologram's acquisition can be performed in a single shot, while allowing satisfactory reconstruction results [41]. The first and second properties are also common for Gabor holography.

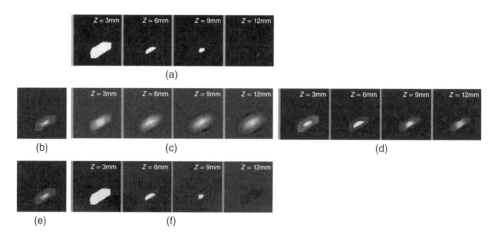

Figure 4.14   (a) Three depth planes of a 3D object. (b) Noisy hologram, (c) depth planes reconstructed using conventional field back-propagation, (d) CS reconstruction results from (b), (e) noisy hologram with reference wave intensity four times larger than in (d), (f) CS reconstructions from (e)

The third property, which differs from Gabor (in-line) holography, is that the SEOL holography behaves as a heterodyne system. This enables the recording of digital holograms with an improved *signal-to-noise ratio* (SNR) by proper control of the amplitude partition of the reference and object arms. The SNR is a figure of merit in DH, and generally its improvement yields enhanced lateral object resolution details [43–45]. Combining this property of the SEOL setup, while introducing the CS formulation, has yielded improved axial resolution with comparison to compressive reconstruction from a Gabor holographic recording [43]. This is demonstrated in Fig. 4.14 where the results of a numerical experiment demonstrates improved axial resolution (3D sectioning) by approaching SEOL in a CS framework together with proper intensity partition on the reference and object beams [43]. In this simulation, the reference beam was set to be four times larger than the object's beam intensity. Comparing the reconstruction results from of the standard back-propagation, compressive holography reconstruction with the hologram recorded in Gabor setup, and compressive SEOL holography reconstruction, it can be witnessed that the latter gives the most accurate reconstruction results.

We may conclude that the SEOL digital holographic setup, combined with the compressive sensing framework can be considered an almost ideal 3D object inference digital holographic based system; it offers the high resolution and large FOV associated with Gabor holography, and provides the robustness and sensitivity associated with off-axis holography setup, along with rapid frame acquisition rate associated with Gabor and off-axis configurations.

4.5   Conclusion

In this chapter we have demonstrated the usefulness of compressive sensing theory in the realm of digital holography. The basic compressive sensing theory was presented along with theoretical analysis of implementation aspects for digital holography. The analysis accounted for the imaging geometry, type of illumination, sensor size and resolution, object spatial size and resolution, sparsity of the object, and sparsifying transform used for representing the object.

A brief survey of compressive digital holographic sensing applications was presented. We have also demonstrated that CS together with appropriate modeling can help in recovering objects imaged through a complex random and/or partially opaque media. Furthermore, inference of an object tomography from its hologram was discussed. Improvement of the 3D resolution with incoherent multiple view projection holography, and with in-line SEOL holography, were demonstrated.

In summary, we have shown that CS can drastically improve the extraction of the information captured with holographic setups. We believe that future mutual synergies between digital holography and compressive sensing will demonstrate further improvements in 3D imaging performance, will yield new optical setup designs, and enable new applications.

Acknowledgments

This work was supported by the Israel Ministry of Science and Technology (MOST) to YR, AS and JR, and by The Israel Science Foundation (ISF) to AS and JR.

References

[1] Kreis T., *Handbook of Holographic Interferometry*, 1st edn, Wiley-VCH Verlag, Weinheim), Chap. 3 (2004).
[2] Donoho D., "Compressed sensing", *IEEE Trans. on Information Theory*, **52**(4), 1289–1306 (2006).
[3] Candès E. J., J. K. Romberg, and T. Tao, "Stable signal recovery from incomplete and inaccurate measurements", *Communications on Pure and Applied Mathematics* **59**(8), 1207–1223 (2006).
[4] Eldar Y. C. and G. Kutyniok, *Compressed Sensing: Theory and Applications*, Cambridge University Press (2012).
[5] Bruckstein A. M., D. L. Donoho, and M. Elad, "From Sparse Solutions of Systems of Equations to Sparse Modeling of Signals and Images", *SIAM Review* **51**(1), 34–81 (2009).
[6] Stern A., "Method and system for compressed imaging", US patent app. Nr. **12**/605866 (2008).
[7] Denis L., D. Lorenz, E. Thiébaut, C. Fournier, and D. Trede, "Inline hologram reconstruction with sparsity constraints," *Opt. Lett.* **34**, 3475–3477 (2009).
[8] Brady D. J., K. Choi, D. L. Marks, R. Horisaki, and S. Lim, "Compressive Holography," *Opt. Express* **17**, 13040–13049 (2009).
[9] Rivenson Y., A. Stern, and B. Javidi, "Compressive Fresnel holography," *Display Technology, Journal of* **6**(10), 506–509 (2010).
[10] Marim M. M., M. Atlan, E. Angelini, and J.-C. Olivo-Marin, "Compressed sensing with off-axis frequency-shifting holography," *Opt. Lett.* **35**, 871–873 (2010).
[11] Rivenson Y. and A. Stern, "Conditions for practicing compressive Fresnel holography," *Opt. Lett.* **36**, 3365–3367 (2011).
[12] Fernandez Cull C., D. A. Wikner, J. N. Mait, M. Mattheiss, and D. J. Brady, "Millimeter-wave compressive holography," *Appl. Opt.* **49**, E67–E82 (2010).
[13] Choi K., R. Horisaki, J. Hahn, S. Lim, D. L. Marks, T. J. Schulz, and D. J. Brady, "Compressive holography of diffuse objects," *Appl. Opt.* **49**, H1–H10 (2010).
[14] Zhang X. and E. Y. Lam, "Edge-preserving sectional image reconstruction in optical scanning holography," *J. Opt. Soc. Am. A* **27**, 1630–1637 (2010).
[15] Coskun A. F., I. Sencan, T.-W. Su, and A. Ozcan, "Lensless wide-field fluorescent imaging on a chip using compressive decoding of sparse objects," *Opt. Express* **18**, 10510–10523 (2010).

[16] Greenbaum A., W. Luo, T. Su, Z. Göröcs, L. Xue, S. O. Isikman, *et al.*, Imaging without lenses: Achievements and remaining challenges of wide-field on-chip microscopy. *Nature Methods* **9**(9), pp. 889–895 (2012).

[17] Marim M., E. Angelini, J.-C. Olivo-Marin, and M. Atlan, "Off-axis compressed holographic microscopy in low-light conditions," *Opt. Lett.* **36**, 79–81 (2011).

[18] Rivenson Y., A. Stern, and J. Rosen, "Compressive multiple view projection incoherent holography," *Opt. Express* **19**, 6109–6118 (2011).

[19] Hahn J., S. Lim, K. Choi, R. Horisaki, and D. J. Brady, "Video-rate compressive holographic microscopic tomography," *Opt. Express* **19**, 7289–7298 (2011).

[20] Rivenson Y., A. Rot, S. Balber, A. Stern, and J. Rosen, "Recovery of partially occluded objects by applying compressive Fresnel holography," *Opt. Lett.* **37**, 1757–1759 (2012).

[21] Horisaki R., J. Tanida, A. Stern, and B. Javidi, "Multidimensional imaging using compressive Fresnel holography," *Opt. Lett.* **37**, 2013–2015 (2012).

[22] Liu Y., L. Tian, J. Lee, H. Huang, M. Triantafyllou, and G. Barbastathis, "Scanning-free compressive holography for object localization with subpixel accuracy," *Opt. Lett.* **37**, 3357–3359 (2012).

[23] Nehmetallah G. and P. Banerjee, "Applications of digital and analog holography in three-dimensional imaging," *Adv. Opt. Photon.* **4**, 472–553 (2012).

[24] Williams L., G. Nehmetallah, and P. Banerjee, "Digital tomographic compressive holographic reconstruction of 3D objects in transmissive and reflective geometries," *App. Opt.* **52**, 1702–1710 (2013).

[25] Lim S., D. Marks, and D. Brady, "Sampling and processing for compressive holography," *Appl. Opt.* **50**, H75–H86 (2011).

[26] Rivenson Y., A. Stern, and B. Javidi, "Overview of compressive sensing techniques applied in holography," *Appl. Opt.* **52**, A423–A432 (2013).

[27] Stern A. and B. Javidi, "Random projections imaging with extended space-bandwidth product," *Journal of Display Technology* **3**(3), 315–320 (2007).

[28] Donoho D. L. and M. Elad, "Optimally sparse representation in general (nonorthogonal) dictionaries via ℓ_1 minimization," *Proc. Nat. Acad. Sci. USA*, **100**(5), 2197–2202 (2003).

[29] Candès E. J. and Y. Plan, A probabilistic and RIPless theory of compressed sensing. *IEEE Transactions on Information Theory* **57**(11), 7235–7254 (2011).

[30] Rudin L., S. Osher and E. Fatemi, "Nonlinear total variation based noise removal algorithm," *Physica D* **60**, 259–268 (1992).

[31] Needell D. and R. Ward, "Stable image reconstruction using total variation minimization," *SIAM Journal on Imaging Sciences* **6**(2), 1035–1058 (2013).

[32] Lustig M., D. L. Donoho, J. M. Santos and J. M. Pauly, "Compressed Sensing MRI," *IEEE Signal Processing Magazine* **25**(2), 72–82 (2008).

[33] Candès E. J. and J. Romberg, "Sparsity and incoherence in compressive sampling," *Inverse Problems* **23**, 969–985 (2006).

[34] Mas D., J. Garcia, C. Ferreira, L. M. Bernardo and F. Marinho, "Fast algorithms for free-space diffraction patterns calculation," *Opt. Comm.* **164**, 233–245 (1999).

[35] Rivenson Y. and A. Stern, "Compressed imaging with separable sensing operator," *IEEE Signal Processing Letters* **16**(6), 449–452 (2009).

[36] Van der Lught A., "Optimum sampling of Fresnel transforms," *Applied Optics* **29**, 3352–3361 (1990).

[37] Shaked N. T., B. Katz, and J. Rosen, "Review of three-dimensional holographic imaging by multiple-viewpoint-projection based methods," *Appl. Opt.* **48**, H120–H136 (2009).

[38] Rivenson Y., A. Stern, and J. Rosen, "Reconstruction guarantees for compressive tomographic holography," *Opt. Lett.* **38**, 2509–2511 (2013).

[39] Javidi B., I. Moon, S. Yeom and E. Carapezza, "Three-dimensional imaging and recognition of microorganism using single-exposure on-line (SEOL) digital holography," *Opt. Express* **13**, 4402–4506 (2005).

[40] Yeom S. and B. Javidi, "Automatic identification of biological microorganisms using three-dimensional complex morphology," *J. Biomed. Opt.* **11**, 024017 (2006).

[41] Stern A. and B. Javidi, "Theoretical analysis of three-dimensional imaging and recognition of micro-organisms with a single-exposure on-line holographic microscope," *J. Opt. Soc. Am. A* **24**, 163–168 (2007).

[42] Moon I., M. Daneshpanah, B. Javidi, and A. Stern, "Automated three-dimensional imaging, identification and tracking of micro/nano biological organisms by holographic microscopy," *Proc. IEEE* **97**, 990–1010 (2009).

[43] Rivenson Y., A. Stern, and B. Javidi, "Improved three-dimensional resolution by single exposure in-line compressive holography," *Appl. Opt.* **52**, A223–A231 (2013).

[44] Charrière F., T. Colomb, F. Montfort, E. Cuche, P. Marquet, and C. Depeursinge, "Shot-noise influence on the reconstructed phase image signal-to-noise ratio in digital holographic microscopy," *Appl. Opt.* **45**, 7667–7673 (2006).

[45] Tippie A. and J. Fienup, "Weak-object image reconstructions with single-shot digital holography," in *Digital Holography and Three-Dimensional Imaging*, OSA Technical Digest (Optical Society of America, 2012), paper DM4C.5.

5

Dispersion Compensation in Holograms Reconstructed by Femtosecond Light Pulses

Omel Mendoza-Yero[1,2], Jorge Pérez-Vizcaíno[1,2], Lluís Martínez-León[1,2], Gladys Mínguez-Vega[1,2], Vicent Climent[1,2], Jesús Lancis[1,2] and Pedro Andrés[3]

[1]*INIT – Institut de Noves Tecnologies de la Imatge, Universitat Jaume I, Spain*
[2]*GROC·UJI – Grup de Recerca d'Òptica, Dept. de Física, Universitat Jaume I, Spain*
[3]*Department d'Òptica, Universitat de València, Spain*

5.1 Introduction

The generation of a desired diffraction pattern when employing a femtosecond laser pulse source is a matter of great interest in several hot topics of research, such as high-speed microprocessing [1], multiphoton microscopy [2], optical trapping of particles [3], and the generation of optical vortices [4]. Among other possible approaches, computer generated holograms (CGHs) are diffractive optical elements (DOEs) designed to shape an optical beam into a user-defined intensity distribution in one shot. Usually, CGHs are implemented onto voltage-addressed flat displays working as phase-only spatial light modulators (SLMs). This implementation allows for dynamic applications since the intensity and phase pattern delivered onto the target can be changed in real time.

It is worth remarking that when a pulsed beam passes through a DOE two relevant modifications appear: *angular dispersion*, due to the strong dependence of the diffraction phenomenon with the wavelength, and *temporal stretching*, originated from the propagation time difference (PTD) in free space. Consequently, DOEs enable precise spatiotemporal control of the femtosecond pulsed beam only in the long-pulse duration regime (above 100 fs), as reported in the pioneer paper by Nolte *et al.* [5]. For shorter pulse durations, the significant bandwidth of ultrashort pulses results in a coupling of the spatiotemporal effects causing the degradation

Multi-dimensional Imaging, First Edition. Edited by Bahram Javidi, Enrique Tajahuerce and Pedro Andrés.
© 2014 John Wiley & Sons, Ltd. Published 2014 by John Wiley & Sons, Ltd.

of the output image and the stretching of the pulse duration. Thus, dispersion compensation techniques are required to minimize these effects at the target.

Several efforts have been conducted in the past few years to compensate for the angular dispersion in DOE-based systems working with femtosecond light beams, with temporal stretching remaining uncorrected. In the paper by Amako *et al.* [6], the femtosecond beam is split by means of a diffraction grating and the lateral walk-off among the different spectral components of diffraction maxima is corrected by focusing the pulsed light with a diffractive lens (DL). The focal length of a DL is proportional to the wavenumber of the incoming radiation, which is the key point to attain dispersion compensation. The pulse is temporally stretched at the focal plane, which leads to a considerable reduction of the laser peak power in the output plane. A system consisting of a Dammann grating and a subsequent *m*-time-density grating, arranged in a conventional double grating mounting, has been proposed to reduce the angular dispersion associated with the *m*th-order beam generated by the Dammann element [7]. Here, pulse broadening due to diffraction is not as large as that produced by the grating pair used in other techniques because of the low period of the Dammann grating: 10 lines/mm. Kuroiwa *et al.* have shown that the large chromatic dispersion effects induced by DOEs can be reduced when the focalization of the pulse is achieved with a refractive lens (RL), instead of being included in the CGH design [8]. Finally, and in the field of micromachining, Jesacher *et al.* have proposed a way to fabricate 3D structures with high spatial quality by doing machining at different depths of a crystal, but close to the optical axis to avoid chromatic aberrations [9].

This chapter describes how the spatiotemporal dispersion associated with the diffraction of broadband femtosecond light pulses through CGHs [10] can be compensated to a first order with a properly designed dispersion compensation module (DCM). We have demonstrated that at least two DLs are needed to compensate for both the spatial and the temporal distortion associated with angular dispersion [11]. The DCM is made up of a hybrid diffractive–refractive lens triplet [11–13]. It compensates not only for spatial elongations caused by chromatic aberrations but also for the pulse stretching effect. Some experimental results in one-shot second harmonic (SH) generation, wide-field two-photon microscopy, and parallel micromachining, that corroborate the strength of the proposed device, are shown.

5.2   Fundamental Features of the DCM

This section is organized as follows. In Section 5.2.1 the basic equation for light transport of an ultrashort pulse through a misaligned optical setup within the framework of the Fresnel–Kirchhoff diffraction formula is derived. The analysis is carried out in one step by means of generalized ray matrices. In Section 5.2.2, this general formulation is evaluated under a second order analysis. In Section 5.2.3, these results are specifically applied to the conventional beam-splitting problem where multibeams diffracted by a low-frequency diffraction grating are gathered with an achromatic lens doublet. Corresponding results achieved with the hybrid, diffractive-refractive, lens triplet are discussed in Section 5.2.4, where the residual spatial and temporal stretching is emphasized. In Section 5.2.5, results of computer simulations are provided, illustrating the dispersion compensation capabilities of our proposal. Finally, Section 5.2.6 presents experimental results confirming the capability of our device for dispersion compensation.

5.2.1 Theory of Propagation of Diffracted Femtosecond Pulses

An analytical model to calculate the spatiotemporal dependence of the electrical field of an ultrashort pulse at the output plane of a misaligned optical setup, within the framework of the Fresnel–Kirchhoff diffraction formula is developed. According to Reference [14], the optical device is described by means of a 3×3 ray transfer matrix ABCDEF. To simplify the mathematical analysis, and without loss of generality, only one transverse coordinate is considered. The extension to the 2D case is straightforward.

To study light propagation, we first consider a pulsed beam with carrier frequency ω_o, with spectral components of frequency ω contained within its spectrum. Figure 5.1 shows the geometry to which we are referring. A monochromatic wave with amplitude $U_{in}(x; \omega)$ is assumed to be incident at the input plane of a misaligned optical device consisting of a low-frequency diffraction grating, with spatial period p, cascaded with a rotationally symmetric, but otherwise arbitrary, ABCD focusing setup. The grating splits the pulsed radiation into several diffracted beams. These beams are gathered by the collecting optics and focused on to an array of light spots. For the nth-order diffracted beam, the light amplitude at the output plane can be evaluated through the equation:

$$U_{out}(x; \omega) = \sqrt{\frac{\omega}{2\pi cBi}} \exp\left[i\frac{\omega L}{c}\right] \exp\left[i\frac{D}{B}\frac{\omega}{2c}x^2\right]$$

$$\times \int_{-\infty}^{\infty} U_{in}(x'; \omega) \exp\left[i\frac{A}{B}\frac{\omega}{2c}x'^2\right] \exp\left[-i\frac{\omega}{cB}x'(x-E)\right] dx'. \quad (5.1)$$

Here, the symbol L stands for the on-axis optical path length between the input and the output planes and c is the speed of the light. Only the case of a low-frequency diffraction grating is considered so that the paraxial approximation remains valid, but the full frequency-dependence of the grating equation is retained. In general, the matrix coefficients A, B, C, and D are wavelength-dependent. Throughout this chapter, we will refer to the value

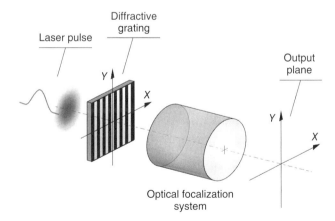

Figure 5.1 Schematic diagram of a Fourier beam splitter. The system, which provides the Fourier transform of the input grating in the output plane, is fully described through the *ABCDEF* ray transfer matrix. *Source*: Mínguez-Vega, G., Tajahuerce, E., Fernández-Alonso, M., Climent, V., Lancis, J., Caraquitena, J., Andrés, P. (2007). Figure 1. Reproduced with permission from The Optical Society

of any wavelength-dependent parameter evaluated at the carrier frequency by employing the subscript o, for example, $A_o = A(\omega_o)$. For the case of the optical device sketched in Fig. 5.1, the spatial-shift coefficient E is given, for the nth-order diffraction maxima, as

$$E = \frac{nB2\pi c}{\omega p}.$$
(5.2)

The output instantaneous irradiance distribution $I_{out}(x;t)$ is obtained as the square modulus of the inverse (temporal) Fourier transform of Eq. (5.1), that is,

$$I_{out}(x;t) = \left| \int_{-\infty}^{\infty} U_{out}(x,\omega)\exp[-i\tilde{\omega}t]\,d\tilde{\omega} \right|^2,$$
(5.3)

with $\tilde{\omega} = \omega - \omega_o$. Generally, for complex optical systems, Eqs. (5.1) and (5.3) do not have an analytical solution and they must be solved numerically [15].

We further proceed by assuming a transform-limited Gaussian-shaped, both spatial and temporal, input pulse. In mathematical terms, the input electric field is given by $U_{in}(x;t) = \exp[-x^2/4\sigma_x^2]\exp[-t^2/4\sigma_t^2]$, where σ_x and σ_t denote the root-mean-square (rms) width of the spatial and temporal irradiance profiles, respectively. Recall that, for a Gaussian pulse, the full width at $1/e^2$ of the intensity profile is four times the rms width. For this waveform, the input field in the spectral domain is $U_{in}(x;\omega) = \exp[-x^2/4\sigma_x^2]\exp[-\sigma_t^2\omega^2]$. After introduction of this waveform into Eq. (5.1), we obtain

$$U_{out}(x;\omega) = \sqrt{\frac{\omega}{2\pi cBi}}\exp\left[i\frac{\omega L}{c}\right]\exp[-\sigma_t^2\tilde{\omega}^2]$$

$$\times \exp\left[i\frac{\omega}{B2c}\left(Dx^2 - \frac{\sigma_x^2 A(x-E)^2}{4\sigma_{x\omega}'^2}\right)\right]\exp\left[-\frac{(x-E)^2}{4\sigma_{x\omega}'^2}\right],$$
(5.4)

with $\sigma_{x\omega}'^2 = \sigma_x^2 A^2 + (c^2 B^2/4\omega^2\sigma_x^2)$. Note that the last exponential term is a spatial Gaussian function peaked at $x = E$ and with spatial extent $\sigma_{x\omega}'$. Generally, for a fixed diffraction order of the grating, each temporal frequency of the pulse provides a different value of E, as indicated in Eq. (5.2), causing an angular dispersion.

5.2.2 Second Order Analysis

Let us consider the factor ω/B in front of the exponential functions. It has been shown that, for a singlet lens, the approximation $\omega/B \cong \omega_o/B_o$ presents a negligible error for pulse durations longer than 15 fs in the focal region [16]. Of course, the error is even smaller for wavelength-corrected focusing systems.

Equation (5.4) is further simplified by expanding in a Taylor series the argument inside the phase exponentials up to the second order around ω_o. Namely,

$$\frac{\omega L}{c} \cong \alpha_o + \alpha_1\tilde{\omega} + \frac{\alpha_2}{2}\tilde{\omega}^2 \quad \text{and}$$

$$\frac{\omega}{B2c}\left(Dx^2 - \frac{\sigma_x^2 A(x-E)^2}{4\sigma_{x\omega}'^2}\right) \cong \beta_o(x) + \beta_1(x)\tilde{\omega} + \frac{\beta_2(x)}{2}\tilde{\omega}^2,$$
(5.5)

where, of course,

$$\alpha_i = \frac{\partial^i}{\partial \omega^i}\left(\frac{\omega L}{c}\right)\Bigg|_{\omega=\omega_o} \text{ and } \beta_i(x) = \frac{\partial^i}{\partial \omega^i}\left(\frac{\omega}{B2c}\left\{Dx^2 - \frac{\sigma_x^2 A(x-E)^2}{4\sigma' x_\omega^2}\right\}\right)\Bigg|_{\omega=\omega_o}. \tag{5.6}$$

Whenever the bandwidth of the pulse is only a small fraction of the carrier frequency, a linear approximation for the spatial-shift term E can be adopted

$$E \cong \frac{2\pi c n}{p\omega_o}B_o\left[1 + \tilde{\omega}\left(\frac{1}{B_o}\frac{\partial B}{\partial \omega}\Bigg|_{\omega=\omega_o} - \frac{1}{\omega_o}\right)\right]. \tag{5.7}$$

Equation (5.7) can be written in a compact way as $E = E_o + E_1\tilde{\omega}$. We also note that the wavelength-dependence of coefficient E is the leading mechanism responsible for chromatic distortion of the output field. Thus, we assume that $\sigma' x_\omega(\omega) \cong \sigma' x_\omega(\omega_o)$. Finally, Eqs. (5.5) and (5.7) are introduced into Eq. (5.4) and we find

$$U_{out}(x;\omega) = \sqrt{\frac{\omega_o}{2\pi cB_o i}}\exp\left[i\left(\alpha_o + \beta_o(x)\right)\right]\ \exp\left[i\left(\alpha_1 + \beta_1(x)\right)\tilde{\omega}\right]$$

$$\exp\left[i\left(\alpha_2 + \beta_2(x)\right)\frac{\tilde{\omega}^2}{2}\right]\exp\left[-\left(\sigma_t^2 + \frac{E_1^2}{4\sigma' x_\omega^2(\omega_o)}\right)\tilde{\omega}^2\right]\exp\left[-\frac{(x-E_o)^2}{4\sigma' x_\omega^2(\omega_o)}\right] \tag{5.8}$$

$$\exp\left[-\frac{(x-E_o)^2}{4\sigma' x_\omega^2(\omega_o)}\right]\exp\left[\frac{2(x-E_o)E_1\tilde{\omega}}{4\sigma' x_\omega^2(\omega_o)}\right].$$

Equation (5.8) is the basis of the following discussion. It describes, in the spectral domain and up to second order, the transformation of a short light pulse by the system of Fig. 5.1. The phase term that contains the terms α_o and $\beta_o(x)$ does not contribute to the field irradiance and will be omitted for further calculations. The exponential term linear with $\tilde{\omega}$ provides the group delay, GD, also called PTD. Specifically, coefficient α_1 produces a non-disturbing shift in the arrival time of the pulse and the presence of $\beta_1(x)$ leads to the x-dependent temporal pulse distortion. The group-velocity dispersion, GVD, caused by the lenses' material and diffraction and responsible for pulse stretching is included in α_2 and $\beta_2(x)$, respectively. Furthermore, the spectral narrowing of the spectrum of the pulse caused by the angular dispersion is controlled by the quantity $E_1/2\sigma'_{x\omega}(\omega_o)$. Obviously, in the temporal domain it results in the stretching of the pulse. Equation (5.8) also accounts for a spatial Gaussian function peaked at E_o and scaled the factor $\sigma'_{x\omega}(\omega_o)$. Finally, the last exponential term is related to the coupling between spatial and temporal coordinates.

5.2.3 Conventional Refractive Lens System

A schematic representation of the setup is shown in Fig. 5.2. The paraxial focal length of the achromatic doublet L_1 is $f(\omega)$, with $\partial f(\omega)/\partial \omega|_{w=w_o} = 0$, and the output plane is located at a distance f_o ahead of the lens that corresponds to the focal plane for ω_o. The doublet has a wavelength-dependent on-axis optical path length, $L = n_1(\omega)d_1 + n_2(\omega)d_2$, where $n_{1,2}$ refers to the refractive indexes and $d_{1,2}$ to the central thickness of each material. The thickness of the glass substrate of the diffractive grating has been neglected in this analysis, as it is usually small compared to L. The ABCD matrix corresponding to light propagation between the grating and the output plane is

$$\begin{bmatrix} A & B \\ C & D \end{bmatrix} = \begin{bmatrix} 1 - \frac{f_o}{f(\omega)} & f_o\left(2 - \frac{f_o}{f(\omega)}\right) \\ -\frac{1}{f(\omega)} & 1 - \frac{f_o}{f(\omega)} \end{bmatrix}. \tag{5.9}$$

Taking into account Eqs. (5.2) and (5.4), we find

$$E = \frac{2cn\beta f_o}{p\omega_o}\left(1 - \frac{\tilde{\omega}}{\omega_o}\right) \text{ and } \sigma'_{x\omega}(\omega_o) = \frac{1}{2}\frac{cf_o}{\sigma_x\omega_o}. \tag{5.10}$$

The achromatic requirement for the doublet leads to $\beta_1(x) = 0$. According to the paper by Kempe et $al.$ [17], we also assume that $\partial^2 n(\omega)/\partial\omega^2|_{w=w_o} = C^{(2)}\partial n(\omega)/\partial\omega|_{w=w_o}/\omega_o$ where $C^{(2)}$ results from the Sellmeier equation and typically it is of order unity [17]. In this way $\partial^2 f(\omega)/\partial\omega^2|_{w=w_o} = 0$ and $\beta_2(x) = 0$.

The narrowing of the spectrum of the pulse at the diffraction maxima E_o is the most important cause for temporal broadening. The GVD of a conventional lens, provided by α_2, is a second-order effect that causes a change in the phase of the different spectral components. For the optical device in Fig. 5.2, GVD effects are at least two orders of magnitude smaller than angular dispersion effects. Consequently, the output spatiotemporal intensity is provided by

$$I_{out}(x';t) = \exp\left[-\frac{(x - E_o)^2}{2\sigma_x^2}\right]\exp\left[-\frac{(t - \alpha_1)^2}{2\sigma_t'^2}\right], \tag{5.11}$$

Where

$$\sigma'^2_t = \sigma_t^2\left(1 + \frac{4n^2\pi^2\sigma_x^2}{p^2\omega_o^2\sigma_t^2}\right) \text{ and } \sigma'^2_x = \sigma'^2_{x\omega}(\omega_o)\left(1 + \frac{4n^2\pi^2\sigma_x^2}{p^2\omega_o^2\sigma_t^2}\right). \tag{5.12}$$

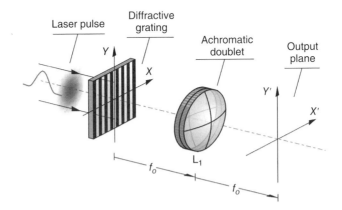

Figure 5.2 Sketch of the conventional grating based multifocal device. *Source*: Mínguez-Vega, G., Tajahuerce, E., Fernández-Alonso, M., Climent, V., Lancis, J., Caraquitena, J., Andrés, P. (2007). Figure 2. Reproduced with permission from The Optical Society

Equation (5.11) demonstrates that the output irradiance is simply the product of two expanded Gaussian functions. In this way, for the nth-order diffraction maximum, we obtain a relative stretching in the temporal domain, σ'^2_t/σ^2_t, which is just the same expression as the relative broadening for the spatial domain $\sigma'^2_x/\sigma'^2_{x\omega}(\omega_o)$. Note that the spatial extent increases with the diffraction order n. Thus, as diffraction order grows, diffraction spots become more and more elliptically distorted.

5.2.4 The Dispersion Compensation Module

Let us focus our attention on the system shown in Fig. 5.3. It is constituted by an achromatic doublet L_1 and two kinoform diffractive lenses, DL_1 and DL_2, with image focal lengths, $Z = Z_o\omega/\omega_o$ and $Z' = Z'_o\omega/\omega_o$, respectively. The input grating, located at a distance z from L, is illuminated by the Gaussian-shaped input pulse. Axial distances f_o, d, and d' denote arbitrary but fixed lengths between the different elements of the system. According to [11], a set of angular dispersion-compensated focal spots is achieved at the output plane when matrix coefficients satisfy $A(\omega_o) = 0$, $\partial(B/\omega)/\partial\omega|_{\omega_o} = 0$, and $\partial A/\partial\omega|_{\omega_o} = 0$. These conditions are fulfilled when $d^2 = -Z'_o Z_o$ and $d' = -d^2/(d + 2Z_o)$ [11].

To proceed with the wave optics analysis, the overall wave matrix ABCD corresponding to light propagation between the grating plane and the output plane needs to be computed. The product of wave matrices associated either with free-space propagation or with a passage through a lens, multiplied in the opposite order to that in which the operations are encountered,

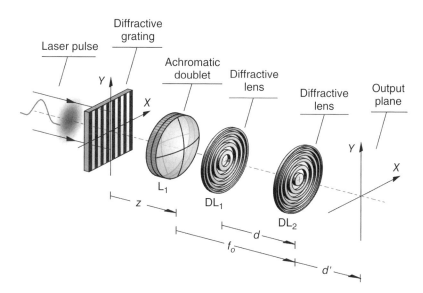

Figure 5.3 Representation of the dispersion-compensation hybrid device for spot array generation. *Source*: Mínguez-Vega, G., Tajahuerce, E., Fernández-Alonso, M., Climent, V., Lancis, J., Caraquitena, J., Andrés, P. (2007). Figure 3. Reproduced with permission from The Optical Society

results in

$$
\begin{bmatrix} A & B \\ C & D \end{bmatrix} = \begin{bmatrix} 1 & d' \\ 0 & 1 \end{bmatrix} \begin{bmatrix} 1 & 0 \\ -\omega_o/Z_o'\omega & 1 \end{bmatrix} \begin{bmatrix} 1 & d \\ 0 & 1 \end{bmatrix} \begin{bmatrix} 1 & 0 \\ -\omega_o/Z_o\omega & 1 \end{bmatrix}
$$
$$
\begin{bmatrix} 1 & f_o - d \\ 0 & 1 \end{bmatrix} \begin{bmatrix} 1 & 0 \\ -1/f(\omega) & 1 \end{bmatrix} \begin{bmatrix} 1 & z \\ 0 & 1 \end{bmatrix}. \tag{5.13}
$$

It is a straightforward matter to show that

$$
E = E_o = \frac{2\pi n c Z_o f_o}{p\omega_o(d + 2Z_o)}, \text{ and } \sigma'_{x\omega}(\omega_o) = \frac{cf_o Z_o}{2\omega_o \sigma_x(d + 2Z_o)}. \tag{5.14}
$$

Thus, $E_1 = 0$ and, from Eq. (5.8), the output spatiotemporal intensity is given by

$$
I_{out}(x; t) = \exp\left[-\frac{(x - E_o)^2}{2\sigma'^2_{x\omega}(\omega_o)}\right] \exp\left[-\frac{(t - \alpha_1)^2}{2\sigma'^2_t}\right], \tag{5.15}
$$

where $\sigma'^2_t = \sigma^2_t + (\alpha_2 + \beta_2(x)/2\sigma_t)^2$. To reach Eq. (5.15), several facts must be taken into account. On the one hand, we consider light intensity only at the vicinity of the different diffraction maxima, $x \cong E_o$. Thus, $\beta_1(x) \cong \beta_1(E_o)$, which involves a non-disturbing temporal delay at the focal spot, and $\beta_2(x) \cong \beta_2(E_o)$. Furthermore, from Eq. (5.5), we deduce that $|\beta_1(x)| << \alpha_1$. Note that the GVD introduced by diffraction, $\beta_2(x)$, shows an anomalous dispersion that is larger for off-axis spots. Equation (5.15) indicates that the spatial width of the nth-order diffraction maximum is essentially the one achieved for continuous wave (CW) illumination for the frequency ω_o. Dispersion compensation drastically reduces the lateral walk-off between the different spectral components at the light spots. As a result, the available bandwidth is improved and, thus, the temporal broadening of the energy is negligible. In fact, the temporal width of the output pulse is limited, in a first-order approximation, by the total GVD. This description is completely general and not only restricted to Gaussian beams.

5.2.5 Comparative Numerical Simulations

Computer simulations have been performed to verify our theoretical approach. In the numerical simulation, a 100 lines/inch diffraction grating is illuminated by a Ti:Sapphire laser producing femtosecond Gaussian pulses centered at $\lambda_o = 2\pi c/\omega_o = 800$ nm, with an input beam size of $\sigma_x = 5$ mm. The following parameters have been accounted for in the devices shown in Figs. 5.2 and 5.3: The achromatic doublet is a biconvex lens of $f_o = 80$ mm, made of BK7 (crown) and SF5 (flint), with central thicknesses of $d_{BK7} = 5$ and $d_{SF5} = 2.5$ mm. The consideration of the dispersion formula for these materials leads to $\alpha_1 = 39.5$ ps and $\alpha_2 = 470$ fs^2. For the elements in Fig. 5.3, the DLs have an image focal length of $Z_o' = -Z_o = 70$ mm, and the axial distances are $d' = d = 70$ and $z = 80$ mm.

The behavior of both multifocal generators in the Fresnel approximation is simulated by means of Eqs. (5.3) and (5.4). It is necessary to emphasize that no approximation is made in order to compute the instantaneous intensity. In Fig. 5.4(a) the relative spatial broadening, $\sigma_x'/\sigma_{x\omega}'(\omega_o)$, is plotted against σ_t. The fifth-order diffraction maximum is considered. In the graph, with a different scaling, the same ratio for the setup of Fig. 5.2 is also shown. A noteworthy improvement of the spatial resolution is apparent. Analog results are obtained for the relative temporal broadening, as shown in Fig. 5.4(b). In both plots, for the conventional grating-based spot array generator, there is a perfect matching between the numerical simulations and the results that would be obtained through approximated Eq. (5.12). In the case of the DOE-based beam splitter, the approximated Eq. (5.15) is no longer valid for pulses shorter than 50 fs, as can be seen from Fig. 5.4(a). For this short-pulse duration, there is not a simple analytical solution.

The integrated intensity has been calculated as

$$I(t) = \int\limits_{-\infty}^{\infty} I_{out}(x, t)dt. \tag{5.16}$$

This integrated intensity is represented in Fig. 5.5. Again, the fifth diffraction order and the following pulse durations are considered: (a) $\sigma_t = 50$ fs and (b) $\sigma_t = 21.24$ fs (which corresponds to intensity FWHM of 50 fs). For comparison, the output Gaussian beam obtained with a monochromatic laser emitting at 800 nm is also drawn. For the 50 fs pulse duration, the output beams obtained with our proposal and with the CW illumination are the same, so lines overlap. For shorter pulse durations a slight difference between both cases can be observed. A clear improvement in the spatial spot size elongation in comparison with the conventional setup is evident.

Finally, Fig. 5.6 (Plate 8) shows the tridimensional distribution of the output spatiotemporal irradiance for the fifth diffraction maximum when the input pulse width is of $\sigma_t = 50$ fs. The output profile, calculated numerically by using Eqs. (5.3) and (5.4), is shown in the right part of the figure for the achromatic doublet and in the left part for the CGH-based system. The size and duration of the output pulse is clearly reduced for short pulses when the system shown in Fig. 5.3 is used. Consequently, due to the spatiotemporal concentration of power, the proposed setup seems a suitable instrument for the generation of user-defined patterns with broadband ultrafast laser pulsed light.

5.2.6 Experimental Results

Once it has been shown how the simulations validate the system, experiments confirming these results will be presented. The optical setup in Fig. 5.3 was constructed with $f_o = 200$ mm and $Z'_o = -Z_o = 150$ mm and a numerical aperture (NA) of 0.02. A Ronchi grating with fundamental frequency 11.8 lp/mm as a diffractive optical element was also used. A picture of the DCM of the system is shown in Fig. 5.7 and the sketch of the complete experimental setup is shown in Fig. 5.8.

The measurement of the complex spatiotemporal light field at the multifocal pattern was carried out by using a recently reported technique that is based on the spatiotemporal amplitude and phase reconstruction by Fourier transform of the interference spectra of the optical beams (STARFISH) [18,19]. This reconstruction technique consists in spatially resolved spectral interferometry using a fiber optic coupler as interferometer. The reference and the test beam are collected by each fiber input port. The test arm spatially scans the transverse profile and the reference arm controls the relative delay between the pulses required for interferometry (between 2–3 ps in the experimental conditions of the present study). Both pulses are

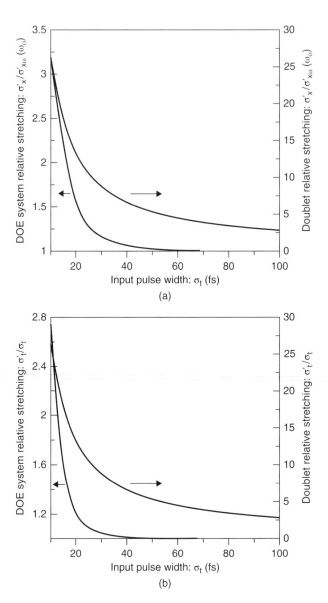

Figure 5.4 Relative stretching versus the input pulse width for the fifth-order diffraction maximum ($n = 5$) after focusing with an achromatic doublet (dashed line) and the system proposed in Fig. 5.3 (continuous line) for: (a) the spatial domain and (b) the temporal domain. *Source*: Mínguez-Vega, G., Tajahuerce, E., Fernández-Alonso, M., Climent, V., Lancis, J., Caraquitena, J., Andrés, P. (2007). Figure 4. Reproduced with permission from The Optical Society

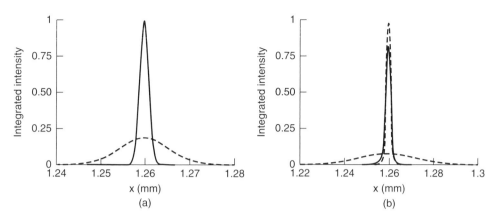

Figure 5.5 Integrated intensity profiles of the fifth diffraction order of a 100 lines/inch diffractive grating after focusing with an achromatic doublet (dashed line), the system of Fig. 5.3 (continuous line) and a monochromatic wave of 800 nm (dash/dot line) for an input pulse duration of: (a) $\sigma_t = 50$ fs and (b) $\sigma_t = 21.24$ fs. *Source*: Mínguez-Vega, G., Tajahuerce, E., Fernández-Alonso, M., Climent, V., Lancis, J., Caraquitena, J., Andrés, P. (2007). Figure 5. Reproduced with permission from The Optical Society

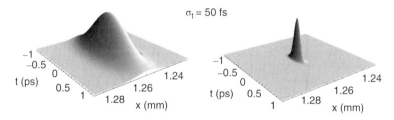

Figure 5.6 (Plate 8) Spatiotemporal profiles of the fifth diffraction order of a 100 lines/inch diffractive grating for an input pulse width of $\sigma_t = 50$ fs. The right part of the figure is obtained after focusing with an achromatic lens doublet and the left part by focusing with the DOE-based system. Note that the time origin is chosen arbitrarily. *Source*: Mínguez-Vega, G., Tajahuerce, E., Fernández-Alonso, M., Climent, V., Lancis, J., Caraquitena, J., Andrés, P. (2007). Figure 6. Reproduced with permission from The Optical Society. *See plate section for the color version*

combined inside the fiber coupler and leave it through the output common port that is directly connected to a standard spectrometer, where the interference spectrum is measured. Notice that this scheme does not require collinear test and reference, avoiding the precise alignment of the beams. Also, the reference beam is not scanned and its spectral phase is obtained on-axis just by means of a direct measurement, as explained next. The fiber coupler must be single mode for the whole spectral range in order to prevent pulse distortions. In our case, it means a 4-μm mode field diameter, which gives us a high spatial resolution. The two arms of the fiber coupler must have equal lengths. Slight phase differences were calibrated and taken into account through a single spectral interferometry measurement using the same input pulse at both fiber ports.

To reconstruct the test pulse, the information of the spectral interference is used to obtain the spectral phase difference between both pulses (reference and test) by applying the fringes

Figure 5.7   Experimental setup

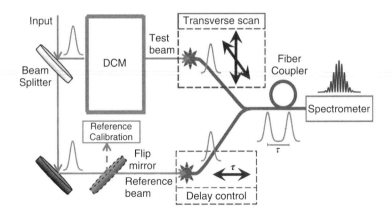

Figure 5.8   Sketch of the experimental setup: a replica of the laser pulse is used as the reference and the other replica passes through the DCM (test beam). The pulses are collected by the arms of the fiber coupler. The reference one controls the relative delay, whereas the test one spatially scans the test beam. The spatially resolved spectral interference is measured after the fiber coupler in the spectrometer

inversion algorithm known as *Fourier-Transform Spectral Interferometry* [20]. Since the reference is calibrated with SPIDER [21], it allows obtaining the spectral phase of the test pulse. This information is combined with the test spectrum measurement, collecting only the test pulse, and it is equivalent to the knowledge of the temporal amplitude and phase by doing the inverse Fourier transform. The extension to spatial domain is achieved by the transverse scan of the test fiber port, thus constituting the full characterization of the spatiotemporal (and spatio-spectral) amplitude and phase of the beam at certain propagation distance. The spatiotemporal coupling is preserved in the measurement because the reference is kept constant, which is what allows measuring test pulse front structure.

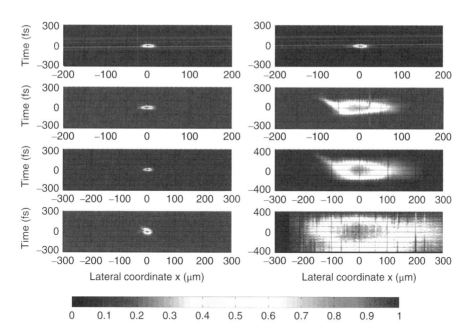

Figure 5.9 (Plate 9) Normalized spatiotemporal light intensity after low NA focusing of the beamlets coming from a diffractive grating (0th, +1st, +2nd, and +3rd diffraction orders from *top to bottom*) with (*left column*) and without (*right column*) DCM. Measurements were captured using STARFISH [18]. The maximum frequency component, for the third diffraction order, is 35.4 lp/mm. *Source*: Martínez-Cuenca, R., Mendoza-Yero, O., Alonso, B., Sola, Í. J., Mínguez-Vega, G., Lancis, J. (2012). Figure 2. Reproduced with permission from The Optical Society. *See plate section for the color version*

The experimental results are shown in Fig. 5.9 (Plate 9). To show dispersive effects, the transverse coordinate at the output plane normal to the grating ruling is displayed along the horizontal axis. The scan is performed through the line with the maximum irradiance for the diffractive foci. The temporal coordinate is displayed along the vertical axis. Right and left columns show the spatiotemporal light distribution corresponding to the 0th, +1st, +2nd, and +3rd diffraction orders with and without DCM, respectively. Each plot is centered in the vicinity of the position of the spectral spot for the center wavelength λ_0. For the plot, light intensity was normalized to the same value for the different maxima. The capability of the DCM to compensate for dispersive stretching is noticeable at increasing frequency components of the grating that are located at outer regions of the output plane.

In order to better observe the quality of the compensation performed with the DCM, the spatial and temporal profiles for the zero values of delay and lateral coordinate are plotted in Figs. 5.10 and 5.11, respectively. The rms widths for the spatial and temporal coordinates, respectively, of the compensated foci are 9.0 μm and 13.4 fs (0th order), 9.3 μm and 13.7 fs (+1st order), 9.3 μm and 16 fs (+2nd order), and 9.6 μm and 31.6 fs (+3rd order). The widths remain nearly constant up to frequency components of about 30 lp/mm. We also notice some ripples in the temporal profile that are due to non-compensated third-order dispersion in the glass components. Also radial GDD effects are apparent for the higher spatial frequencies. The foci generated without DCM suffer from a higher spatiotemporal stretching (even one order

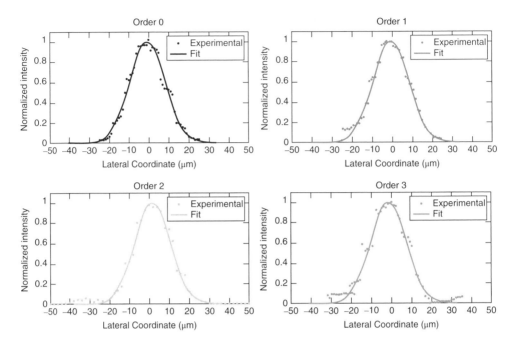

Figure 5.10   Spatial profile of the different diffraction orders obtained with the DCM

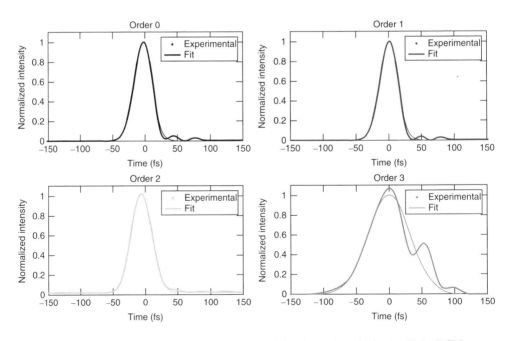

Figure 5.11   Temporal profile of the different diffraction orders obtained with the DCM

of magnitude for the higher spatial frequencies). Specifically only 10%, 2.5%, and 0.5% of the available peak-power for the compensated foci is delivered to the spatiotemporal smeared diffraction orders +1st, +2nd, and +3rd, respectively.

5.3 Holographic Applications of the DCM with Ultrafast Light Pulses

Current high-gain femtosecond amplifiers running at kHz repetition rate provide output pulse energies at the mJ level that overpass by a huge amount the required energy for some applications. The use of multiple beams in parallel to take advantage of the full power of the amplifier system has been proposed and demonstrated in fields as material processing and multiphoton microscopy. Such methods rely on the use of DOEs or microlens arrays to divide the laser beam into several beamlets that, after focusing, scan the sample simultaneously. CGHs are usually the choice as they allow dynamic codification. However, as shown in the previous section, the use of DOEs is problematic for laser pulses in the femtosecond regime due to the increase of the eccentricity of the focused spots and the pulse stretching in time. Under this scenario, the use of the proposed DCM is a valuable tool.

This section is organized as follows. In Section 5.3.1, the application of the DCM to generate arbitrary irradiance patterns for the SH signal is shown [13]. In Section 5.3.2, we demonstrate efficient generation of wide-field fluorescence signals in two-photon microscopy [22]. Finally, in Section 5.3.3, multi-beam high spatial resolution laser micromachining with femtosecond pulses is presented [23].

5.3.1 Single Shot Second Harmonic Signals

Arbitrary irradiance patterns at the SH signal have been experimentally demonstrated. To excite the SH signal, the DCM described in Section 5.2.6 was used, though a telescope composed of a RL of focal length 100 mm and a 10× microscope objective (160 mm conjugated, 0.25 NA) was added. In this way, a reduced image of the foci plane onto the sample was obtained. In our case, the sample holds for an uncoated Type-I β-BaB$_2$O$_4$ crystal (10 × 10 × 0.02 mm, $\theta = 29.1°$, $\varphi = 0°$). Additional relay optics introduces extra group velocity dispersion (GDD) and third-order dispersion (TOD). Although GDD was compensated for by tuning the prism pair located at the last section of the femtosecond laser to maximize the SH yield, TOD prevents from achieving a transform-limited 28 fs pulse at the sample. Spatial resolution was fixed by the NA of the microscope and was estimated to be of 1.53 mm. Field curvature of the microscope objective increases the beam diameter slightly in the low spatial frequency region while harmonic components higher than 20 lp/mm are more seriously affected by the residual chromatism. To observe the SH signal, the BBO crystal is imaged onto a conventional CCD sensor (Ueye UI-1540M) by means of a 10× microscope objective also conjugated at 160 mm. We placed a suited filter (BG39-Schott crystal) before the CCD camera to prevent the infrared signal. We used a Fourier CGH codifying the complex Fourier transform of a "smiling face" in amplitude as an off-axis binary computer generated hologram. The computer reconstruction of the Fourier CGH shows a set of more than 100 diffractive foci. The spatial spectrum of the sample spreads from around 5 to about 40 lp/mm. Irradiance uniformity of the foci is estimated to be 65%. Results of the recorded SH signals are shown in Fig. 5.12 without (a) and with (b)

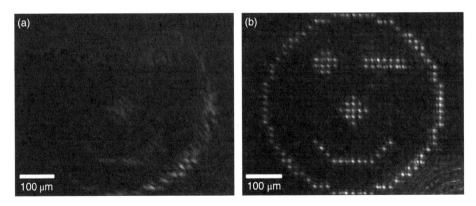

Figure 5.12 SH signal from a BBO crystal illuminated with a multispot pattern originated from a CGH without (a) and with (b) the DCM. *Source*: Martínez-Cuenca, R., Mendoza-Yero, O., Alonso, B., Sola, Í. J., Mínguez-Vega, G., Lancis, J. (2012). Figure 3. Reproduced with permission from The Optical Society

the DCM. For the uncompensated system only the lower spatial frequencies are able to excite SH signal at the incoming light energy.

5.3.2 Wide-Field Fluorescence Signals in Two-Photon Microscopy

Two- and three-dimensional images of living tissue are widely demanded in life sciences [24]. In this context, multiphoton absorption processes [25] allow for a selective excitation of fluorescence signals, providing a convenient way for acquiring images with good quality. The probability of these nonlinear processes is extremely low because they rely on the absorption of two or more photons to excite a molecule to a certain vibrational state from which a fluorescence signal is emitted. Hence, the fluorescence emission is restricted to a very small region of the laser focus, allowing not only high spatial resolution imaging deep into tissue [26], but also natural axial resolution.

In this context, since the laser focus is traditionally scanned across the live sample that originates two-photon imaging, signal acquisition suffers from time limitations [27]. In fact, the time required to acquire a complete two-dimensional image of the whole field of view depends, among other factors, on the response time of scanners. Although this drawback has been overcome by using different strategies to increase the scanning speed, which include resonant mirror systems [28], acousto-optic devices [29], or polygon-mirror scanners [30], such strategies can reduce the dwell time per pixel and compromise signal-to-noise ratio. In addition, by selectively scanning only points of interest rather than all pixels of the sample, very fast pixel-by-pixel scans processes have been reported [31,32].

In the last years, an alternative approach to spatially multiplex the laser focus, and then carry out multibeam scanning simultaneously, has been introduced. This allows for a more efficient use of the available laser power to either speed up imaging or to increase sampling per frame. The generation of an array of laser foci can be performed by multiple beam splitters [33], spinning-disks [34], or DOEs [35]. However, arbitrary diffraction patterns created by CGHs in SLMs clearly offer more flexibility and customizability than imprinted fixed elements.

For instance, they can be used for simultaneously photo-manipulation of living samples in a variable and temporally dynamic manner and/or for the stimulation or inhibition of multiple neurons at once to achieve specific excitatory/inhibitory effects [27,36].

However, for short pulses under 100 fs, the irradiance patterns generated with CGHs appear naturally blurred. Apart from possible distortion effects due to wavefront aberrations or light scattering, this unwanted physical phenomenon is mainly attributed to the linear dependence on the scale of Fraunhofer irradiance patterns on the wavelength of light, usually known as *spatial chirp*. In fact, spatial chirp causes an increasing of the eccentricity of the focused spots in applications, such as real time two-photon absorption microscopy [37]. In addition, the pulse-front tilt due to the PTD between pulses coming from the CGH at different transverse positions also stretches the pulse in time as commented in previous sections.

Here, we experimentally show two-photon absorption and fluorescence emission in Rhodamine B (RB) with spatially and temporally well-resolved irradiance patterns. To create arbitrary irradiance patterns in RB, CGHs were encoded onto a phase only SLM. This allows for a dynamical and programmable optical setup with a great deal of flexibility and customizability. Spatial chirp and pulse-front tilt effects were corrected at first order, by using a hybrid diffractive-refractive optical system that was experimentally tested in the reconstruction of dynamical CGHs [38].

Figure 5.13 shows the proposed setup. A mode-locked Ti:sapphire laser (Femtosource, Femtolaser) provided the excitation source. The temporal width of the pulses was 30 fs intensity FWHM, and the central wavelength of the corresponding spectra was 800 nm with a bandwidth of approximately 50 nm. The maximum energy per pulse reached 0.8 mJ at 1 kHz repetition rate. The average energy per pulse was adjusted with a half-wave plate and a polarizer and the spatial beam width was fixed using a reflective beam expander. The ultrashort light pulses from the Ti:sapphire laser passed before exit through a post-compression stage based on fused silica Brewster prisms. Hence, negative dispersion could be introduced to later compensate for the positive material dispersion in the beam delivery path to the observation plane.

In the setup, the pulsed laser beam propagates through a beam splitter onto a Fourier CGH encoded into a phase-only SLM. In Fig. 5.13, the dashed box corresponds to the DCM, which images the reconstructed hologram onto its back focal plane (FP). The CGHs are calculated by using the well-known Gerchberg–Saxton iterative Fourier transform algorithm, but carried out in two stages as proposed by Wyrowski [39]. The DCM has been constructed with the

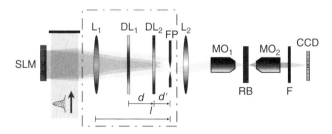

Figure 5.13 Schematic of the diffractive-refractive optical system used to improve two-photon absorption and fluorescence emission in Rhodamine B. *Source*: Pérez-Vizcaíno, J., Mendoza-Yero, O., Mínguez-Vega, G., Martínez-Cuenca, R., Andrés, P., Lancis, J. (2013). Figure 1. Reproduced with permission from The Optical Society

Figure 5.14 Details of the experimental setup used to focus the light pulses onto the RB and record the fluorescence emission

following lens parameters: $f_o = 300\,\text{mm}$, $Z'_o = -Z_o = 150\,\text{mm}$, and the axial distances are d' = d = 150 and l = 300 mm.

In order to excite the fluorescence signal in the RB, we used a telescope composed of the RL L_2 of focal length $f_2 = 100\,\text{mm}$, and a 20× microscope objective MO_1 of focal length 10 mm. This telescope produces a reduced image of the irradiance pattern at the RB, which is contained within a cubic glass box of 10 mm thickness. In Fig. 5.14 there is a picture of this part of the setup. Note that the additional relay optics preserves the spatiotemporal light distribution of the irradiance pattern except for the reduced spatial scale. To observe the fluorescence signal, the plane of RB is imaged onto a conventional CCD sensor (Ueye UI-1540M, 1280 × 1024 pixel resolution and 5.2 pixel pitch) by means of a 50× microscope objective MO_2. We placed a suited filter F (BG39-Schott crystal) before the CCD camera to prevent the infrared signal.

In Fig. 5.15, results of the recorded fluorescence signal without (central column) and with (right column) the DCM are shown. In our experiment, the ability of the optical system to generate arbitrary irradiance distributions is illustrated with two examples, representing a bicycle in Fig. 5.15(a)–(c), and a spiral in Fig. 5.15(d)–(f). In the first column of Fig. 5.15, computer reconstruction of the CGHs corresponding to the bicycle (a) and the spiral (d) are shown. The computer reconstruction of the holograms spreads over an area A_0 of more than 400 × 400 pixels (or 2.08 × 2.08 mm²). The spatial spectrum of the samples ranges from frequencies of around 25 to about 38 lp/mm.

In order to ensure wide-field fluorescence signaling all over the sample plane for the repetition rate and the pulse width of our laser system, the average power was adjusted to 3 mW. As explained earlier, patterning was accomplished through diffractive beam delivery. However, as reported in [40] we found out that spatial and temporal broadening of the pulse at the sample plane makes the use of CGH unsuitable, recording no signal. The fluorescence signal was subsequently recovered when the DCM was employed (see Fig. 5.15c and f). We

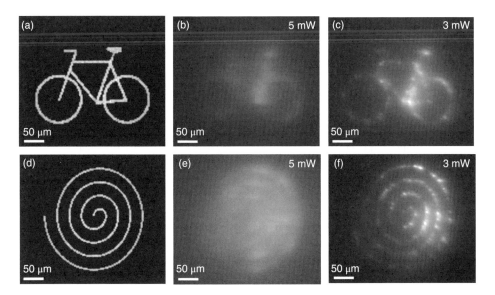

Figure 5.15 Arbitrary fluorescence irradiance patterns obtained in RB without compensation (b and e) and with DCM (c and f). Corresponding reconstructed CGHs for the central wavelength of the laser (a and d). *Source*: Pérez-Vizcaíno, J., Mendoza-Yero, O., Mínguez-Vega, G., Martínez-Cuenca, R., Andrés, P., Lancis, J. (2013). Figure 2. Reproduced with permission from The Optical Society

attribute this effect to the dispersion compensation capability of the DCM that preserves temporal width of the laser pulse at the sample plane, thus increasing the fluorescence signal with respect to the uncompensated situation. Note that the number of emitted photons at the sample depends inversely on the pulse width. In addition, due to spatial chirp compensation, transverse spatial resolution for the CGH reconstruction is also maintained. Finally, we check that diffraction-driven fluorescence signaling is possible without the DCM, but at the expense of using an additional 2 mW average power to compensate for temporal broadening. This is not recommended when using biological samples because high levels of excitation intensity may cause photodamage and photobleaching. The uncompensated spatial chirp leads to a blurred signal, which prevents correct hologram reconstruction and thus irradiance patterning, as shown in Fig. 5.15(b) and (e).

5.3.3   High Speed Parallel Micromachining

High precision micro- and nano-structuring of materials with femtosecond laser pulses is usually performed under low fluence regime (about a few µJ). In this regime, when focusing an ultrashort pulse onto a given material surface, a small area of it can be ionized. Hence, the ablation of the material due to the Coulomb explosion causes minimal thermal or mechanical damages in the surrounding of the processed area. This attractive physical phenomenon, together with several well-established techniques for the generation of user-defined irradiance patterns makes femtosecond laser processing a very promising tool for industrial applications.

In particular, parallel processing by means of complex irradiance patterns allows overcoming extensive attenuation required for current regenerative or multipass amplifier systems, providing pulse energies in the range of the mJ. Furthermore, it reduces the long fabrication time characteristic of sequentially dot-by-dot scans over a sample. In this context, optical techniques for parallel processing have been used, among other tasks, to generate desired focal patterns with the help of microlens arrays [41–43], perform temporal focusing of pulsed beams [44,45], achieve multibeam interference of femtosecond beams [46,47] or carry out holographic patterning for material microstructuring by using CGHs [8,48–51]. In addition, holographic femtosecond laser processing assisted by a SLM can be also regarded as a dynamic method for arbitrary irradiance patterning. Note that microstructured surfaces could be useful for the fabrication of microfluidic devices or integrated photonics, among other applications.

For non-thermal micromachining long pulses can be used (up to the range of ~ 10 ps depending on the material) [52]. However, optimal energy coupling with the help of suitably shaped temporal pulses gives us the possibility to guide the ablation towards user-defined directions, offering extended flexibility for high-quality material processing. Some experiments have demonstrated the influence of femtosecond shaped pulses in material processing. Just to mention a few cases, asymmetric shaped pulses with third-order dispersion can produce holes smaller than the diffraction limit in dielectrics [53]; with doubled pulses delayed a specific time, it is possible to control the size of the transferred spots done in metal oxides [54], and the shape of ablation channels in fused silica can be modified by changing the pulse duration [55].

Then, to overcome both chromatic aberrations due to the strong dependence of the diffraction phenomenon with the wavelength when using CGHs and the temporal stretching originated by propagation the time difference in free space, we propose to accomplish femtosecond micromachining with the DCM.

The experimental setup followed the model shown in Fig. 5.13 but the SLM was replaced by a static CGH designed to provide an array of 8×8 spots in the Fourier plane and instead of the RB a stainless steel sample was used. To see the processed focal spots in a wide spatial field, we used an optical microscope. In these conditions, 52 blind holes were ablated onto the sample when the DCM was used (please note that some spots of the DOE were out of the pupil of the microscope objective). This number was reduced to 16 when a conventional setup was employed. Owing to the dispersion effects, in the latter case only focal spots corresponding to lower spatial frequencies had enough fluence to make micromachining in the material surface. In our setup, the highest spatial frequency marked on the metal surface without the DCM was 33 lp/mm, while with the DCM we achieved ablation at 50 lp/mm, which implies increasing more than three times the ablation area. Details of the processed material are shown in Fig. 5.16.

In Fig. 5.17 the drilled holes obtained for the same spatial frequencies with and without the DCM are compared. It is clear that the shape of the processed spots affected by chromatic aberrations departs to a great extent from the ideal circular form. This fact prevents from the use of certain DOEs for micromachining under ultrashort pulsed illumination. However, when the DCM was employed the resulted spots retrieved the desired circular shape.

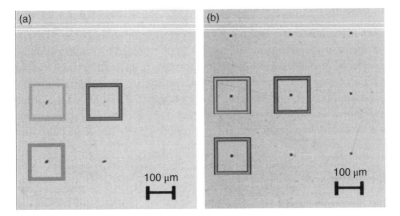

Figure 5.16   Details of a region of the surface of the ablated sample observed by optical microscopy (a) with a conventional setup (b) with the DCM

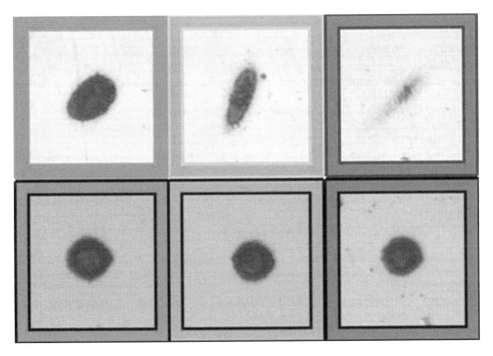

Figure 5.17   Images of the holes corresponding to different diffraction orders, with a conventional setup (*top*) and with the DCM (*bottom*)

5.4 Conclusion

We have shown that a suited dispersion-reversed optical setup should be used to correct for the spatial and temporal dispersions induced by CGHs illuminated by ultrashort pulsed light. On this regard, we have introduced a simple DCM (formed by a RL and two DLs) able to provide spatiotemporal irradiance patterns of very good quality, in comparison with those patterns achieved by conventional setups. In order to corroborate the validity of the proposed DCM, we have shown three different dispersion-compensated experiments ranging from wide field two-photon microscopy to parallel laser micromachining with ultrashort pulsed light.

Acknowledgments

This work has been partly funded by Generalitat Valenciana through the Excellence Net about Medical Imaging (project ISIC/2012/013) and the Prometeo Excellence Programme (project PROMETEO/2012/021). Partial support from Universitat Jaume I through the projects P1-1B2012-55 and P1·1B2013-53 is also acknowledged.

References

[1] Chichkov, B.N., Momma, C., Nolte, S., VonAlvensleben, F., Tunnermann, A. (1996) Femtosecond, picosecond and nanosecond laser ablation of solids. *Applied Physics A* **63**, 109–115.

[2] Denk, W., Strickler, J.H., Webb, W.W. (1990) 2-photon laser scanning fluorescence microscopy. *Science* **248**, 73–76.

[3] Grier D.G. (2003) A revolution in optical manipulation. *Nature* **424**, 810–816.

[4] Bezuhanov K., Dreischuh A., Paulus G.G., Schätzel M.G., Walther H. (2004) Vortices in femtosecond laser fields. *Optics Letters* **29**, 1942–1944.

[5] Nolte, S. (2003) Micromachining. In: *Ultrafast Optics: Technology and Application*, Fermann M.E., Galvanauskas A. and Sucha G. (eds). Marcel Dekker, New York.

[6] Amako, J., Nagasaka, K., Kazuhiro, N. (2002) Chromatic-distortion compensation in splitting and focusing of femtosecond pulses by use of a pair of diffractive optical elements. *Optics Letters* **27**, 969–971.

[7] Li, G., Zhou, C., Dai, E. (2005) Splitting of femtosecond laser pulses by using a Dammann grating and compensation gratings. *Journal of the Optical Society of America A* **4**, 767–-772.

[8] Kuroiwa, Y., Takeshima, N., Narita, Y., Tanaka, S., Hirao, K. (2004) Arbitrary micropatterning method in femtosecond laser microprocessing using diffractive optical elements. *Optics Express* **12**, 1908–1915.

[9] Jesacher, A., Booth, M.J. (2010) Parallel direct laser writing in three dimensions with spatially dependent aberration correction. *Optics Express* **18**, 21090–21099.

[10] Lancis, J., Mínguez-Vega, G., Tajahuerce, E., Climent, V., Andrés, P., Caraquitena, J. (2004) Chromatic compensation of broadband light diffraction: ABCD-matrix approach. *Journal of the Optical Society of America A* **21**, 1875–1885.

[11] Mínguez-Vega, G., Lancis, J., Caraquitena, J., Torres-Company, V., Andrés, P. (2006) High spatiotemporal resolution in multifocal processing with femtosecond laser pulses. *Optics Letters* **31**, 2631–2633.

[12] Mínguez-Vega, G., Tajahuerce, E., Fernández-Alonso, M., Climent, V., Lancis, J., Caraquitena, J., Andrés, P. (2007) Dispersion-compensated beam-splitting of femtosecond light pulses: Wave optics analysis. *Optics Express* **15**, 278–288.

[13] Martínez-Cuenca, R. , Mendoza-Yero, O., Alonso, B., Sola, Í. J., Mínguez-Vega, G., Lancis, J. (2012) Multibeam second-harmonic generation by spatiotemporal shaping of femtosecond pulses. *Optics Letters* **37**, 957–959.

[14] Martínez, O.E. (1988) Matrix formalism for pulse compressors. *IEEE Journal of Quantum Electrononics* **24**, 2530–2536.

[15] Fuchs, U., Zeitner, U.D., Tünnermann, A. (2005) Ultra-short pulse propagation in complex optical systems. *Optics Express* **13**, 3852–3861.

[16] Kempe, M., Rudolph, W. (1993) Femtosecond pulses in the focal region of lenses. *Physical Review A* **48**, 4721–4729.

[17] Kempe, M., Stamm, U., Wilhelmi, B., Rudolph, W. (1992) Spatial and temporal transformation of femtosecond laser pulses by lenses and lens systems. *Journal of the Optical Society of America B* **9**, 1158–1165.

[18] Alonso, B. Sola, I.J., Varela, O., Hernández-Toro, J., Méndez, C., San Román, J., *et al.* (2010) Spatiotemporal amplitude-and-phase reconstruction by Fourier-transform of interference spectra of high-complex-beams. *Journal of the Optical Society of America B* **27**, 933–940.

[19] Mendoza-Yero, O., Alonso, B., Varela, O., Mínguez-Vega, G., Sola, I.J., Lancis, J., *et al.* (2010) Spatiotemporal characterization of ultrashort pulses diffracted by circularly symmetric hard-edge apertures: theory and experiment. *Optics Express* **18**, 20900–20911.

[20] Lepetit, L., Cheriaux, G., Joffre, M. (1995) Linear techniques of phase measurement by femtosecond spectral interferometry for applications in spectroscopy. *Journal of the Optical Society of America B* **12**, 2467–2474.

[21] Iaconis, C., Walmsley, I.A. (1998) Spectral phase interferometry for direct electric-field reconstruction of ultrashort optical pulses. *Optics Letters* **23**, 792–794.

[22] Pérez-Vizcaíno, J., Mendoza-Yero, O., Mínguez-Vega, G., Martínez-Cuenca, R., Andrés, P., Lancis, J. (2013) Dispersion management in two-photon microscopy by using diffractive optical elements. *Optics Letters* **38**, 440–442.

[23] Torres-Peiró, S., González-Ausejo, J., Mendoza-Yero, O., Mínguez-Vega, G., Andrés, P., Lancis, J., Parallel laser micromachining based on diffractive optical elements with dispersion compensated femtosecond pulses (submitted).

[24] Zipfel, W.R., Williams, R.M., Webb, W.W. (2003) Nonlinear magic: multiphoton microscopy in the biosciences. *Nature Biotechnology* **21**, 1369–1377.

[25] Masters, B.R., So, P.T.C. (2008) *Handbook of Biomedical Nonlinear Optical Microscopy*. Oxford University Press, New York.

[26] Denk, W., Delaney, K.R., Gelperin, A., Kleinfeld, D., Strowbridge, B.W., Tank, D.W., Yuste, R. (1994) Anatomical and functional imaging of neurons using 2-photon laser scanning microscopy. *Journal of Neuroscience Methods* **54**, 151–162.

[27] Nikolenko, V., Watson, B.O., Araya, R., Woodruff, A., Peterka, D.S., Yuste, R. (2008) SLM microscopy: scanless two-photon imaging and photostimulation with spatial light modulators. *Frontiers in Neural Circuits* **2**, 1–14.

[28] Rochefort, N.L., Garaschuk, O., Milos, R.-I., Narushima, M., Marandi, N., Pichler, B., *et al.* (2009) Sparsification of neuronal activity in the visual cortex at eye-opening. *Proceeding of the National Academy of Science* **106**, 15049–15054.

[29] Otsua, Y., Bormutha, V., Wonga, J., Mathieub, B., Duguéa, G.P., Feltz, A., Dieudonnéa, S. (2008) Optical monitoring of neuronal activity at high frame rate with a digital random-access multiphoton (RAMP) microscope. *Journal of Neuroscience Methods* **173**, 259–270.

[30] Warger, W.C., Laevsky, G.S., Townsend, D.J., Rajadhyaksha, M., DiMarzio, Ch.A. (2007) Multimodal optical microscope for detecting viability of mouse embryos in vitro. *Journal of Biomedical Optics* **12**, 044006.

[31] Göbel, W., Helmchen, F. (2007) New angles on neuronal dendrites in vivo. *Journal of Neurophysiology* **98**, 3770–3779.

[32] Lillis, K.P., Enga, A., White, J.A., Mertz, J. (2008) Two-photon imaging of spatially extended neuronal network dynamics with high temporal resolution. *Journal of Neuroscience Methods* **172**, 178–184.

[33] Nielsen, T., Fricke, M., Hellweg, D., Andresen, P. (2001) High efficiency beam splitter for multifocal multiphoton microscopy. *Journal of Microscopy* **201**, 368–376.

[34] Bewersdorf, J., Pick, R., Hell, S.W. (1998) Multifocal multiphoton microscopy. *Optics Letters.* **23**, 655–657.

[35] O'Shea, D.C., Suleski, T.J., Kathman, A.D., Prather, D.W. (2004) *Diffractive Optics: Design, Fabrication, and Test.* SPIE Press Book.

[36] Watson, B.O., Nikolenko, V., Araya, R., Peterka, D.S., Woodruff, A., Yuste, R. (2010) Two-photon microscopy with diffractive optical elements and spatial light modulators. *Frontiers in Neuroscience* **4**, 1–8.

[37] Buist, A.H., Müller, M., Squer, J., Brakenhoff, G.J. (1998) Real time two-photon absorption microscopy using multi point excitation. *Journal of Microscopy* **192**, 217–226.

[38] Martínez-León, L., Clemente, P., Tajahuerce, E., Mínguez-Vega, G., Mendoza-Yero, O., Fernández-Alonso, M., *et al.* (2009) Spatial-chirp compensation in dynamical holograms reconstructed with ultrafast lasers. *Applied Physics Letters* **94**, 011104.

[39] Wyrowski, F. (1990) Diffractive optical elements: iterative calculation of quantized, blazed phase structures. *Journal of the Optical Society of America A* **7**, 961–969.

[40] Sacconi, L., Froner, E., Antolini, R., Taghizadeh, M.R., Choudhury, A., Pavone, F.S. (2003) Multiphoton multifocal microscopy exploiting a diffractive optical element. *Optics Letters* **28**, 1918–1920.

[41] Kato, J., Takeyasu, N., Adachi, Y., Sun, H., Kawata, S. (2005) Multiple-spot parallel processing for laser micronanofabrication. *Applied Physics Letters* **86**, 044102.

[42] Matsuo, S., Juodkazis, S., Misawa, H. (2004) Femtosecond laser microfabrication of periodic structures using a microlens array. *Applied Physics A* **80**, 683–685.

[43] Salter, P.S., Booth, M.J. (2011) Addressable microlens array for parallel laser microfabrication. *Optics Letters* **36**, 2302–2304.

[44] Kim, D., So, P.T.C. (2010) *High-throughput three-dimensional lithographic microfabrication Optics Letters* **35**, 1602–1604.

[45] Vitek, D.N., Adams, D.E., Johnson, A., Tsai, P.S., Backus, S., Durfee, C.G., *et al.* (2010) Temporally focused femtosecond laser pulses for low numerical aperture micromachining through optically transparent materials. *Optics Express* **18**, 18086–18094.

[46] Shoji, S., Kawata, S. (2000) Photofabrication of three-dimensional photonic crystals by multibeam laser interference into a photopolymerizable resin. *Applied Physics Letters* **76**, 2668–2670.

[47] Kondo, T., Matsuo, S., Juodkazis, S., Mizeikis, V., Misawa, H. (2003) Multiphoton fabrication of periodic structures by multibeam interference of femtosecond pulses. *Applied Physics Letters* **82**, 2758–2760.

[48] Hayasaki, Y., Sugimoto, T., Takita, A., Nishida, N. (2005), *Variable holographic femtosecond laser processing by use of a spatial light modulator Applied Physics Letters* **87**, 031101.

[49] Hasegawa, S., Hayasaki, Y. (2009) Adaptive optimization of a hologram in holographic femtosecond laser processing system. *Optics Letters* **34**, 22–24.

[50] Kuang, Z., Perrie, W., Leach, J., Sharp, M., Edwardson, S.P., Padgett, M., *et al.* (2008) High throughput diffractive multi-beam femtosecond laser processing using a spatial light modulator. *Applied Surface Science* **255**, 2284–2289.

[51] Kuang, Z., Liu, D., Perrie, W., Edwardson, S., Sharp, M., Fearon, E., *et al.* (2009) Fast parallel diffractive multi-beam femtosecond laser surface micro-structuring. *Applied Surface Science* **255**, 6582–6588.

[52] Stuart, B.C., Feit, M.D., Herman, S., Rubenchik, A.M., Shore, B.W., Perry, M.D. (1996), Nanosecond-to-femtosecond laser-induced breakdown in dielectrics. *Physical Review B* **53**, 1749–1761.

[53] Englert, L., Wollenhaupt, M., Haag, L., Sarpe-Tudoran, C., Rethfeld, B., Baumert, T. (2008) Material processing of dielectrics with temporally asymmetric shaped femtosecond laser pulses on the nanometer scale. *Applied Physics A* **92**, 749–753.

[54] Papadopouloua, E.L., Axentea, E., Magoulakisa, E., Fotakisa, C., Loukakosa, P.A. (2010), Laser induced forward transfer of metal oxides using femtosecond double pulses. *Applied Surface Science* **257**, 508–511.

[55] Vázquez de Aldana, J.R., Méndez, C., Roso, L. (2005) Saturation of ablation channels micro-machined in fused silica with many femtosecond laser pulses. *Optics Express* **14**, 1329–1338.

Part Two

Biomedical Applications and Microscopy

6

Advanced Digital Holographic Microscopy for Life Science Applications

Frank Dubois, Ahmed El Mallahi, Christophe Minetti and
Catherine Yourassowsky
Microgravity Research Centre, Université Libre de Bruxelles, Belgium

6.1 Introduction

A *digital holographic microscope* (DHM) is a powerful tool for recording full interferometric information of objects. It provides the capability to record sequences of holograms of dynamic phenomena at high rates and with short exposure times. The analyses are based on the digital holographic refocusing of both the optical intensities and quantitative optical phase maps. In this chapter, we present the developments we have realized on these instrumentations and holographic information processing. Our instrumental developments are based on the use of optical illumination sources of reduced coherences in order to improve the optical quality of the both phase maps and intensities. Although the reduction of the spatial coherence is most significant with a transmission DHM, the reduction in temporal coherence is also relevant to reducing the noise with high scattering samples. In this way, we recently developed a microscope with multi-wavelength sources of both spatial and temporal reduced coherences enabling one-shot recording. The DHM configurations are detailed in Section 6.2. The processing of holograms has been developed in several applications. In Sections 6.3 and 6.4, they are illustrated in water monitoring and the study of dynamical behavior of red blood cells. Some concluding remarks are provided in Section 6.5.

Multi-dimensional Imaging, First Edition. Edited by Bahram Javidi, Enrique Tajahuerce and Pedro Andrés.
© 2014 John Wiley & Sons, Ltd. Published 2014 by John Wiley & Sons, Ltd.

6.2 DHM Configurations

6.2.1 Phase Stepper DHM

The first DHM we developed was a phase stepper configuration with a partially coherent source realized from a spatially filtered *light emitting diode* (LED) (Fig. 6.1) [1]. The partially coherent illumination improves the quality of the holographic recording by decreasing the coherent artifact noise. With this DHM, the image quality is excellent, even considering the intensities, and is comparable to those obtained with classical optical microscopes (Fig. 6.2). According to the extraction of the complex amplitude by phase stepping [2], the partially coherent illumination allows retaining the two major capabilities of the DHM: (1) refocusing of objects recorded out of focus; (2) performing quantitative phase contrast imaging, even in high-scattering media [3–6].

We mainly used this microscope configuration for adherent cell culture applications.

6.2.2 Fast Off-Axis DHM

With the phase stepper configuration, the angle between the reference and the object beams incident on the camera sensor is as small as possible. Therefore, the method requests the sequential recording of several interferometric images that limits acquisition speed due to the camera frame rate. It means that the object has to remain static during the complete acquisition that takes the time of recording several frames. To overcome this limitation, we developed an off-axis fast DHM [7] with a partial coherent source for which the phase map and intensity method are obtained from each recorded hologram thanks to the Fourier method (Fig. 6.3) [8,9].

In this configuration, an angle is introduced between the object beam and the reference beam in order to obtain spatial heterodyne fringes (off-axis configuration). With respect to the phase stepper configuration, the full one-shot recording of the holographic information is a decisive advantage in the analysis of rapidly varying phenomena. An example obtained with the DHM set up to perform time lapse on fresh water flux in micro-channels is shown by Fig. 6.4.

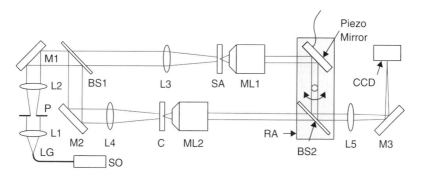

Figure 6.1 DHM microscope with LED illumination and phase stepper. SO: The LED module, LG: liquid light guide, L1–L5: Achromatic lenses, P: Aperture, M1–M3: Mirrors, BS1 and BS2: Beamsplitters, ML1, ML2: Microscope lenses, SA: Sample, C: Optical path compensator, RA: rotation assembly, CCD: camera, Piezo Mirror: mirror mounted on a piezoelectric transducer

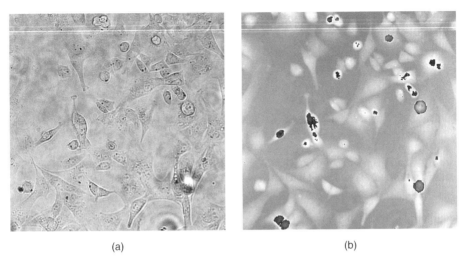

(a)   (b)

Figure 6.2 Adherent cell culture. (a) Intensity, (b) quantitative phase image obtained with the phase stepper DHM with LED illumination. FOV = 400 × 400 μm. *Source*: Dubois *et al*. 2011. Reproduced with permission from Springer

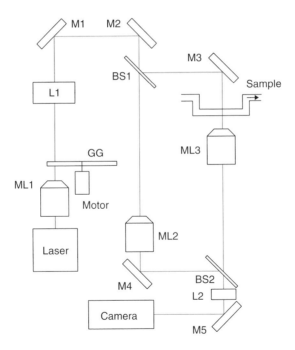

Figure 6.3 Fast DHM configuration with a reduced spatial coherence source created from a laser beam

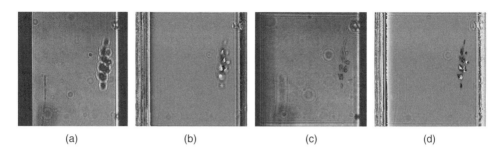

(a)   (b)   (c)   (d)

Figure 6.4   An alga in a micro-channel. FOV $= 370 \times 370$ μm. (a) out of focus intensity image, (b) out of focus phase image, (c) refocused intensity image, (d) refocused phase image

Each particle or organism can be refocused and analyzed thanks to the quantitative phase contrast imaging capability, allowing a high throughput monitoring of fresh water. However, the off-axis configuration requests an optical source of high temporal coherence. Otherwise, the fringe modulation is not constant due to variable optical delays over the field of view between the object and reference beams. If the path length differences between the reference beam and the object beam become larger locally than the coherence length of the incident beam, no interference can be observed and the phase information is lost. This means that, with temporally partially coherent light, the difference between the path lengths at different positions in the recording plane can be sufficient to disrupt coherency, so interferences will only be observed in part of the recording plane in which the coherency is maintained.

As shown in the Fig. 6.3, the optical source of high temporal coherence with reduced partial spatial coherence can be obtained with a laser beam focused close to a moving ground glass. For a given position of the ground glass, the transmitted light through the sample is a speckle field. When the ground glass is moving, and assuming that the exposure time is long enough to obtain an averaging effect, this type of source is equivalent to a spatial partial coherent light with a spatial coherence width equal to the average speckle field. However, illumination fluctuations arise when short exposure times are requested. When very fast acquisition is required, it is difficult to achieve fast enough moving ground glass.

6.2.3   Color DHM

In order to overcome the limitation of the fast off-axis DHM described here previously, we developed a new DHM configuration, which enables the use of the off-axis configuration with a partial temporal and spatial coherent light beam. It is a significant improvement as it allows operation of the microscope in fast mode without the disadvantage of fluctuations resulting from the configuration with a laser. Moreover, this implementation enables the use of low cost sources such as LED and gives the possibility of simultaneously recording red-green-blue high quality holograms to provide full color digital holographic microscopy at very low noise levels [10]. In this new DHM configuration, the coherence plane of the reference light beam is not perpendicular to the propagation direction in the vicinity of the recording plane. This results in the ability of this light beam to interfere with object beam, whose coherence plane is perpendicular to the propagation direction, and to produce a fringe contrast independent of the position on the recording plane. This permits the recording of off-axis interfering fringes

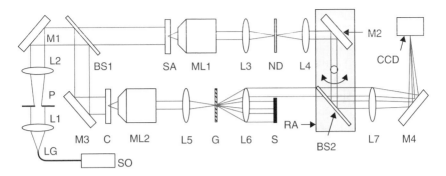

Figure 6.5 The optical setup. SO: The RGB LED module, LG: liquid light guide, L1–L7: Achromatic lenses, P: Aperture, M1–M4: Mirrors, BS1 and BS2: Beamsplitters, ML1, ML2: Microscope lenses, SA: Sample, C: Optical path compensator, G: Grating, S: Optical stop, RA: rotation assembly, CCD: Camera. ND: Neutral density filter to adjust the beam intensity ratio. *Source*: Dubois 2012. Reproduced with permission from the Optical Society

(spatially heterodyne fringes) even in the case of light with limited coherence length, such as the light produced by a LED, a gas discharge lamp. Those configurations take advantage of the particular properties of diffraction gratings for producing an off-axis reference beam without disrupting the temporal coherence of interfering beams in the recording plane. The transmission DHM configuration based on that concept is shown by Fig. 6.5.

The DHM setup is based on a Mach–Zehnder interferometer. The light source SO is realized with three LEDs forming a RGB source with wavelengths 470, 530, and 630 nm. A liquid light guide LG (diameter of 3 mm) homogenizes the illumination. The light beam, collimated by L1, is filtered by the aperture P in order to increase the spatial coherence. The incoming light beam in the DHM is split by the beamsplitter BS1 into object and reference beams. The object beam, transmitted by BS1, illuminates the sample (SA) in transmission. The microscope lens (ML1) and lenses L3, L4, and L7 make the image of one plane of the sample on the CCD sensor. The back focal plane of L3, where a neutral density filter ND is located, is in the front focal plane of L4. The sensor is in the back focal plane of L7. The neutral density filter achieves a suitable beam ratio between the object and the reference beam.

The lenses ML1, L3, and L4 of the object beam have their equivalent counterparts in the reference path, respectively, the lenses ML2, L5, and L6, to achieve equivalent optical paths for the reference and object beams. In the reference path, there is an additional grating G and an optical stop S located, respectively, at the back focal planes of L5 and L6. The grating plane is conjugated with this sensor plane. In the front focal plane of the ML2 lens, an optical path compensator C is placed to roughly compensate for the optical thickness of the sample SA. One of the first diffraction orders created by the grating is not blocked by the optical stop and reaches the sensor plane with an incidence angle that allows the off-axis configuration [11]. The object and the reference beam paths are accurately adjusted by rotating the rotation assembly RA around an axis perpendicular to the optical axes.

As the grating G is conjugated to the sensor plane, we observe that the diffraction does not affect the alignment for spatial coherence superposition.

Let us now consider the partial temporal coherence influence on the fringe contrast. We assume first that the complex amplitude emerging out of the grating plane is quasi-monochromatic of optical frequency . It is imaged on the sensor plane where it is expressed by:

$$g_o(x, y, v) = A \exp\left\{jk2\left(f_6 + f_7\right)\right\} \exp\{jKx\} \, g_i \left(-\frac{f_6}{f_7}x, -\frac{f_6}{f_7}x\right) s(v) \qquad (6.1)$$

where g_o and g_i are, respectively, the complex amplitudes in the sensor plane and the back focal plane of the lens L6, is a constant, $k = 2\pi/\lambda$, f_6 and f_7 are the focal lengths of the lenses L6 and L7, K is the slant factor introduced by the grating on the incident beam on the sensor and $s(v)$ the spectral distribution. We consider now the partially temporal coherent illumination. To optimize the fringe contrast, the propagation times of the object and reference channels between BS1 and the sensor plane have to be equal as far as possible. The temporal behavior is obtained by considering the propagation of a Dirac optical pulse $\delta(t)$. For that purpose, we assume an infinitely broad constant spectrum and we compute the Fourier transformation of Eq. (6.1) on the frequency variable v to obtain the temporal representation. Assuming that $s(v)$ is constant in the plane of g_i, it results in:

$$g_o(x, y, t) = A'\delta\left(t - \frac{2\left(f_6 + f_7\right)}{c}\right) \exp\{jKx\} \, g_i \left(-\frac{f_6}{f_7}x, -\frac{f_6}{f_7}x\right). \qquad (6.2)$$

Equation (6.2) expresses that an optical Dirac pulse in the back focal plane of L6 takes an identical time $t = 2(f_6 + f_7)/c$, to reach every point of the sensor plane regardless of position, where c is the speed of light in the vacuum. Therefore, the grating does not introduce any disturbance to performing global path equalization between the reference and object beams in the off-axis configuration.

For the first color DHM tests, we used a monochromatic CCD camera. Therefore, we successively recorded the three holograms with the red, green, and blue LEDs switched on. The holograms are individually processed by the Fourier method to obtain, in the three colors, the complex amplitudes and the corresponding intensities and optical phases. Digital refocusing is achieved by using the reconstruction algorithm described in [1]. The color intensities are recombined to obtain a composite color image.

Figure 6.6 (Plate 10) shows results obtained on the alga *Odontella* sp., colored with neutral red dye. This figure shows the intensities of the recombined RGB channels for out of focus

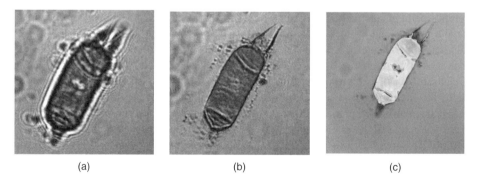

(a)   (b)   (c)

Figure 6.6 (Plate 10)   (a) Out of focus color intensity of the alga *Odontella* sp., (b) refocused intensity, (c) composite phase image of the RGB channels. *See plate section for the color version*

recorded holograms (Fig. 6.6a), the refocused ones (Fig. 6.6b), and the combined phase image (Fig. 6.6c). The RGB LED source gives very low noise and the intensity of image quality is excellent and fully exploitable for microscopy purposes. We observed some small chromatic aberrations. There are small lateral shifts between the three chromatic channels and they are not simultaneously refocused. We measured a total drift of about 10 μm between the blue and red channels along the optical axis. To reduce those effects, we shifted laterally the three chromatic images and we applied slightly different digital holographic refocusing distances to simultaneously obtain the refocused RGB channel images.

6.3   Automated 3D Holographic Analysis

6.3.1   *Extraction of the Full Interferometric Information*

Once a digital hologram is recorded, the complex amplitude is first computed using the Fourier method [8,9]. The intensity and the phase are then computed. This process is illustrated in Fig. 6.7, where Fig. 6.7(a) represents a digital hologram of a green alga, *Pediastrium* sp. The computed intensity and phase information is illustrated respectively in Fig. 6.7(b) and (c). The non-uniform background present on the phase map is due to the experimental cell, which is not completely flat, and to small misalignment of the interferometer. Therefore it is necessary to implement a phase map correction that subtracts the background phase. A method based on the phase map derivative is developed and described in [12]. The resulting compensated phase map is illustrated in Fig. 6.8(a). Thanks to this aberration, compensation of the optical phase, the optical thickness of the alga is measured quantitatively. Figure 6.8(b) illustrates the pseudo-3D representation of this algae from which quantitative measurement can be performed.

6.3.2   *Automated 3D Detection of Organisms*

This part describes the methodology adopted to automatically detect all particles in the experimental volume, recently proposed in [13]. Holograms are first recorded at a frame rate of 24 frames per second with an exposure time of 200 μs. The recorded focus plane is set to the

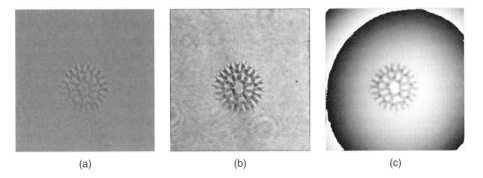

| (a) | (b) | (c) |

Figure 6.7   Extraction of the full interferometric information. (a) Recorded hologram of the alga *Pediastium* sp. (b) Intensity image. (c) Phase image

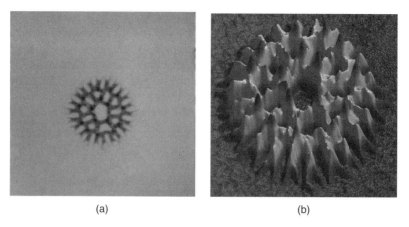

(a)   (b)

Figure 6.8 (a) Compensation of the phase. (b) Pseudo-3D representation of the algae (with inverted contrast)

middle of channel thickness. Once the full interferometric information is extracted from the digital holograms, automated 3D detection is performed.

The 3D detection of the particles is then split into two steps: (1) detection in the X-Y plane is performed, and (2) the position on the Z-axis is determined. X and Y are respectively the horizontal axis and the vertical axis, while the Z-axis is the depth axis. The determination of the X-Y position of particles can be done using a classical thresholding method based on a global threshold of the intensity or the compensated phase, depending on the species' nature. For large concentrations this method is not robust enough. Therefore a new approach based on propagating matrices has been investigated.

6.3.2.1   Using the Usual Thresholding Method

To detect the 3D position of each organism, a classical procedure consists in "thresholding" the intensity or the phase map depending on the nature of the objects under investigation. For opaque particles, the detection is made on the intensity image, while for transparent particles it is performed on the compensated phase images. The organisms under study in the frame of water quality project are algae and can be considered as phase objects. Therefore X-Y detection is achieved by analyzing the phase fluctuations with respect to the background values. The compensation of the phase images shifts the background close to the zero gray level. The particles are then detected by thresholding the compensated phase map to provide a rough detection of particles present in the X-Y plane field of view. This process uses the Otsu method [14] that chooses the threshold level to minimize the intra-class variance of the black and white pixels. The computation of the threshold is then used to convert the compensated phase into a binary image. The centroid of the segmented particle is then computed and a first set for the position in the X-Y plane is obtained. A filtering process is also implemented in order to remove all small particles that are not significant; for example, bacteria, dust, and other very small defects. The binary images are filtered by selecting the range of areas or sizes (major and minor axis) of the particles of interest. The binary image is then cleaned and only contains the particles of interest.

Once the particle is detected in the *X-Y* plane, the *Z* position of the particle has to be determined. This position is established by numerically investigating the micro-channel volume (refocusing of the different slices). To determine the optimal focus plane, a refocusing criterion [15] whose robustness has been determined recently [16] is used. It has been shown that this refocusing criterion is maximal for phase objects as is the case with this experiment. The refocusing criterion is computed inside a region of interest (ROI) that is established around the detected particle [17]. This ROI is determined on the basis of the enlarged segmented area of each object in such a way that the diffraction patterns created by the object are also involved in the ROI.

When the *Z* position has been established, the particle is re-segmented in its focus plane to obtain a precise delimitation of the object. Indeed, the first segmentation (performed in the out of focus recorded plane) only provides a rough estimation of the size of the object. This first estimation is sufficient to determine the ROI for the *Z* scanning but cannot provide any precise information of the object for the classification. By segmenting in the focus plane, the precise shape and the *X-Y* position of the object can be computed, and thereby the 3D position of each object.

This method has been quantified by processing thousands holograms of different organisms, illustrated in Fig. 6.9. These three different organisms are typically present in water and were investigated for testing the detection software: cysts of the parasitic protozoan *Giardia lamblia*, as well as algae *Scenedesmus dimorphus*, and *Chlorella autotrophica*. The classification of these three organisms is presented in Section 6.4.

Applying the developed automated 3D detection algorithm, giving a mean detection score of 80.4% (standard deviation: 5.2%). This mean rate has been computed by comparing a manual and an automated count. This score is not high enough and can be explained by the fact that the detection is based on a simple global thresholding of the compensated phase in the recorded plane. For large concentrations, organisms can be recorded with very different focus depths and thus can lead to different optical phase fluctuations inside the same phase image. All particles then cannot be detected with the same threshold. In Fig. 6.10, an example of concentrated *Chlorella autotrophica* samples are shown where we can see in Fig. 6.10(b) some

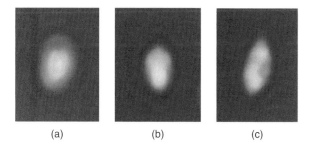

| | | |
| (a) | (b) | (c) |

Figure 6.9 Cropped studied organisms: Compensated phase images (a) *Giardia lamblia*, (b) *Chlorella autotrophica*, (c) *Scenedesmus dimorphus*. *Source*: A. El Mallahi, C. Minetti and F. Dubois 2013. Reproduced with permission from the Optical Society

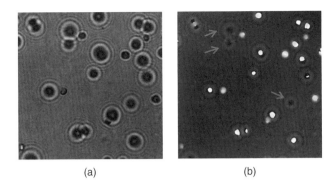

(a)   (b)

Figure 6.10   (a) Intensity image of an algae *Chlorella autotrophica*. (b) Compensated phase map where we can see that some algae (with the arrows) cannot be detected because they are far from the recorded plane. *Source*: A. El Mallahi, C. Minetti and F. Dubois 2013. Reproduced with permission from the Optical Society

algae (with arrows) that will not be detected by a global threshold. Therefore, a new procedure based on the propagation operator has been investigated to overcome this limitation.

6.3.2.2   Using Robust Propagating Matrices

In the case of large concentrations, the previous more usual method is not robust enough to detect particles in the volume with a detection level higher than 80%. Therefore, a new robust method for the 3D automated detection based on the free propagation operator is investigated.

Let us consider the complex amplitude distribution $u_d(x', y')$ refocused at a distance d along the optical axis by computing the Kirchhoff–Fresnel propagation integral in the paraxial approximation:

$$u_d\left(x', y'\right) = \exp\left(jkd\right) F_{x',y'}^{-1} \exp\left(\frac{-jkd\lambda^2}{2}\left(v_x^2 + v_y^2\right)\right) F_{v_x,v_y}^{+1} u_0\left(x, y\right) \tag{6.3}$$

where $u_0(x, y)$ is the complex optical field in the focus plane, $u_d(x', y')$ is the complex amplitude field propagated at a distance d along the optical axis, $k = 2\pi/\lambda$, (x, y) and (x', y') are the spatial variables respectively in the focus plane and in the reconstructed plane, (v_x, v_y) are the spatial frequencies, $j = \sqrt{-1}$ and $F^{\pm 1}$ denotes the direct and inverse 2D continuous Fourier transformations. The Eq. (6.3) is implemented in a discrete form:

$$u_d(s'\Delta, t'\Delta) = \exp(jkd) F_{s',t'}^{-1} \exp\left(\frac{-jkd\lambda^2}{2N^2\Delta^2}\left(U^2 + V^2\right)\right) F_{U^2,V^2}^{+1} u_0(s\Delta, t\Delta) \tag{6.4}$$

where N is the number of pixels in both directions (to match the Fast Fourier transform computations), s, t, s', t', U, and V are integers varying from 0 to $N - 1$, and $F^{\pm 1}$ is now the direct and inverse discrete Fourier transformations.

The complex amplitude $u_d(s'\Delta, t'\Delta)$ is computed for each refocused distance d inside the experimental volume, from the lower to the upper slides of the micro-channel. Then we compute two 3D matrices containing respectively the intensity I and the optical phase P for all the refocused distances:

$$I\left(s'\Delta, t'\Delta, d\right) = \mathrm{Re}\left(u_d\left(s'\Delta, t'\Delta\right)\right)^2 + \mathrm{Im}\left(u_d\left(s'\Delta, t'\Delta\right)\right)^2$$

$$P\left(s'\Delta, t'\Delta, d\right) = \tan^{-1}\left(\frac{\mathrm{Im}\left(u_d\left(s'\Delta, t'\Delta\right)\right)}{\mathrm{Re}\left(u_d\left(s'\Delta, t'\Delta\right)\right)}\right) \tag{6.5}$$

where Re and Im are respectively the real and imaginary parts. Note here that each plane corresponding to a distance d of the 3D P-matrix can be compensated for, as described in this section, in order to remove the background.

Four matrices can now be computed for respectively the minimum and the maximum along the d dimension of the 3D matrices I and P. For these two 3D matrices, we investigate all the d planes and we keep the minimum and the maximum; giving four new matrices:

$$I_{\min}\left(s'\Delta, t'\Delta\right) = \min_d\left(I\left(s'\Delta, t'\Delta, d\right)\right)$$

$$I_{\max}\left(s'\Delta, t'\Delta\right) = \max_d\left(I\left(s'\Delta, t'\Delta, d\right)\right)$$

$$P_{\min}\left(s'\Delta, t'\Delta\right) = \min_d\left(P\left(s'\Delta, t'\Delta, d\right)\right)$$

$$P_{\max}\left(s'\Delta, t'\Delta\right) = \max_d\left(P\left(s'\Delta, t'\Delta, d\right)\right) \tag{6.6}$$

where $\min_d$ and $\max_d$ are, respectively, the minimum and the maximum along all d planes. It results in four bi-dimensional matrices that can be used simultaneously for the 3D detection of particles. Fig. 6.11(a–d) illustrates the four matrices of Eq. (6.6) based on the intensity and phase image in Fig. 6.10.

Each of these four matrices can then be automatically thresholded using the Otsu method, providing four thresholding parameters, $T_{I_{\min}}, T_{I_{\max}}, T_{P_{\min}}$, and $T_{P_{\max}}$. On the basis of the four thresholded images, a simple logical operation (logical AND) provides the binary matrix Fig. 6.11(e) and robust 2D detection becomes possible. The scanning along the z-axis is performed to detect the z position of each detected particle as described in the previous section.

Applying the same methodology as in the previous section, the mean detection rate now gives 95.1% (standard deviation: 2.3%) of correct detections thanks to the propagation operator that easily detects out-of-focus objects. This score obviously depends on the sample nature. Indeed, samples with a lot of aggregates or clogging organisms can perturb the detection rate. To overcome this limitation, a new method, based on a complete analysis of the evolution of the focus planes, has been recently developed to separate aggregates of overlapped particles [18].

6.4 Applications

The developed automated 3D detection process has been used for different applications. We present two of them here. The first one is dedicated to the classification of micro-organisms while the second one looks at the dynamics of red blood cells.

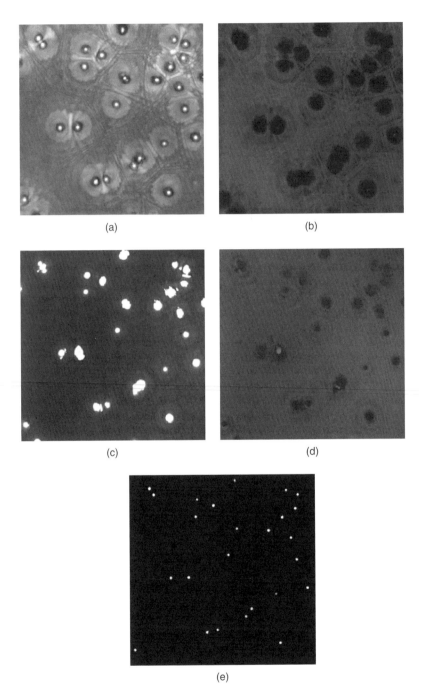

(a)

(b)

(c)

(d)

(e)

Figure 6.11 Propagating matrices (a) I_{max}, (b) I_{min}, (c) P_{max} (+ compensation), (d) P_{min} (+ compensation), (e) by thresholding these four matrices and then combining them, a binary matrix can be computed, on which robust 2D detection can be done. *Source*: A. El Mallahi, C. Minetti and F. Dubois 2013. Reproduced with permission from the Optical Society

6.4.1   Holographic Classification of Micro-Organisms

Object classification by digital holography is a promising approach offering a large number of applications. The development of automated procedures to detect and recognize living organisms using digital holography are of significant benefit. Using single-exposure on-line holographic microscope (SEOL), Javidi *et al.* have investigated 3D imaging and recognition of micro-organisms in [19–22] where the Fresnel field of the object illuminated by coherent light is recorded. In [23], 3D phase objects (triangle, rectangle, and circle) have been also recognized and classified by digital Fresnel holography. The identification of microorganisms has been addressed in [24] by the use of microscopic integral imaging. It is based on recording the multi-view directional information of a 3D scene with incoherent illumination. Another class of pattern recognition problem based on photon counting imagery has been investigated in [25], where the effect of noise has been taken into account to recognize a target in a 3D scene. Recently in [26], Liu *et al.* have applied statistical clustering algorithms to recognize and classify two classes of red blood cells (RBCs). Also recently, an automated statistical procedure to quantify the morphology of RBCs using DHM was presented in [27].

We have developed a classification procedure based on the full interferometric information of species (intensity + phase). For this reason, it makes sense to use the term *holographic classification*. This procedure has been applied to several applications. One of them concerns the monitoring of drinking water resources. Over the past three decades, serious health crises associated with the parasitic protozoan *Giardia lamblia* have triggered major efforts worldwide for its detection in the water resources used to produce drinking water [28,29]. This flagellated protozoan is responsible for giardiasis [30], a widespread diarrheal disease affecting around two hundred million symptomatic individuals in the world [28]. Humans can be infected through the following transmission routes: person-to-person transmission, direct contact with contaminated mammals, or ingestion of infected food and/or water. *Giardia lamblia* is able to live in very harsh environments thanks to its vegetative transmission form, the cyst, whose size varies between $11–14\,\mu m$ in length and $6–10\,\mu m$ in width. In particular, the cyst is able to resist the levels of chlorine traditionally used in water disinfection. Considering the lack of efficient preventive means against *Giardia* cysts, they must be detected as early as possible in water resources [31]. Usual detection methods are too time consuming as they require the use of different stains prior to detection and recognition.

A new classification procedure has been developed and applied to the monitoring of drinking water. Three different organisms (typically present in water) were investigated in this study: cysts of *Giardia lamblia*, as well as the algae *Scenedesmus dimorphus*, and *Chlorella autotrophica*. *Giardia lamblia* is a parasitic flagellated protozoan that was obtained in cyst form. *Scenedesmus dimorphus* and *Chlorella autotrophica* are both types of green algae and were selected for their high similarities in size range and shape to *Giardia lamblia* cysts [32].

These three species are illustrated in Fig. 6.9. A three-class classification is then investigated. Once the 3D position of each particle is detected, a set of features is computed to perform their classification. Two types of feature can be extracted from holographic information. In the first step, a set of morphological features can be obtained from the segmentation of the compensated phase image. In the second step, as both intensity and optical phase are extracted from holograms, textural features based on these two physical quantities are computed.

6.4.1.1 Using Morphological Features

A set of classical morphological features can be first extracted from the segmented binary image of the detected organisms. This set of region descriptors includes the following: *area*, *perimeter*, *major* and *minor axes*, *eccentricity*, and *equivalent circle diameter*. Area is calculated as the number of pixels in the segmented region, the perimeter is the number of pixels in the boundary of the segmented particles. The major and minor axes are the length in pixels of the major and minor axes of the ellipse. Eccentricity is computed as the ratio of the distance between the foci of the ellipse surrounding the segmented particle and its major axis length. The value of the feature is between 0 and 1 (an ellipse whose eccentricity is 0 is a circle, while an ellipse whose eccentricity is 1 is a line segment). The *equivalent circle diameter* is a number specifying the diameter of a circle with the same area as the region. Each value is then converted into µm, given by the field of view (FOV) of the camera and the magnification of the objective lenses used (63× objective lenses, given a FOV of 115 × 115 µm).

To perform a classification process, the best discriminants among this set of morphological features have to be determined. As the three species have very similar shapes and sizes, this list of features is not sufficient. Moreover, during the particle's travel in the micro-channel, it can rotate around its own axis and any of these out-of-plane orientations can be recorded on the holograms. It results in a wide range of different morphological values that can be captured for organisms of the same species.

Consequently, in our multi-species classification experiment, the use of those morphological features only would lead to substantial regions of overlap among similar species in the feature space. Those features are, therefore, not sufficient to adequately distinguish between species used in our experiments. For a robust classification, other features have to be introduced and added to the morphological features. Given the advantage of capturing both intensity as well as phase information with digital holographic microscopy, we choose to introduce textural features to the classification procedure. As we use the full interferometric information, the term *holographic classification* is used.

6.4.1.2 Using Textural Features

Digital holography gives information about both the intensity and phase of the object. When an object is focused, the intensity gives information about the absorbance of the organism while the phase is a measure of the optical thickness.

Thanks to these two physical quantities, a textural analysis based on statistical measures can be computed. A typical category of such measures is based on statistical moments whose expression of the *n*th is given by:

$$\mu_n = \sum_{i=0}^{L-1} \left(q_i - m \right)^n p(q_i) \tag{6.7}$$

where q_i is the intensity, $p(q)$ is the histogram of the intensity levels in a region, L is the number of possible intensity levels (in this case for gray level images, $L = 256$) and $m = \sum_{i=0}^{L-1} q_i p(q_i)$ is the mean intensity.

These measures are based on the histogram of the pixel intensity value. For each image, a set of six texture measures is computed, which are: *average gray level* and *average contrast*, *smoothness*, *skewness*, *uniformity*, and *entropy*. The mathematical expression of those six measures can be found in [33]. All these textural features are measured in the domain given by the segmentation of the particle. The first two features, average gray level and average contrast, give the mean values in a gray level scale with its standard deviation. The smoothness (R) is measured inside each particle domain and is dependent on the variance (which is normalized to the range [0, 1] by dividing it by $(L-1)^2$). The smoothness equals 0 for the region of constant intensity and increases with fluctuations in pixel value. It ranges from $0-1$. The third statistical moment measures the skewness of the histogram. It gives 0 for symmetric histograms, a positive value for histograms skewed to the right above the mean and negative value for histograms skewed to the left. This moment is also normalized to 1 by dividing it by $(L-1)^2$. The uniformity texture measure is at its maximum when all gray levels are equal. The entropy of the object region is also implemented to quantify the randomness of the texture.

These six texture features are computed for all detected organisms both on the intensity and the compensated phase images. A set of 12 textural features is then available for each organism to be classified. Figure 6.12 shows the pseudo-3D representation of the compensated phase of organisms illustrated in Fig. 6.9. We can see that even if there is a high similarity between the different species, thanks to the optical phase it becomes possible to distinguish and discriminate between them. Indeed, in spite of having the same shapes, these organisms have different optical thicknesses that are exploited thanks to the interferometric information. Figure 6.13 (Plate 11) illustrates two feature spaces, one using textural features based on intensity information (Fig. 6.13a) and the other using textural features based on compensated phase information (Fig. 6.13b). We observe that the use of quantitative optical phase information can lead to high separability between the three species. It has to be pointed out that the use of a partially coherent source provides phase maps of high quality with a very low noise level [34] permitting an accurate textural analysis.

6.4.1.3   Robust SVMs Classifier

After a graphic analysis of the dataset, a classification process is performed. For this, support vector machines (SVMs) [35–37] are investigated to optimally separate the dataset into the three categories. The basic idea of this supervised learning model is the construction of a set of $(N-1)$-dimensional hyperplanes (N = number of features = 18 in our case) that optimally separates the data classes in the N-dimensional feature space. For each hyperplane, the best separation is obtained by the hyperplane that has the largest distance to the nearest training data point of any class. The multi-class classification is performed by training one SVM per class using a one-against-all method [38]. The use of a radial basis function (RBF) kernel is chosen as a majority of features can be separated by a nonlinear region. This kernel function maps the data into a different space where the hyperplanes can be used to separate between classes. The parameters of the SVMs (a cost parameter that controls the flexibility of the hyperplanes to avoid overfitting and the parameter of the RBF kernel) are automatically selected based on a grid search. The evaluation is performed by a 10-fold cross-validation. The dataset is randomly partitioned into 10 subsets, performing the training on nine subsets (training set), and validating

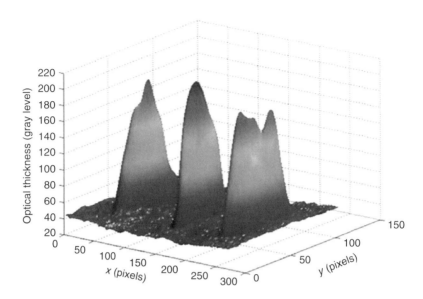

Figure 6.12 Pseudo-3D representation of the compensated phase of the organisms illustrated in Fig. 6.9, where we can see that, even if the three organisms have the similar morphology features, thanks to the optical phase, it becomes possible to distinguish between them. Textural features are therefore selected to perform the classification process rather than morphological ones. The z-axis is the optical thickness. *Source*: A. El Mallahi, C. Minetti and F. Dubois 2013. Reproduced with permission from the Optical Society

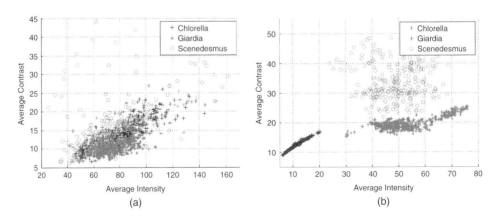

Figure 6.13 (Plate 11) Feature space representation using (a) intensity information of the detected particles of the three species, and (b) compensated phase of detected particles of the three species *See plate section for the color version*

Table 6.1   Performance of the classifier (%) with the corresponding standard deviation: Confusion matrix to compare the truth labels and the decisions of the classifier

Actual Classes	Predicted Classes		
	Chlorella	Giardia	Scenedesmus
Chlorella	**98.1 ± 2.3**	1.1 ± 0.2	0.8 ± 0.3
Giardia	0.8 ± 0.2	**98.6 ± 1.7**	0.6 ± 0.1
Scenedesmus	1.1 ± 0.5	1.4 ± 0.4	**97.5 ± 2.4**

the classifier on the last subset (testing set). The cross validation is then repeated 10 times, with each of the 10 subsets used only once as the testing set. The 10 results are then averaged to produce a single robust mean estimation, with its corresponding standard deviation.

In Table 6.1, the confusion matrix of the classifier is shown to illustrate the performance of the algorithm. It contains information about the actual and predicted classification of the three tested species. The diagonal of the confusion matrix stores the number of correctly classified organisms, while the non-diagonal elements refer to the misclassified objects. Thanks to the complete list of features (morphological and textural) resulting from the full interferometric information, a robust classifier that distinguishes the three species with more than 97% is built. (There are approximately 500 samples of each class.) For the *Giardia lamblia* cysts, the proportion of organisms that were correctly identified is 98.6% while 0.8% of all the detected *Giardia* has been classified as *Chlorella autotrophica* alga and 0.6% as *Scenedesmus dimorphus* alga. The accuracy of the built classifier can be computed as the total number of correct predictions and gives a level of 98.1%.

6.4.2   Dynamics of Red Blood Cells (RBCs)

Blood is a complex biological fluid made up of more than 50% with red blood cells (RBC). Those cells, with a disk diameter of approximately 6.2–8.2 µm, a thickness at the thickest point of 2–2.5 µm, and a minimum thickness in the center of 0.8–1 µm, are much smaller than most other human cells. They are deformable and can adopt various shapes depending on the channel they are flowing in. They are the principal means of delivering oxygen to body tissues via blood flow through the circulatory system. They take up oxygen in the lungs or gills and release it while squeezing through the body's capillaries. These cells' cytoplasm is rich in hemoglobin, an iron-containing biomolecule that can bind to oxygen and is responsible for blood's red color.

The distribution of RBCs in a confined flow is inhomogeneous and shows a marked depletion near the walls, due to competition between migration away from the walls and shear-induced diffusion resulting from interactions between particles [39].

The migration of blood cells forms the physical basis of the formation of a cell free plasma layer near vessel walls in the microcirculation. Furthermore, with the very high concentration of RBCs in human blood, strong hydrodynamic interactions take place. Consequently, blood is a complex flow from a rheological point of view leading to complex flow patterns in microcirculation networks where the diameter of blood vessels becomes comparable to the cell size.

Those inhomogeneities have important consequences on the transport of oxygen and may lead to situations where one branch of a capillary bifurcation is not irrigated by RBCs.

In this part, we report the investigation and quantification of the lift force that pushes cells away from the vessel walls when a shear flow is applied. To cope with the DHM requirements in terms of concentration and limit the hydrodynamic interactions between pairs of cells, we work with highly diluted and washed blood. The resulting fluid is made of RBCs in an external solution (BSA – bovine serum albumin and PBS – phosphate buffer saline) alone or in combination with Dextran (typically 1%). With the addition of Dextran, one can increase the viscosity of the external fluid, and easily play on lift velocity. To avoid any screening of the lift by sedimentation, we perform these experiments in microgravity during parabolic flight campaigns onboard the Airbus A300 Zero-G from Novespace.

6.4.2.1 Experimental Protocol

Experiments are performed in a shear flow chamber made of two glass plates with a gap of about 170 μm (see Fig. 6.14). The bottom plate is fixed while the top one rotates creating a linear velocity profile of the fluid inside the chamber. The suspension of RBCs is inserted via the inlet in the center of the bottom plate and the DHM monitors the suspension at a distance of 7 cm with respect to the center of the cell.

During a parabolic maneuver creating a period of 22 s of microgravity (10^{-2} g), we first undergo a hyper gravity period of 1.8 g. We take the benefit of this period to stop the shear flow and let the cells sediment on the bottom plate of the cell. When entering the microgravity period, the shear flow is applied and the DHM monitors the 22 s period of microgravity. The RBC, initially lying on the bottom plate will lift away from the wall and the 3D position of

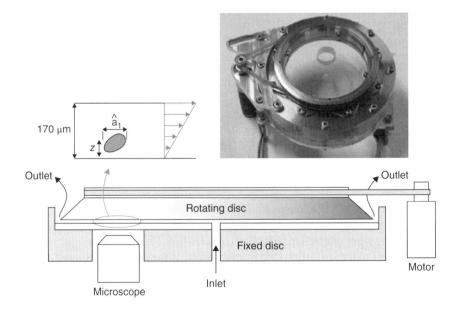

Figure 6.14   Shear flow chamber used in parabolic flights

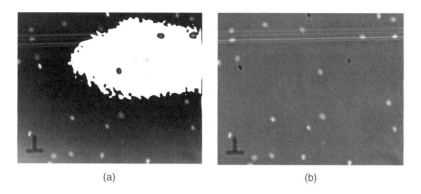

(a)   (b)

Figure 6.15   (a) Example of phase maps with RBCs, and (b) compensated phase image

each RBC will be retrieved thanks to the interferometric information obtained with the DHM. This zero-g period is then followed by another 1.8 g period where the shear flow is stopped again to let the sample sediment. In this way, we have a very simple and reproducible initial condition.

6.4.2.2   Hologram Analysis

Holograms are acquired at a frequency of 24 holograms per second. The field of view is $270 \times 215\,\mu m$. Depending on the concentration of the sample, between 10 and 100 cells are present in the field of view. Thanks to an adequate adjustment of the interferometer, we obtain very flat phase maps that can easily be compensated [12] resulting in bright spots (RBC) on an almost constant background (see Fig. 6.15). The X-Y detection of the cells is made on a simple threshold of the compensated phase map. When the refractive index difference between cells and the external solution is not sufficient to detect the X-Y position with a simple threshold, we use the propagation matrices (described earlier) that give more robust and reliable results.

Thanks to the mono dispersed size of RBCs, we determine the $\underline{Z}$ position by locally propagating around the X-Y position of each RBC with a fixed region of interest (ROI) of 120×120 pixels. Red blood cells behave like phase objects but have a non-negligible absorption. They cannot be considered either as pure phase objects or as pure amplitude objects. Consequentially, the best focus is determined by propagating the ROI along the different Z values and by computing, for each plane, the integrated gradient inside the ROI. The curve exhibits a well-defined and narrow minimum at the best focus plane of the object. This procedure is repeated for each RBC inside each hologram and provides the evolution of the suspension as a function of time.

6.4.2.3   Experimental Results

The lift force of deformable objects has been subject to a theoretical study by Olla [40,41] and a predictive law has been determined, which gives the distance from object to wall as a function of the shear rate (γ), the radius of the object (R), the time (t), and a parameter (U) describing the lift velocity and containing the viscosity ratio (between inner and outer fluid).

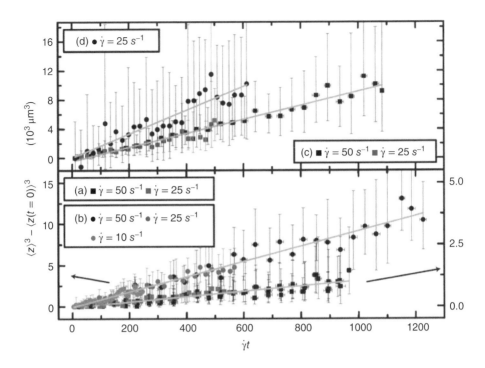

Figure 6.16 RBC: wall distance $<z^3>$ vs. γt for different outer viscosities: (a) 1.4 mPa.s; (b) 6.1 mPa.s; (c) 9.3 mPa.s; (d) 13 mPa.s. Bold lines indicate fit to Olla's prediction, with $UR^3 = 0.36, 3.1, 3.2, 5.4\,\mu m^3$, respectively. *Source*: G. Coupier, A. Srivastav, C. Minetti, and T. Podgorski 2013. Reproduced with permission from the American Physical Society

The law can be reduced to the following expression:

$$\frac{\langle Z^3 \rangle}{R^3} = 3U\gamma t \qquad (6.8)$$

exhibiting a linear dependency between the rescaled mean position of objects to the power of 3 and the shear rate multiplied by time. Experimental results confirm this predictive law very well by exhibiting linear curves with different lift velocities for the different viscosity ratios (Fig. 6.16). By symmetry, a tumbling rigid object should not migrate on average. The non-zero lift suggests that RBC deformability allows symmetry breaking: It is stretched when oriented in the direction of the elongational component of the flow, while it is compressed when orthogonal, resulting in an averaged asymmetric shape, leading to a migration law similar to the one known for a fixed shape and orientation. By increasing the external viscosity, the stresses on the RBC membrane are higher and lead to increased deformation, which in turn enhances the lift.

These results clearly demonstrate the capability of digital holographic microscopy for monitoring RBC suspension in micro-flow circulation.

6.5 Conclusion

In this chapter we showed developments on DHM working with partially coherent illumination that allows us to drastically reduce coherence noise. The several described instruments have their specific advantages with respect to the applications. In this way, it is possible to achieve high quality phase shifting or off-axis DHMs. More recently, we proposed the implementation of a color off-axis DHM making possible the use of weak spatial and temporal coherences sources. The improved image quality leads to the implementation of powerful processing up to the automated classification of living organisms. With respect to the classification, the additional phase information constitutes a huge improvement. The DHMs working with partial coherence sources are demonstrated on applications that are currently developed in our laboratory.

Acknowledgments

The authors thank l'Institut Bruxellois pour la Recherche et l'Innovation (INNOVIRIS), the Walloon Region and the PRODEX office for supports in the frame of the HoloFlow, DECISIV and BIOMICS projects.

The authors also thank Eva-Maria Zetsche from ANCH laboratory, Vrije Universiteit Brussel, for providing us with algae samples.

References

[1] Dubois, F., Joannes, L., and Legros, J.C., "Improved three-dimensional imaging with a digital holography microscope with a source of partial spatial coherence," *Appl. Optics*, **38**, 7085–7094 (1999).

[2] Zhang, T., and Yamaguchi, I., "Three-dimensional microscopy with phase-shifting digital holography," *Opt. Lett.* **23**, 1221–1223 (1998).

[3] Dubois, F., Yourassowsky, C., and Monnom, O., "Microscopie en holographie digitale avec une source partiellement cohérente," in *Imagerie et Photonique pour les Sciences du Vivant et la Médecine*, M. Faupel, P. Smigielski, and R. Grzymala, (eds) Fontis Media (2004).

[4] Marquet, P., Rappaz, B.J., Magistretti, P.J., Cuche, E., Emery, Y., Colomb, T., and Depeursinge, C., "Digital holographic microscopy: a noninvasive contrast imaging technique allowing quantitative visualization of living cells with subwavelength axial accuracy," *Opt. Lett.* **30**, 468–470 (2005).

[5] Dubois, F., Yourassowsky, C., Monnom, O., Legros J.C., Debeir, O., Van Ham, P., *et al.*, "Digital holographic microscopy for the three-dimensional dynamic analysis of in vitro cancer cell migration," *J. Biomed. Opt.* **11–5**, 054032 (2006).

[6] Carl D., Kemper B., Wernicke, G., and von Bally, G., "Parameter-optimized digital holographic microscope for high-resolution living-cell analysis," *Appl. Opt.* **43**, 6536–6544 (2004).

[7] Dubois, F., N. Callens, C. Yourassowsky, M. Hoyos, P. Kurowski and O. Monnom, "Digital holographic microscopy with reduced spatial coherence for 3D particle flow analysis," *Applied Optics*, **45**, No. 5, 964–961 (2006).

[8] Kreis, T., "Digital holographic interference-phase measurement using the Fourier-transform method," *J. Opt. Soc. Am. A* **3**, 847–855 (1986).

[9] Takeda, M., Ina, H., and Kobayashi, S., "Fourier-transform method of fringe-pattern analysis for computer-based topography and interferometry," *J. Opt. Soc. Am.* **72**, 156–160 (1982).

[10] Dubois, F. and Yourassowsky, C., "Full off-axis red-green-blue digital holographic microscope with LED illumination," *Optics Letters* **37**(12), 2190–2192 (2012).

[11] Kolman, P. and Chmelík, R., "Coherence-controlled holographic microscope", *Opt. Express* **18**, 21990–22003 (2010).

[12] Minetti, C., Callens, N., Coupier, G., Podgorski, T., and Dubois, F., "Fast measurements of concentration profiles inside deformable objects in microflows with reduced spatial coherence digital holography," *Appl. Op.* **47**, 5305–5314 (2008).

[13] El Mallahi, A., Minetti, C., and Dubois, F., "Automated three-dimensional detection and classification of living organisms using digital holographic microscopy with partial spatial coherent source: application to the monitoring of drinking water resources," *Appl. Optics* **52**, A62–A80 (2013).

[14] Sezgin, M. and Sankur, B., "Survey over image thresholding techniques and quantitative performance evaluation," *Journal of Electronic Imaging* **13**(1), 146–165 (2004).

[15] Dubois, F., Schockaert, C., Callens, N., and Yourassowsky, C., "Focus plane detection criteria in digital holography microscopy by amplitude analysis," *Opt. Exp.* **14**, 5895–5908 (2006).

[16] El Mallahi, A. and Dubois, F., "Dependency and precision of the refocusing criterion based on amplitude analysis in digital holographic microscopy," *Opt. Exp.* **19**, 6684–6698 (2011).

[17] Antkowiak, M., Callens, N., Yourassowski, C. and Dubois, F., "Extended focusing imaging of a microparticle field with digital holographic microscope," *Opt. Exp.* **33**, 1626–1628 (2008).

[18] El Mallahi, A. and Dubois, F., "Separation of overlapped particles in digital holographic microscopy," *Opt. Exp.* **21**, 6466–6479 (2013).

[19] Javidi B., Moon, I., Yeom, S., and Carapezz, E., "Three-dimensional imaging and recognition of microorganisms using single-exposure on-line (SEOL) digital holography," *Opt. Exp.* **13**, 4492–4506 (2005).

[20] Stern, A. and Javidi, B., "Theoretical analysis of three-dimensional imaging and recognition of micro-organisms with a single-exposure on-line holographic microscope," *J. Opt. Soc. Am. A* **24**, 163–168 (2007).

[21] Moon, I. and Javidi, B., "Shape tolerant three-dimensional recognition of biological microorganisms using digital holography," *Opt. Exp.* **13**, 9612–9622 (2005).

[22] Javidi, B., Yeom, S., Moon, I., and DaneshPanah, M., "Real-time automated 3D sensing, detection, and recognition of dynamic biological micro-organic events," *Opt. Exp.* **14**, 3806–3826 (2006).

[23] Nelleri, A., Joseph, J., and Singh, K., "Recognition and classification of three-dimensional phase objects by digital Fresnel holography," *Appl. Opt.* **45**, 4046–4053 (2006).

[24] Javidi, B., Moon, I., and Yeom, S., "Three-dimensional identification of biological microorganism using integral imaging," *Opt. Exp.* **14**, 12096–12108 (2006).

[25] DaneshPanah, M., Javidi, B., and Watsin, E.A., "Three dimensional object recognition with photon counting imagery in the presence of noise," *Opt. Exp.* **18**, 26450–26460 (2010).

[26] Liu, R., Dey, D.K., Boss, D., Marquet, O., and Javidi, B., "Recognition and classification of red blood cells using digital holographic microscopy and data clustering with discriminant analysis," *J. Opt. Soc. Am. A* **28**, 1204–1210 (2011).

[27] Moon, I., Javidi, B., Yi, F., Boss, D., and Marquet, P., "Automated statistical quantification of three-dimensional morphology and mean corpuscular hemoglobin of multiple red blood cells," *Opt. Exp.* **20**, 10295–10309 (2012).

[28] World Health Organization (WHO), "Guidelines for drinking-water quality Volume 1 Recommendations," Geneva (2006).

[29] Huang, D.B. and White, A.C., "An updated review on Cryptosporidium and Giardia," *Gastroenterol. Clin. N. Am.* **35**, 291–314 (2006).

[30] Craun, G.F., "Waterborne giardiasis," *Human Parasitic Diseases* **3**, 267–293 (1990).

[31] Bouzid, M., Steverding, D., and Tyler, K.M., "Detection and surveillance of waterborne protozoan parasites," *Cur. Opin. Biotechnol.* **19**, 302–306 (2008).

[32] Rodgers, M.R., Flanigan, D.J., and Jakubowski, W., "Identification of algae which interfere with the detection of *Giardia* cysts and *Cryptosporidium* oocysts and a method for alleviating this interference," *Appl. Environ. Microbio.* **61**, 3759–3763 (1995).

[33] Gonzalez, R.C. and Woods, R.E., *Digital Image Processing*, Prentice Hall, Upper Saddle River, NY (2002).

[34] Dubois, F., Novella Requena, M.-L., Minetti, C., Monnom, O., and Istasse, E., "Partial coherence effects in digital holographic microscopy with a laser source," *Appl. Opt.* **43**, 1131–1139 (2004).

[35] Cortes, C. and Vapnik, V., "Support vector network," *Machine Learning* **20**, 273–297 (1995).

[36] van der Heijden, F., Duin, R.P.W., de Ridder, D., and Tax, D.M.L., *Classification*, Parameter Estimation and State Estimation, John Wiley & Sons, Ltd, Chichester (2004).

[37] Webb, A., *Statistical Pattern Recognition*, John Wiley & Sons, Ltd, Chichester (2002).

[38] Chang, C.-C. and Lin, C.-J., "LIBSVM: A library for support vector machines," *ACM Trans. Intell. Syst. Technol.* **2**, 1–27 (2011).

[39] Grandchamp, X., Coupier, G., Srivastav, A., Minetti, C., and Podgorski, T., "Lift and down-gradient shear-induced diffusion in red blood cell suspensions," *Phys. Rev. Lett.* **110**, 108101 (2013).

[40] Olla P., "The lift on a tank-treading ellipsoidal cell in a shear flow," *J. Phys. II*, **7**, 1533 (1997).

[41] Callens, N., Minetti, C., Coupier, G., Mader, M., Dubois, F., Misbah, C., and Podgorski, T., "Hydrodynamic lift of vesicles under shear flow in microgravity," *Europhys. Lett.* **83**(2), 24002 (2008).

7

Programmable Microscopy

Tobias Haist[1], Malte Hasler[1], Wolfang Osten[1] and Michal Baranek[2]

[1]*Institute für Technische Optik, Universität Stuttgart, Germany*
[2]*Department of Optics, Palacky University Olomouc, Czech Republic*

7.1 Introduction

Today, a huge number of imaging methods for the investigation of microscopic specimens are available and most of these methods have some parameters that will strongly affect imaging. For microscope users, it is–at least most of the time–impossible to chose the perfect combination for the imaging task at hand.

Programmable microscopy is a technique that allows one to easily switch at high speed between these different imaging methods. If we denote the imaging method by p_0 and its parameters by $p_1, p_2, \ldots$ "optimized imaging" means that the overall set of parameters $p(p_0, p_1, p_2, \ldots)$ leads to the "best" suited image to a given task.

In a multi-image context we can use multiple images obtained with different parameters p in order to find a good, or even perfect, imaging result. Moreover, we might use the images obtained with different parameters in order to generate one optimized image by digital post-processing.

Up until now, only a few very rudimentary examples of such approaches have been realized but we think that this methodology might lead to the next generation of powerful microscopes. In order to realize such methods, fortunately, not much hardware is necessary. It is enough to incorporate programmable elements into the optical train of the microscope.

The preferred approach is to use spatial light modulators (SLMs) in the illumination and/or imaging path of the microscope. This has been proposed in the past for a lot of different, interesting applications. Particularly within the last 10 years a lot of methods have been investigated, mainly due to the availability of high quality spatial light modulators. Important methods are optical micromanipulation (Eriksen *et al.* 2002; Grier 2003; Hayasaki *et al.* 1999; Reicherter *et al.* 1999), multi-photon microscopy (Nikolenko *et al.* 2008; Peterka *et al.* 2010; Qin *et al.* 2012), structured-light illumination (Chang *et al.* 2009; Choi and Kim 2012; Hussain and Campos 2013; Wang *et al.* 2009), Raman and coherent anti-Stokes

Multi-dimensional Imaging, First Edition. Edited by Bahram Javidi, Enrique Tajahuerce and Pedro Andrés.
© 2014 John Wiley & Sons, Ltd. Published 2014 by John Wiley & Sons, Ltd.

Raman imaging (Jesacher *et al.* 2011), point-spread function engineering (Kenny *et al.* 2012), interference microscopy (Schausberger *et al.* 2010), microscopic spectroscopy (Pham *et al.* 2012), stereo microscopy (Hasler *et al.* 2012; Lee *et al.* 2013), holographic microscopy (Mico *et al.* 2010; Valencia and Moliner 2010), phase contrast imaging (Bernet *et al.* 2006; Glückstad and Mogensen 2001; Khan *et al.* 2011; Kim and Popescu 2011; Maurer *et al.* 2011; McIntyre *et al.* 2009a), confocal imaging (Heintzmann *et al.* 2001), lithography (Gittard *et al.* 2011; Haist *et al.* 1999; Jesacher and Booth 2010; Zhou *et al.* 2013), multispot-imaging and spectroscopy (Nikolenko *et al.* 2010; Pham *et al.* 2012; Shao *et al.* 2012), and aberration correction (Débarre *et al.* 2009; Haist *et al.* 2008; Reicherter *et al.* 2004; Scrimgeour and Curtis 2012).

The main advantage of this sort of programmable microscopy is that we can change from one imaging method to the next very fast, always using the same image sensor. Therefore, different information about a specimen can be obtained and the multiple images can be fruitfully combined into one final image.

In this contribution we concentrate on approaches that use a spatial light modulator in the imaging path. We explain the optical design considerations that have to be taken into account when implementing programmable microscopy and we describe the current most important applications, aberration correction, and phase contrast imaging.

7.2 Optical Design Considerations and Some Typical Setups

A sketch of a typical setup used to manipulate the imaging path of a microscope is shown in Fig. 7.1. The spatial light modulator is ideally located in a plane conjugate to the exit pupil of the microscope objective lens (MO). For a conventional MO the exit pupil is inside the MO, therefore, some sort of imaging is necessary. Most often a telescopic imaging system is employed because it preserves the wavefront curvature (a plane wave on the SLM leads to a plane wave in the pupil of the MO). Also, it can easily be integrated into the collimated path of most modern research microscopes. With such an approach the SLM is located in a Fourier plane of the object, therefore, a filter written into the SLM can be regarded as a Fourier hologram.

For this case of telescopic imaging one chooses the magnification of the telescope $|\beta'| = f_2/f_1 = D_{SLM}/D_{MO}$ based on the effective aperture of the SLM D_{SLM} and the diameter of the pupil of the MO D_{MO}. If the system is used with different objective lenses, a zoom system can be employed or one has to find a compromise concerning the usable number of pixels and the usable numerical aperture of the microscope. This necessary number of pixels depends on the details of the application, especially on the carrier frequency that is used.

In practice, light modulators are never perfect. Deviations in modulation characteristics lead to unwanted diffraction orders and, therefore, to a loss of contrast in the image plane due to the overlay of multiple, modified, and shifted copies of the object. Therefore, it is advantageous to superimpose a carrier frequency to the filter. This separates the filtered image (positive first diffraction order) from other components (e.g., zeroth diffraction order). This means that the Fourier holograms written into the SLM are off-axis holograms. The basic idea is shown in Fig. 7.2.

On the other hand, a high carrier frequency leads to a reduced diffraction efficiency and increased fringing field effects (see Section 7.3). Additionally, as the response curve of the SLM depends strongly on the wavelength, the display has to be characterized carefully to

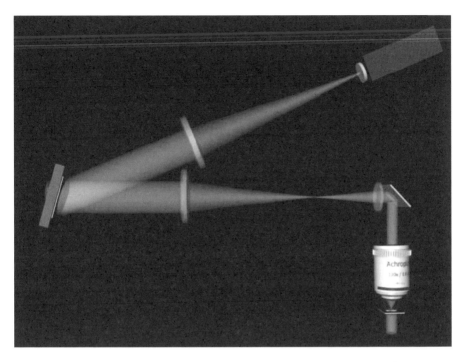

Figure 7.1 Basic setup for using a spatial light modulator (SLM) in the imaging path of a programmable microscope. The pupil of the microscope objective lens is imaged onto the SLM using a Kepler telescope

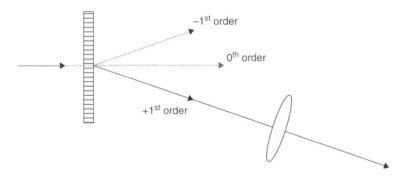

Figure 7.2 A carrier frequency approach that can be employed in order to seperate unwanted diffraction orders of the modulator from the desired order (here: positive first order)

minimize the loss of light. In practice, a carrier frequency of four gray levels is a good compromise but the usable object field is reduced when SLMs with a limited space bandwidth are used because the angular separation of the different orders due to the carrier frequency effectively limits the field of view.

One should keep in mind that such an approach is only necessary due to the non-ideal modulation behavior of the SLM (compare Section 7.3).

It is also possible to try to eliminate the images due to unwanted diffraction orders by post-processing of multiple images. We explain this idea based on the unwanted zeroth order of the hologram. Without special optimization precautions (Liang *et al.* 2012) this order will be present when using commercial SLMs. Two images are recorded with different filters displayed on the SLM. One image with the SLM set to a constant phase and the other incorporating a filter, for example, one with a small carrier frequency. If one subtracts the (correctly normalized) images it is possible to eliminate the disturbing order (see Fig. 7.3).

For this sort of imaging without strong carrier frequency, incoherent light has to be used because otherwise, of course, the behavior is nonlinear in intensity and, therefore, the subtraction would not eliminate the unwanted images.

Typically, most applications that have been published up until now use coherent light for imaging. However, it is well known that in microscopy, coherence leads to problems most of the time, namely speckles, other interference-related disturbances and artifacts (e.g., problems due to dust on an optical surface). It is possible to use averaging approaches, for example, rotating diffusers or vibrating modulators, or even multiple recordings using different SLM addressings, in order to avoid these problems.

Of course, the more straightforward approach is to reduce the coherence of the illumination. To this end, in most of our current programmable imaging setups we use LEDs in combination with bandpass filters. The bandpass is necessary because the dispersion at the carrier frequency of the gratings written into the SLM leads to chromatic aberrations in the image plane. The tolerable spectral width (and therefore the coherence length) has to be computed based on the geometry of the imaging system (magnification, pixel size of the sensor, SLM position) and the filters or gratings that will be written into the SLM. An interesting alternative is to use an additional grating in order to cancel the dispersion as described by Steiger *et al.* (2012).

It is also possible to vary the position of the SLM so that it is not located conjugate to the pupil of the MO. Depending on the diameters in the system and the location of the MO pupil this might make it possible to strongly reduce the overall size of the system. In this case, however, one has to carefully analyze the system with respect to vignetting (vignetting starts if

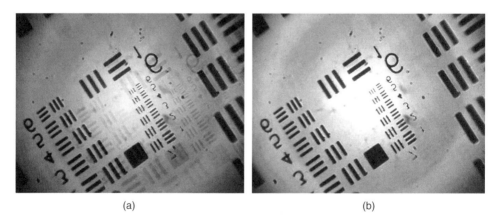

(a)   (b)

Figure 7.3 Subtraction of three images which have been obtained with and without filters written into the SLM leads to elimination of the unwanted diffraction order. (a) single image, (b) image obtained by subtracting a zeroth-order image and a minus first order image

the marginal rays of an on-axis object point fall onto the edge of the SLM) and the illumination might be modified.

For most phase-contrast methods (e.g., Zernike-based methods or dark field, see Section 7.5) it is necessary that the light source is conjugate to the SLM. For a typical Köhler-type illumination and having the SLM located in the Fourier plane of the specimen this is automatically the case. But shorter designs are possible if one does not use a perfect Köhler geometry. Of course, the illumination should be more or less homogeneous in the object plane and the illumination source should be imaged onto the SLM. The numerical aperture of the illumination is limited but this is only a restriction if extremely good resolutions are to be achieved.

Figure 7.4 shows one such design. Like all of our current setups, it employs a HDTV LCOS SLM (Holoeye Pluto, 1920×1080 pixels, 8 μm pixel pitch, compare Section 7.3) and a 1/3'' CCD camera (SVS-Vistek eco204, 1024×768 pixels). For the SLM to work optimally it is necessary to ensure polarization of the incoming beam.

The 16:9 format of the SLM is only used within a circular pupil for this setup. For applications that are very demanding concerning field of view (typically proportional to the space-bandwidth product of the modulator) an anamorphotic imaging system can be used.

In Fig. 7.5 a more complex setup is shown that is not optimized concerning size. Compared to the basic setup this system allows us to chose three different types of illumination (transmission homogeneous, transmission structured, reflection homogeneous (EPI)). The homogeneous transmission illumination employs a classical Köhler illumination based on a high power LED (OSRAM Diamond Dragon, $\lambda = 632$ nm with a 1 nm bandwidth filter and a Zeiss $10 \times NA = 0.22$ lens). The Köhler illumination in reflection is basically the same, except for the different wavelength of $\lambda = 532$ nm. The wavelength filter is of course necessary in this case as well. Additionally a beamsplitter will have to divide illumination and imaging pathways.

The third option employs a slightly more complicated illumination using a laser. Since the wavefronts on the SLM generally are circular, and the SLM has a format of 16:9, a large portion will, in single use, remain unoccupied. This part is used for controlling the laser illumination.

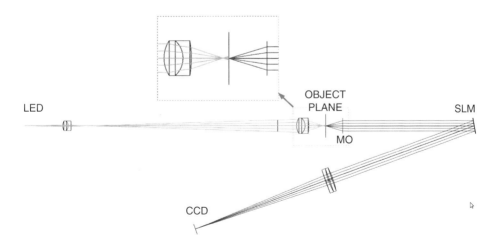

Figure 7.4 Principle of compact setup with a non-Köehler illumination. The illumination source is imaged onto the SLM. The pupil of the MO is not imaged onto the SLM

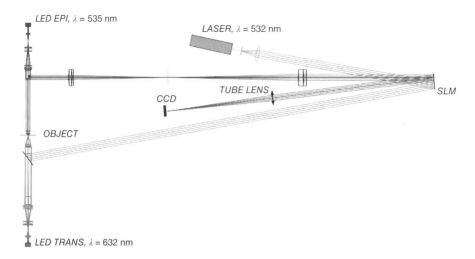

Figure 7.5 Extended setup with programmable imaging and illumination. Three different illuminations are present. The center path shows the imaging

The laser is expanded and propagates via the SLM and by means of a Kepler telescope to the illumination optics.

7.3　Liquid Crystal Spatial Light Modulator

Of course, the core element of a programmable microscope is the spatial light modulator. For the most common applications (the ones that we show in this chapter), it is beneficial to locate this modulator in a plane conjugate to the pupil or in a plane that is at least separated from planes conjugate to the object. And in this case it is then an advantage to be able to modulate the phase.

For applications where only the correction of aberrations is required, non-pixelated elements (most often: membrane mirrors: Paterson *et al.* 2000; Sherman *et al.* 2002) are the elements of choice. In this case a relatively broad spectral range can be used, no unwanted diffraction orders will occur and the light efficiency is close to 100%. Unfortunately, such elements are still quite sensitive to damage and external disturbances, can only be used for correcting low order aberrations with small amplitudes, need high voltages, and the overall system (addressing and modulation) is expensive. A large variety of other specialized AO-modulators like membrane, bimorph, piezoelectric, or electrostatic mirrors, and low-resolution liquid crystal modulators have been and still are being developed: see Tyson (1998) and Olivier (2007) for an overview. But the cost of these modulators is considerably above what can be employed in a commercial research microscope.

For applications where more complicated modulation patterns are necessary, pixelated elements are used. Amplitude-modulation certainly is possible but leads to symmetric diffraction orders and a strong loss of light. Therefore, the element of choice for most applications is a liquid crystal phase modulator with a high fill factor. Most often reflective parallel aligned liquid-crystal on silicon (LCoS) modulators are used because they offer large space-bandwidth-products in combination with high efficiency at a reasonable price.

When using such an element one should be aware of their behavior when modulating light. We do not want to go into detail here (see Lazarev *et al.* 2012; Zwick *et al.* 2010) but want to point out the most important things that might have to be considered:

- Modulation properties (variation of amplitude, phase, and polarization depending on gray level),
- number and geometry of the pixels (e.g., fill factor, space-bandwidth product, transfer function of the pixels),
- additional factors that lead to a loss of contrast (scattering, absorption, coatings),
- discritization and quantization,
- fringing field effects (electrical and optical crosstalk),
- variation with time (e.g., due to pixel refresh and PWM-addressing),
- temperature dependence (e.g., temperature increase due to the energy of the incident light),
- dependence on polarization,
- reconstruction geometry, including incidence angle.

To make things even more complicated, many of these factors depend-, in a non-trivial way-, on secondary factors (e.g., characteristic of the graphics board used for addressing the SLM). Figure 7.6 shows a microscopic image of a small part of a Holoeye Pluto modulator when displaying the Windows Desktop with the mouse pointer moving. It is clearly visible that

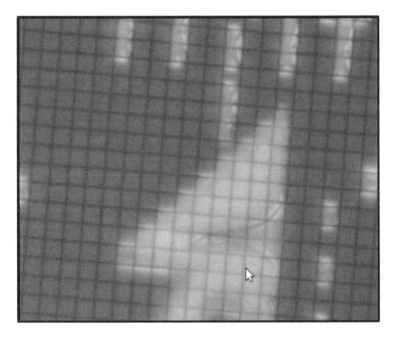

Figure 7.6 Microscopic image of part of a Holoeye Pluto LCoS modulator when displaying the windows desktop with the mouse pointer being moved. Strong fringing field effects as well as effects due to the fuid dynamics of liquid crystal are visible

the individual pixels will not modulate the light in a homogeneous way and that the local behavior of the liquid crystal depends on the local neighborhood as well as on time (see the drag behind the mouse pointer being moved). Especially when used in combination with large carrier frequencies, fringing-field effects become an important source of error. In practically all applications today, such extremely complicated effects are not considered and one lives with the fact that these effects will lead to unwanted diffraction orders and a loss of light in the desired diffraction order (Lingel *et al.* 2013).

7.4 Aberration Correction

Aberration correction is one of the most interesting applications of SLMs in widefield microscopy. The price of the imaging system is strongly correlated with the quality of the employed optics. Furthermore, even if perfect optics are employed, aberrations are often unavoidable in practice due to the specimen itself.

The most important aberration in microscopy (not counting defocus) is spherical aberration (SA) because it is automatically introduced if one focuses into a specimen with the "wrong" refractive index (Booth *et al.* 1998; Booth 2007a; Toeroek *et al.* 1996).

Differernt techniques can be used to reduce or even eliminiate spherical aberration. Most commonly, objectives with correction collars are used. But the adjustment of such collars is a time-consuming manual process (Schwertner *et al.* 2005). Another manual alternative are variable tube-lengths as proposed by Sheppard and Gu (Sheppard and Gu 1991).

But apart from spherical aberration, other aberrations might be present and lead to a deterioration of the image quality. Even small tilts of the cover glass (Arimoto and Murray 2004) have to be considered if the best resolution at high numerical apertures are to be achieved. And, of course, a lot of aberrations are present if one focuses into a thick specimen (Schwertner *et al.* 2004).

7.4.1 Isoplanatic Case

In holographic microscopy (Mico *et al.* 2010; Valencia and Moliner 2010) aberration correction is very simple. Since the complete object wavefront is measured directly (and the image is reconstructed digitally) it is possible to digitally correct the aberrations prior to the reconstruction.

In the following we do not address holographic microscopy any further but concentrate on conventional imaging-based wide-field microscopy. Even there, aberration correction is, in principle, straightforward when using a programmable microscope with the SLM located in a plane conjugate to the pupil. One method for sensing the aberration and one SLM for correcting it is needed. In this case one just writes the conjugate of the aberrated wavefront into the SLM. Therefore, the aberration is canceled and the imaging is improved. Often when implemented with a pixelated light modulator, an additional carrier frequency (phase tilt) as described before, is employed in analogy to conventional off-axis holography. This way the wavefront is coded in the deformation of fringes displayed on the SLM and even binary modulators can be used. In any case, correction is, in principle, very simple and the main challenge is to know the aberration that one has to correct.

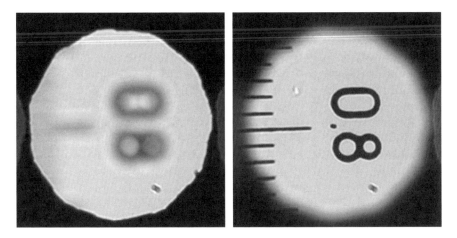

Figure 7.7 Stochastic aberration correction using the method of Warber *et al.* (2010). *Source*: Warber *et al.* 2010. Figure 8. Reproduced with permission from The Optical Society of America (OSA)

Traditionally, standard wavefront sensor concepts are employed but it should be noted that the SLM can also be used to measure aberrations. Different approaches are possible like scene-based Shack–Hartmann sensing (von der Luehe 1988), interference-based methods (Liesener *et al.* 2004c) or direct search-based optimization methods (Liesener *et al.* 2004b). Figure 7.7 shows one example obtained with a system measuring and correcting the aberrations in wide-field microscopy (Warber *et al.* 2010). In combination with microscopic guide stars (Reicherter *et al.* 2004) in principle, a lot of techniques that are also used in astronomic adaptive optics might be employed, for example, Shack–Hartmann sensors, pyramidal sensors (Chamot and Dainty 2006), conoscopic sensors (Buse and Luennemann 2000), curvature sensors (Paterson and Dainty 2000), and phase retrieval methods (Rondeau *et al.* 2007; Teague 1985). Booth *et al.* introduced a modal sensor for microscopy because of the necessity to only correct for lower Zernike orders (Neil *et al.* 2000a,b).

Measurement of the aberrations is quite simple if a single isolated point in the object plane is available. This is the case, for example, in laser scanning microscopy and it becomes possible to completely avoid the wavefront sensor by just optimizing the PSF of the system using an iterative approach. Different algorithms have been proposed to achieve this (Booth 2007b; Liesener *et al.* 2004a; Marsh *et al.* 2003; Sherman *et al.* 2002).

Unfortunately, these methods are not usable for the wide-field microscopic imaging of extended objects because we do not have direct access to the point spread function and only the convolution of the aberrated point spread function with the object to be imaged is detected. The aberration determination is more complicated. Scene-based aberration measurement methods then are to be employed (Bowman *et al.* 2010; Débarre *et al.* 2009; Haist *et al.* 2008; Poyneer 2003; Rimmele *et al.* 2003; von der Luehe 1988; Warber *et al.* 2010).

A simple and straightforward approach is to just correct the aberrations manually (Osten *et al.* 2005; Reicherter *et al.* 2005). For most microscopic applications, which are dominated by only a few low-order terms anyway, this works quite well. Figure 7.8 shows an example of such a correction where diffraction limited imaging could be easily achieved with an SLM microscope, even at high numerical apertures.

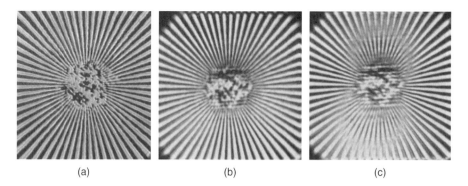

(a)   (b)   (c)

Figure 7.8 Imaging of a Siemens star in transmission with (a) Zeiss Ergoplan (Leitz Wetzlar objective lens, NA = 0.95, l = 540–580 nm) and (b) the SLM microscope (Olympus UmPlanFl objective lens, NA = 0.8, l = 633 nm) with aberration correction, (c) without aberration correction. The half-pitch of the smallest grating structures is 450 nm. *Source*: Hasler *et al.* 2012, Figure 2. Reproduced with permission from The Optical Society of America (OSA)

7.4.2   Field-Dependent Aberrations

A principal problem of aberration correction using the SLM is that the aberration might be field dependent. This is typically the case if the aberration is induced by the specimen itself or by a cheap microscope objective lens. In this case there is not a single wavefront aberration that needs to be corrected so we cannot expect that we will obtain a sharp image over the whole object field.

In a multi-image approach we use different aberration corrections for different field positions, so-called isoplanatic-patches, and the acquired images are then simply combined by digital postprocessing. If the symmetric aberrations are dominating there is no shift in the individual images and therefore no special registration or stitching is necessary to combine the images. One can just copy the corrected areas of different images into one overall image. Figure 7.9 shows an example.

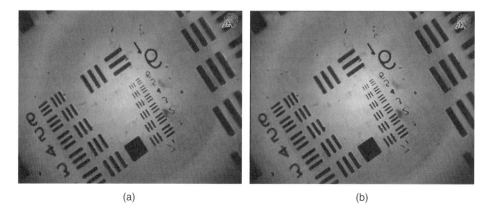

(a)   (b)

Figure 7.9   Widefield aberration correction using multiple images. (a) single image correction, (b) multi-image correction which leads to multiconjugate adaptive optics

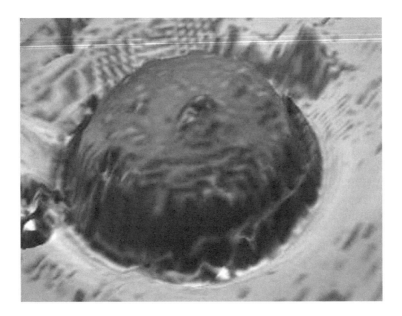

Figure 7.10 Combination of several defocused images allows one to obtain the three-dimensional shape of the specimen

7.4.3 Defocusing

A very special aberration, of course, is defocus. Here, a multi-image approach can be also easily followed using a programmable microscope. We write different zone plates into the SLM and, therefore, different slices of the specimen are sampled. Again, using postprocessing it is possible to combine the images, for example, for obtaining an image with an extended depth of focus, a three-dimensional representation of the object (see e.g., Fig. 7.10), or for visualizing phase objects (Camacho and Zalevsky 2010).

7.5 Phase Contrast Imaging

Most biological (and some important non-biological) specimens are more or less transparent and, therefore, conventional imaging leads to very low contrast. One approach, of course, is to stain the specimen but this might be not convenient and might disturb or even destroy it. Therefore, a lot of different phase contrast techniques have been invented over the years.

In a conventional phase contrast microscope the user has a very limited possibility to change the phase contrast imaging (typically, the phase contrast objective or the bias in differential interference contrast is changed). Programmable microscopy is very much suited to phase contrast imaging, especially in a multi-image context, because one can easily change between different phase contrast methods and their parameters in real time. A good impression of the variance of the images that one obtains by using different phase contrast filters written into the SLM is shown in Fig. 7.11.

Defocusing (compare the previous section), in practice the simplest method of phase contrast filtering, leads to a decrease in lateral resolution. Very small defocusing will not

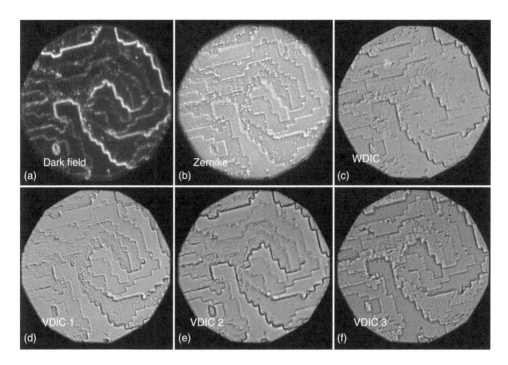

Figure 7.11 Images of a part of a molded computer-generated hologram obtained with different SLM phase contrast settings. *Source*: Warber *et al.* 2011. Figure 6. Reproduced with permission from The Optical Society of America (OSA)

introduce much contrast. But it is an interesting technique when multiple images are used (phase retrieval) and as already described in Section 7.4.3 the defocus can be applied using an SLM (Camacho and Zalevsky 2010).

7.5.1 Dark Field

Conceptually, the most straightforward phase contrast technique is dark field imaging. Despite its simplicity it nevertheless delivers very impressive images for certain applications. The main idea is to block the direct light that will propagate unaffected through the object plane. This means that without a specimen the image on the camera will be dark. To achieve this using a phase modulator, a grating is written into the SLM at all positions that are conjugate to the light source. This way it is possible to direct the unaffected light away from the camera (Fig. 7.12) (Warber *et al.* 2009). Of course, as in conventional microscopy, ring-like or point-like or even more complicated patterns can be employed.

7.5.2 Zernike Phase Contrast

A small variation of the concept leads to Zernike's phase contrast method. In this case the direct light is not blocked but rather shifted in phase. This results in a change of the

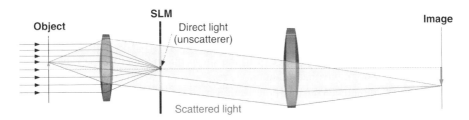

Figure 7.12 Principle of dark field microscopy. Light that is not deflected by structures of the object will be diffracted away from the camera

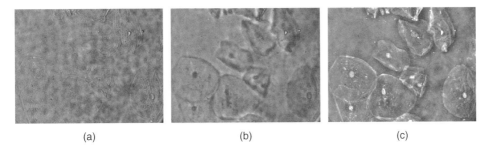

Figure 7.13 Imaging of human mucosa cells using programmable microscopy. (a) bright field image, (b) Zernike setting 1, (c) Zernike setting 2. It is obvious that the character of the image strongly changes if different phase contrast settings are used

interference between the direct light and the light being scattered or diffracted at the object and, therefore, the transparent structures will become visible (Zernike 1955). Compared to conventional Zernike phase contrast the SLM again allows one to change all filter parameters in real time (Glückstad and Mogensen 2001). This allows one to optimize the filter for a given specimen (Fig. 7.13).

As with the dark field technique, one can use more or less arbitrary illumination patterns (Maurer *et al.* 2008). By shifting the phase and recording multiple images it becomes possible to perform even quantitative phase analysis.

7.5.3 Interference Contrast

Apart from the interference between unaffected direct light and scattered light it is also possible to use shearing methods, where the light field coming from the object interferes with a sheared copy of itself. Most common is lateral shearing, which is called *differential interference contrast* (DIC) (Mehta and Sheppard 2008; Pluta 1989). DIC is especially popular in biological research because it can be used even with high numerical aperture (NA) illumination and, therefore, leads to a comparatively good lateral resolution as well as good axial discrimination. Very small features can be imaged with good contrast because the intensity in the image plane is a nonlinear function (approximately a sine squared: Pluta 1989) of the object's phase gradient in the direction of the shear. The shearing in one direction leads to a pseudo-three-dimensional appearance of phase objects due to a "shadow" in one direction. This can be seen as an advantage or a disadvantage. The effect often is visually very pleasing (and partly responsible for the popularity of DIC) but also leads to loss of information in the direction perpendicular to the shear and the microscope user perceives a three-dimensional topography that might not be real.

Classical DIC is implemented using polarization-based shearing, which unfortunately leads to quite complex and expensive setups. If one only needs quasi-monochromatic imaging, grating-based DIC (David *et al.* 2002; Lohmann and Sinzinger 2006) is an interesting alternative. It can be easily implemented using a programmable microscope, for example, by the complex superposition of two gratings (McIntyre *et al.* 2009b, 2010; Warber *et al.* 2009). Again, all parameters (strength of lateral shear, phase bias between the two copies, shear orientation) can be easily changed in real time.

Sometimes it is also beneficial to replace lateral shear by vertical shearing. To this end the interference between the point image and a defocused point image is used. This leads to the so-called vertical differential interference contrast (Warber *et al.* 2011, 2012).

As with conventional DIC, the results depend strongly on the phase retardation, which in this case is the phase difference between the focused and the slightly defocused part of the point-spread-function. For imaging of neighbouring phase defects the behaviour becomes very complex if spatial coherence in the neighbourhood is given. This is in direct analogy to conventional coherent imaging. Again, the resulting intensity distributions depend strongly on nearly all parameters of the filter and the specimen itself. This is generally the case for interference-based phase contrast imaging of small neighboring structures and makes a good interpretation of such structures sometimes difficult. Therefore, it is a good idea to limit the spatial coherence as much as possible by using a large numerical aperture for illumination.

Figure 7.14 (Plate 12) clearly shows this complexity for the coherent imaging of two neighboring point-like defects using different parameters. By employing the right parameters, the neighbouring defects are clearly separated. It is also obvious that contrast reversal can be achieved using different parameter settings. This can be employed to achieve very strong contrasts by subtracting two recorded images with different parameters (see also Section 7.5.4).

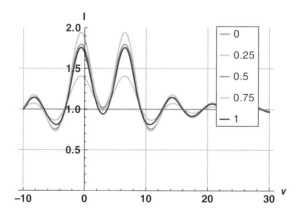

Figure 7.14 (Plate 12)   Imaging of two neighboring point-like phase defects using VDIC with different parameters. *See plate section for the color version*

Compared to conventional DIC, the main advantage of VDIC is the isotropy. This eliminates the pseudo-3D like appearance of the images. The images contain more or less the same information but the appearance is different and as such it is also a viable tool to complement other phase contrast methods. Another interesting extension that leads to isotropy, and that can be easily implemented using programmable microscopy, is spiral phase contrast (Bernet *et al.* 2006; Fürhapter *et al.* 2005).

Apart from the mentioned, perhaps most important methods, a lot of other techniques and an endless number of variations are possible.

7.5.4 Combining Different Phase Contrast Images

We have already seen that the phase contrast images strongly depend on the method, the parameters of the method, and the specimen itself. It is clear that by combining images, which have been obtained using different parameter sets or different methods, this might improve the visibility of structures or make the enclosed information more visible. This approach is technically very attractive because it is so easy to combine the images. The images have been obtained with the same static optical setup and on the same CCD. Therefore, no registration or stitching is necessary. Every object detail stays at exactly the same pixel (at least while the object itself is not moving).

One example is shown in Fig. 7.15 (Plate 13). Here, two images obtained using Zernike's phase contrast and DIC are combined by postprocessing. Zernike's method is sensitive to the

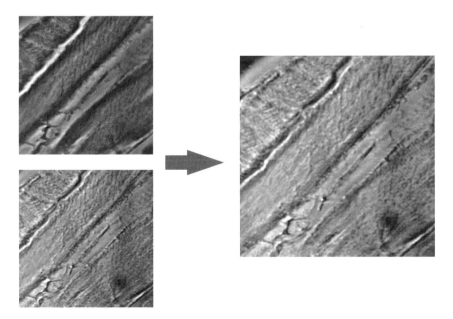

Figure 7.15 (Plate 13) Combination of an image obtained using Zernike phase contrast (hue) and one obtained using DIC (intensity) in order to visualize a phase structure. *See plate section for the color version*

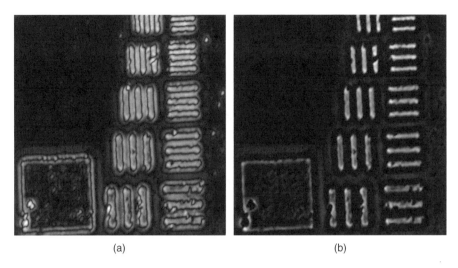

(a)   (b)

Figure 7.16   Combination of two VDIC images when imaging part of a three-bar binary phase target.
(a) The dark contours surounding the bars are precise indicators for the lateral extension of the structures.
(b) Strong contrast improvement for the structure. *Source*: Warber *et al.* 2011. Figure 7. Reproduced with
permission from The Optical Society

phase of the object. For an object consisting of more or less homogeneous material the phase
is proportional to the local thickness. DIC, on the other hand, is sensitive to the gradient of the
phase and is therefore very well suited to enhance small structures. In Fig. 7.15 we therefore
used the Zernike image for the hue and the DIC image for the intensity of a color image. This
way color is related to the local height.

Another example is shown in Fig. 7.16 where an USAF phase target has been imaged using
two different sets of parameters of VDIC. The subtraction of the two images leads to a strongly
increased contrast.

7.6   Stereo Microscopy

In stereo microscopy the specimen is imaged from two different directions. The angle between
the two directions is the so-called triangulation angle. Practically, this can be achieved in programmable microscopy by using two (or even more) different zones of the pupil of the system.
Figure 7.17 depicts the basic idea (Hasler *et al.* 2012).

On the left part of the pupil we use a grating displayed on the left part of the SLM. Therefore, the light which falls onto that side of the pupil will form an image that is shifted with
respect to the image due to the right part of the image. On the camera, we therefore will have
two separated copies of the object. We can use these two images for visualization (e.g., sending them to a 3D display) or for computing the topography of the object using stereo-vision
algorithms.

For transparent objects, unfortunately, we might obtain local contrast reversals. Therefore,
for phase contrast imaging, stereo-based microscopy techniques at the moment do not lead to
satisfactory results.

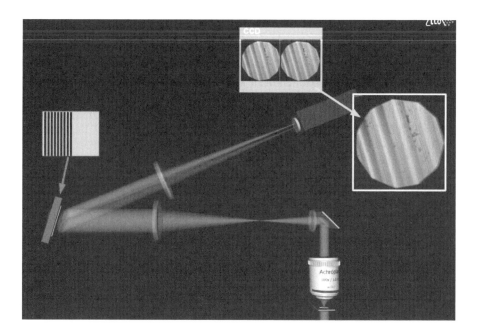

Figure 7.17 Principle of SLM-based stereo microscopy

7.7 Conclusion

We have reviewed different multidimensional imaging methods implemented using SLM-based programmable microscopy. By incorporating a spatial light modulator into the imaging path of a microscope it becomes possible to realize a lot of different imaging methods by software. Due to the possibility of addressing the SLM in video real time one can obtain images obtained with different microscopy methods and different parameters very fast and it is possible to combine these images by postprocessing.

Apart from the optimization of parameters and imaging methods, we have shown multi-image advantages for the optimization of the imaging quality (field-dependent aberrations, multiconjugate adaptive optics), contrast enhancement of phase contrast images, improved visualization of phase specimen, and three-dimensional registration of objects.

At the moment we only scratched at the surface of the possibilities. A wealth of new imaging capabilities will arise if one really exploits the power of postprocessing in a more thorough way. We also only used the spatial light modulator in the imaging path of the microscope. As we have shown in Section 7.2 it is also possible to use spatial light modulation in the illumination path of the microscope. This way, even more imaging methods can be achieved in a flexible way by programming the SLM. Important methods are confocal-like, structured-illumination techniques, and special illumination methods, for example, for Raman- or fluorescence-based imaging.

The disadvantages of SLM-based microscopy are the reduced light efficiency of the microscope and – for most methods – the limitation to quasi-monochromatic light. Still, we think that part of the future of microscopy lies in this sort of multi-image generation and combination by postprocessing, which is enabled by the enormous flexibility of programmable microscopy.

References

Arimoto R and Murray JM 2004 A common aberration with water-immersion objective lenses. *Journal of Microscopy* **216**(1), 49–51.

Bernet S, Jesacher A, Maurer C and Ritsch-Marte M 2006 Quantitative imaging of complex samples by spiral phase contrast microscopy. *Optics Express* **14**(9), 2766–2773.

Booth M, Neil M and Wilson T 1998 Aberration correction for confocal imaging in refractive-index-mismatched media. *Journal of Microscopy* **192**, 90–98.

Booth MJ 2007a Adaptive optics in microscopy. *Philosophical Transactions. Series A, Mathematical, Physical, and Engineering Sciences* **365**(1861), 2829–2843.

Booth MJ 2007b Wavefront sensorless adaptive optics for large aberrations. *Optics lett*, **32**(1), 5–7.

Bowman RW, Wright AJ and Padgett MJ 2010 An SLM-based shack-hartmann wavefront sensor for aberration correction in optical tweezers. *Journal of Optics* **12**(12), 124004.

Buse K and Luennemann M 2000 3D imaging: Wave front sensing utilizing a birefrigent crystal. *Phyical Review Letters* **85**, 3385–3387.

Camacho L and Zalevsky Z 2010 Quantitative phase microscopy using defocusing by means of a spatial light modulator. *Optics Express* **18**(7), 6755–6766.

Chamot S and Dainty C 2006 Adaptive optics for ophtalmic applications using a pyramid wavefront sensor. *Optics Express* **14**, 518–526.

Chang B, Chou L, Chang Y and Chiang S 2009 Isotropic image in structured illumination microscopy patterned with a spatial light modulator. *Optics Express* **17**(17), 8206–8210.

Choi Jr and Kim D 2012 Enhanced image reconstruction of three-dimensional fluorescent assays by subtractive structured-light illumination microscopy. *J Opt Soc Am A Opt Image Sci Vis.* **29**(10), 2165–2173.

David C, Nohammer B, Solak HH and Ziegler E 2002 Differential x-ray phase contrast imaging using a shearing interferometer. *Applied Physics Letters* **81**(17), 3287.

Débarre D, Botcherby EJ, Watanabe T, Srinivas S, Booth MJ and Wilson T 2009 Image-based adaptive optics for two-photon microscopy. *Optics Lett.* **34**(16), 2495–2497.

Eriksen RL, Mogensen PC and Glückstad J 2002 Multiple-beam optical tweezers generated by the generalized phase-contrast method. *Opt. Letters* **27**(4), 267–269.

Fürhapter S, Jesacher A, Bernet S and Ritsch-Marte M 2005 Spiral phase contrast imaging in microscopy. *Optics Express* **13**(3), 689–694.

Gittard SD, Nguyen A, Obata K, Koroleva A, Narayan RJ and Chichkov BN 2011 Fabrication of microscale medical devices by two-photon polymerization with multiple foci via a spatial light modulator. *Optics Express* **2**(11), 267–275.

Glückstad J and Mogensen PC 2001 Optimal phase contrast in common-path interferometry. *Applied Optics* **40**(2), 268–282.

Grier D 2003 A revolution in optical manipulation. *Nature* **424**, 810–816.

Haist T, Hafner J, Warber M and Osten W 2008 Scene-based wavefront correction with spatial light modulators *Proceedings of SPIE*, vol. **7064**, pp. 70640M–70640M–11. SPIE.

Haist T, Wagemann EU and Tiziani HJ 1999 Pulsed-laser ablation using dynamic computer-generated holograms written into a liquid crystal display. *Journal of Optics A: Pure and Applied Optics* **1**(3), 428–430.

Hasler M, Haist T and Osten W 2012 Stereo vision in spatial-light-modulator–based microscopy. *Opt. Lett.* **37**(12), 2238–2240.

Hayasaki Y, Itoh M, Yatagai T and Nishida N 1999 Nonmechanical optical manipulation of microparticle using spatial light modulator. *Optical Review* **6**(1), 24–27.

Heintzmann R, Hanley QS, Arndt-Jovin D and Jovin TM 2001 A dual path programmable array microscope (PAM): simultaneous acquisition of conjugate and non-conjugate images. *Journal of Microscopy* **204**(Pt 2), 119–35.

Hussain A and Campos J 2013 Holographic superresolution using spatial light modulator. *Journal of the European Optical Society - Rapid Publications* 8, DOI: 10.2971/jeos.2013.13007.

Jesacher A and Booth MJ 2010 Parallel direct laser writing in three dimensions with spatially dependent aberration correction. *Optics Express* **18**(20), 132–134.

Jesacher A, Roider C, Khan S, Thalhammer G, Bernet S and Ritsch-Marte M 2011 Contrast enhancement in widefield CARS microscopy by tailored phase matching using a spatial light modulator. *Optics Lett.* **36**(12), 2245–2247.

Kenny F, Lara D and Dainty C 2012 Complete polarization and phase control for focus-shaping in high-NA microscopy. *Optics Express* **20**(13), 2234–2239.

Khan S, Jesacher A, Nussbaumer W, Bernet S and Ritsch-Marte M 2011 Quantitative analysis of shape and volume changes in activated thrombocytes in real time by single-shot spatial light modulator-based differential interference contrast imaging. *Journal of Biophotonics* **4**(9), 600–609.

Kim T and Popescu G 2011 Laplace field microscopy for label-free imaging of dynamic biological structures. *Optics Letters* **36**(23), 4704–4706.

Lazarev G, Hermerschmidt A, Krüger S and Osten S 2012 LCOS Spatial light modulators: trends and applications. *in Optical Imaging and Metrology*, W. Osten, N. Reingang (eds), Springer. pp. 1–23.

Lee MP, Gibson GM, Bowman R, Bernet S, Ritsch-Marte M, Phillips DB and Padgett MJ 2013 A multi-modal stereo microscope based on a spatial light modulator. *Opt. Express* **21**(14), 16541–16551.

Liang J, Wu SY, Fatemi FK and Becker MF 2012 Suppression of the zero-order diffracted beam from a pixelated spatial light modulator by phase compression. *Appl. Opt.* **51**(16), 3294–3304.

Liesener J, Hupfer W, Gehner A and Wallace K 2004a Tests on micromirror arrays for adaptive optics. *Proc. SPIE.*

Liesener J, Reicherter M and Tiziani H 2004b Determination and compensation of aberrations using SLMs. *Optics Communications* **233**, 161–166.

Liesener J, Seifert L, Tiziani H and Osten W 2004c Active wavefront sensing and wavefront control with SLMs. *Proc. SPIE* **5532**(1), 147–158.

Lingel C, Haist T and Osten W 2013 Optimizing the diffraction efficiency of SLM-based holography with respect to the fringing field effect. *Appl. Opt.* **52**(28), 6877–6883.

Lohmann A and Sinzinger S 2006 *Optical Information Processing*. TU Illmenau Universitätsbibliothek.

Marsh P, Burns D and Girkin J 2003 Practical implementation of adaptive optics in multiphoton microscopy. *Opt. Express* **11**, 1123–1130.

Maurer C, Jesacher A, Bernet S and Ritsch-Marte M 2008 Phase contrast microscopy with full numerical aperture illumination. *Optics Express* **16**(24), 19821–19829.

Maurer C, Jesacher A, Bernet S and Ritsch-Marte M 2011 What spatial light modulators can do for optical microscopy. *Laser & Photonics Reviews* **5**(1), 81–101.

McIntyre TJ, Maurer C, Bernet S and Ritsch-Marte M 2009a Differential interference contrast imaging using a spatial light modulator. *Optics Lett.* **34**(19), 2988–2990.

McIntyre TJ, Maurer C, Bernet S and Ritsch-Marte M 2009b Differential interference contrast imaging using a spatial light modulator. *Opt. Lett.* **34**(19), 2988–2990.

McIntyre TJ, Maurer C, Fassl S, Khan S, Bernet S and Ritsch-Marte M 2010 Quantitative SLM-based differential interference contrast imaging. *Optics Express* **18**(13), 14063–14078.

Mehta SB and Sheppard CJR 2008 Partially coherent image formation in differential interference contrast (DIC) microscope. *Optics Express* **16**(24), 19462–19479.

Mico V, Garcia J, Zalevsky Z and Javidi B 2010 Phase-shifting Gabor holographic microscopy. *Journal of Display Technology* **6**(10), 484–489.

Neil Ma, Booth MJ and Wilson T 2000a Closed-loop aberration correction by use of a modal Zernike wave-front sensor. *Optics lett,* **25**(15), 1083–1085.

Neil Ma, Juskaitis R, Booth MJ, Wilson T, Tanaka T and Kawata S 2000b Adaptive aberration correction in a two-photon microscope. *Journal of Microscopy* **200** (Pt 2)(July), 105–108.

Nikolenko V, Peterka DS and Yuste R 2010 A portable laser photostimulation and imaging microscope. *Journal of Neural Engineering* **7**(4), 045001.

Nikolenko V, Watson BO, Araya R, Woodruff A, Peterka DS and Yuste R 2008 SLM microscopy: Scanless two-photon imaging and photostimulation with spatial light modulators. *Front Neural Circuits* **2**, 5.

Olivier S 2007 Adaptive optics. *in "Micro-Opto-Electro-Mechanical Systems"*, ME Motamedi (ed.) SPIE Press Book. pp. 453–475.

Osten W, Kohler C and Liesener J 2005 Evaluation and application of spatial light modulators for optical metrology *a Reunion Espanola de Optoelectronica OPTOEL '05, Elche, Spain.*

Paterson C and Dainty JC 2000 Hybrid curvature and gradient wave-front sensor. *Opt. Lett.* **25**(23), 1687–1689.

Paterson C, Munro I and Dainty J 2000 A low cost adaptive optics system using a membrane mirror. *Optics Express* **6**, 175–185.

Peterka DS, Nikolenko V, Fino E, Araya R, Etchenique R, Yuste R, *et al.* 2010 Fast two-photon neuronal imaging and control using a spatial light modulator and ruthenium compounds. *Library* **7548**, 1–9.

Pham H, Bhaduri B, Ding H and Popescu G 2012 Spectroscopic diffraction phase microscopy. *Optics Lett.* **37**(16), 3438–3440.

Pluta M 1989 *Advanced Light Microscopy Vol. 2*. Elsevier.

Poyneer La 2003 Scene-based Shack–Hartmann wave-front sensing: analysis and simulation. *Applied Optics* **42**(29), 5807–5815.

Qin W, Shao Y, Liu H, Peng X, Niu H and Gao B 2012 Addressable discrete-line-scanning multiphoton microscopy based on a spatial light modulator. *Optics Lett.* **37**(5), 827–829.

Reicherter M, Gorski W, Haist T and Osten W 2004 Dynamic correction of aberrations in microscopic imaging systems using an artificial point source. *SPIE-Int. Soc. Opt. Eng. Proceedings of Spie-the International Society for Optical Engineering* **5462**(1), 68–78.

Reicherter M, Haist T, Wagemann EU and Tiziani HJ 1999 Optical particle trapping with computer-generated holograms written on a liquid-crystal display. *Optics Lett.* **24**(9), 608.

Reicherter M, Haist T, Zwick S, Burla A, Seifert L and Osten W 2005 Fast hologram computation and aberration control for holographic tweezers. *Proceedings of SPIE* **5930**, Optical Trapping and Optical Micromanipulation II, 59301Y.

Rimmele T, Richards K, Hegwer S, Ren D, Fletcher S, Gregory S, *et al.* 2003 Solar adaptive optics: A progress report. *Proc. SPIE.* **4839**, Adaptive Optical System Technologies II, 635

Rondeau X, Thiebaut E, Tallon M and Foy R 2007 Phase retrieval from speckle images. *J. Opt. Soc. Am. A* **24**, 3354–3364.

Schausberger SE, Heise B, Maurer C, Bernet S, Ritsch-Marte M and Stifter D 2010 Flexible contrast for low-coherence interference microscopy by Fourier-plane filtering with a spatial light modulator. *Optics Letters* **35**(24), 4154–4156.

Schwertner M, Booth M and Wilson T 2005 Simple optimization procedure for objective lens correction collar setting. *Journal of Microscopy* **217**, 184–1887.

Schwertner M, Booth MJ and Wilson T 2004 Simulation of specimen-induced aberrations for objects with spherical and cylindrical symmetry. *Journal of Microscopy* **215**(Pt 3), 271–280.

Scrimgeour J and Curtis JE 2012 Aberration correction in wide-field fluorescence microscopy by segmented-pupil image interferometry. *Optics Express* **20**(13), 388–394.

Shao Y, Qin W, Liu H, Qu J, Peng X, Niu H and Gao BZ 2012 Ultrafast, large-field multiphoton microscopy based on an acousto-optic deflector and a spatial light modulator. *Optics Lett.* **37**(13), 2532–2534.

Sheppard CRJ and Gu M 1991 Aberration compensation in confocal microscopy. *Applied Optics* **30**, 3563–3568.

Sherman L, J.Y. Y, Norris A and T.B. N 2002 Adaptive correction of depth-induced aberrations in multiphoton scanning microscopy using a deformable mirror. *Journal of Microscopy* **206**, 65–71.

Steiger R, Bernet S and Ritsch-Marte M 2012 SLM-based off-axis Fourier filtering in microscopy with white light illumination. *Optics Express* **20**(14), 15377–15384.

Teague MR 1985 Image formation in terms of the transport equation. *J. Opt. Soc. Am. A* **2**(11), 1434.

Toeroek P, Sheppard C and Laczik Z 1996 Dark-field and differential phase contrast imaging modes in confocal microscopy using a half-aperture stop. *Optik* **103**(3), 101–106.

Tyson R 1998 *Principles of Adaptive Optics,* Academic Press.

Valencia UD and Moliner D 2010 Common-path phase-shifting lensless holographic microscopy. *Optics Lett.* **35**(23), 3919–3921.

von der Luehe O 1988 Wavefront error measurement technique using extended, incoherent light sources. *Optical Engineering* **27**(12), 1078–1087.

Wang Cc, Lee Kl and Lee Ch 2009 Wide-field optical nanoprofilometry using structured illumination. *Optics Lett.* **34**(22), 3538–3540.

Warber M, Haist T, Hasler M and Osten W 2012 Vertical differential interference contrast. *Optical Engineering* **51**(1), 013204–1–013204–7.

Warber M, Hasler M, Haist T and Osten W 2011 Vertical differential interference contrast using SLMs. *Proc. SPIE* **8086**, 80861E–80861E–10.

Warber M, Maier S, Haist T and Osten W 2010 Combination of scene-based and stochastic measurement for wide-field aberration correction in microscopic imaging. *Applied Optics* **49**(28), 5474–5479.

Warber M, Zwick S, Hasler M, Haist T and Osten W 2009 SLM-based phase-contrast filtering for single and multiple image acquisition. *Proc SPIE* **7442**, Optics and Photonics for Information Processing III, 74420E.

Zernike F 1955 How I discovered phase contrast. *Science* **121**(3141), 345–349.

Zhou Q, Yang W, He F, Stoian R, Hui R and Cheng G 2013 Femtosecond multi-beam interference lithography based on dynamic wavefront engineering. *Optics Express* **21**(8), 53–56.

Zwick S, Haist T, Warber M and Osten W 2010 Dynamic holography using pixelated light modulators. 7.1 *Applied Optics* **49**(25), F47–58.

8

Holographic Three-Dimensional Measurement of an Optically Trapped Nanoparticle

Yoshio Hayasaki

Center for Optical Research and Education (CORE), Utsunomiya University, Japan

8.1   Introduction

The "Optical tweezers" technique [1] is a well-known technique used to trap and manipulate minute objects in a liquid without contact. It has been used for manipulating viruses and bacteria [2], single RNA molecules [3,4], and DNA molecules [5]. It has also been used to study the motion of the biological motor protein kinesin [6] and to sequence DNA [7]. It has been employed in not only biological applications but also for sorting objects in microfluidic systems [8–10], and measuring the mechanical properties [11] and shapes of microstructures [12]. Holographic optical tweezers [13–15] have also been used to create arbitrary three-dimensional structures [16–18].

In these applications, precise position measurement of the trapped object expands the ability to perform mechanical and structural measurements, and several methods have been developed. Position measurement with a quadrant photodiode (QPD) has been performed using spatial changes in light intensity according to the displacement of the trapped object with high temporal resolution and high sensitivity [1,19–23]. However, it is difficult to measure many objects simultaneously and to measure a three-dimensional (3D) position.

The use of a camera was effective for measuring optically trapped objects three-dimensionally [24]. A recently developed high-speed camera provided a temporal resolution comparable to that of a QPD [25]. Measuring the 3D position of trapped particles of

Multi-dimensional Imaging, First Edition. Edited by Bahram Javidi, Enrique Tajahuerce and Pedro Andrés.
© 2014 John Wiley & Sons, Ltd. Published 2014 by John Wiley & Sons, Ltd.

micrometer-order diameters was demonstrated using a holographic method [26]. Three-dimensional position measurement was also demonstrated by using stereomicroscopy [27]. The Brownian motion of a polystyrene sphere with a 0.8 μm diameter was measured with a digital holographic microscope, although this technique did not use optical tweezers [28]. Despite these numerous studies, there has been little research into measuring the 3D position of subwavelength-sized particles (nanoparticles) with diameters smaller than the focal spot diameter used in optical tweezers techniques.

Recently, we demonstrated digital holographic measurement of a nanoparticle's position held in optical tweezers [29,30]. The diameter was smaller than that of nanoparticles used in previous studies [26–28]. The optical tweezers suppressed the movement of the nanoparticle and controlled its position in the use of a measurement probe. Digital holography [31] has been used to detect particles with no relation to optical tweezers [32–34]. Digital focusing on to an object was performed by a pattern matching between a diffraction image and a template image [29], but it has been performed by using features of the focused image, such as an iterative gradient computation [35], the maximization of the intensity local changes [36], a correlation coefficient [37], the sparsity of wavelet coefficients [38], and an integrated amplitude modulus [39,40]. An in-line digital holography setup can be also used for detecting the positions of particles [31,41,42]. In the system, in addition to an in-line digital holography setup, low-coherence light [41,43,44] was used to measure a nanoparticle, because the scattered light from the nanoparticle was very weak and was buried in speckles produced by unwanted scattered light.

The optical tweezers have been applied to manipulate gold nanoparticles when the motion, state, and function of a biological system are measured. The 3D trapping of the gold nanoparticle with a 36 nm diameter was first demonstrated by Svoboda and Block [6]. The optically trapped gold nanoparticle was used as a probe tip of a near-field optical microscope [45–47]. The heat effect of the optically trapped gold nanoparticle has also been discussed [48]. The stable optical trapping of 18–254 nm gold particles has been demonstrated [49]. Optical trappings of silver nanoparticles [50] and gold nanorods [51] has also been demonstrated. Theoretical calculations of the optical trapping force of the metal nanoparticles were described [52–54]. A gold nanoparticle with a 9.5 nm diameter [55] has been successfully trapped in optical tweezers and detected with a quadrant photodiode (QPD) method. These holographic methods were also applied to measure the position of the gold nanoparticle with the heterodyne technique [56,57] and the dark field arrangement [58].

In this chapter, position measurement of an optically trapped nanoparticle using low-coherence, inline-digital holography with a pattern matching method, and a 3D sub-pixel estimation is described. In Section 8.2, the optical setup, the image processing procedure, and sample preparation are described. In Section 8.3, the measurement accuracy of the experimental setup and the experimental results of measuring the Brownian motion of a polystyrene particle with diameters of 200 and 500 nm, while changing the trapping laser power and the 3D position measurement of a gold nanoparticle with 60 nm a diameter, is also demonstrated. In Section 8.4, a method for achieving higher accuracy in 3D position measurement of nanoparticles, using an in-line digital holographic microscope by increasing weak scattered light from a nanoparticle relative to the reference light with a low spatial frequency attenuation filter, is described [59]. In Section 8.5, some concluding remarks are presented.

8.2 Experimental Setup

8.2.1 Optical Tweezers System

Figure 8.1 shows the experimental setup, composed of an optical tweezers system and a low-coherence, in-line digital holographic microscope. This system was placed on a vibration isolation optical table. In the optical tweezers system, a beam generated from an Yb-fiber laser with a wavelength of 1070 nm (IPG Photonics, YLM-10) was collimated and focused in a sample solution using a 60× oil-immersion microscope objective lens (OL) with a numerical aperture of $NA = 1.25$. The irradiation laser power at the sample was controlled with a half-wave plate and a polarization beam splitter (PBS) and calculated as the product of the power measured before introducing the laser beam and the transmittance of the OL.

The samples were suspensions of polystyrene beads (diameter = 202 nm, standard deviation = 10 nm, 2.66% solid suspension, Polysciences, Inc.) diluted 1:48000 with distilled water, polystyrene beads (diameter = 535 nm, standard deviation = 10 nm, 2.69% solids suspension, Polysciences, Inc.) diluted 1:2500 with distilled water, and gold nanoparticles (diameter = 60 nm, 10% deviation, Tanaka Precision Metals) diluted 1:5000 in distilled water.

8.2.2 In-Line Digital Holographic Microscope

8.2.2.1 Optical Setup

In the in-line digital holographic microscope, the light source was a green light emitting diode (LED) (Luxeon, Green Revel Star/O) with a center wavelength of $\lambda_c = 528$ nm and a

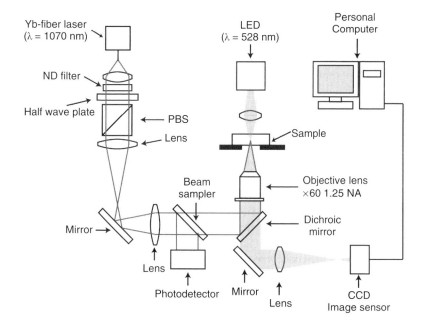

Figure 8.1 Experimental setup

spectral width of $\Delta\lambda = 32\,\text{nm}$ at the full-width half maximum. The green light was focused to obtain stronger illumination with the power of $\sim 6\,\text{mW}$ on the sample. The scattered light from a nanoparticle (the object light) and the straight-through light (the reference light) were magnified by the OL and a lens with a focal length of 300 mm, and interfered on a cooled charge-coupled device (CCD) image sensor (Bitran, BU50LN) with 772×580 pixels, 16-bit resolution, 1.0 ms shutter speed and 8.6 frames/s (fps). The diffraction limit and the Rayleigh length of the microscope calculated from λ_c and NA were $0.61\lambda_c/NA = 258\,\text{nm}$ and $\lambda_c/NA^2 = 338\,\text{nm}$, respectively.

The lateral and axial magnifications of the in-line digital holographic microscope were estimated as follows. In the estimation of the lateral magnification, a 10.0 µm scale bar on the sample plane was measured as 121 pixels on the CCD image sensor, with a pixel spacing of 8.3 µm; therefore the lateral magnification M_L was 1.0×10^2. The corresponding spatial sampling interval on the sample plane was $\Delta x = \Delta y = 83\,\text{nm}$. The axial magnification was estimated by the ratio of the displacements along the axial direction between the object plane and the image plane. In more detail, a 200 nm polystyrene particle fixed on a slide glass was axially moved between 0 and 4000 nm in steps of 500 nm, and the CCD image sensor was moved to the focused point (dark spot). The step of 500 nm along the axial movement of the sample corresponded to the axial movement of 3.9 mm in the focus position of CCD image sensor. Therefore the axial magnification was $M_A = 3.9\,\text{mm}/500\,\text{nm} = 7.8 \times 10^3$. For example, the axial step of 10 µm in the diffraction calculation on the image space corresponded to $\Delta z = 1.3\,\text{nm}$ in the sample space.

8.2.2.2 Diffraction Calculation

In the in-line digital holographic microscope, there was a short path difference between the object light and the reference light, allowing interference between them with low-coherence light. The interference image, $|u_k(x, y, z)|^2$, for wave-number $k = 2\pi/\lambda$ is approximated by the interference between a spherical wave from a point source (scattered light from nanoparticle) and a plane wave (straight-through light):

$$|u_k(x,y,z)|^2 = \left| A_k^{(r)} \exp(ikz) + \frac{A_k^{(s)}}{\sqrt{x^2+y^2}} \exp\left(ik\frac{x^2+y^2}{2z}\right) \right|^2$$

$$= A_k^{(r)2} + \frac{A_k^{(s)2}}{x^2+y^2} + \frac{2A_k^{(r)}A_k^{(s)}}{\sqrt{x^2+y^2}} \cos k\left(z - \frac{x^2+y^2}{2z}\right) \tag{8.1}$$

where $A_k^{(r)}$ and $A_k^{(s)}$ are the amplitudes of the straight-through light and scattered light, respectively. The scattered light from a nanoparticle is very weak [60,61]. Therefore, it will be lost among speckle noise if a high coherence light source such as a laser is used. When a broadband (low-coherence) light source is used, the interference image is described as

$$I_h(x,y,z) = \int_k \left|u_k(x,y,z)\right|^2 dk. \tag{8.2}$$

From the interference pattern, the diffraction calculation was performed on the basis of the angular spectrum method [62]. The object was a single nanoparticle, which was a very simple object, so that the amplitude of the captured interference image was given by the amplitude-only hologram with a constant phase. The complex amplitude of a diffraction image $u(x, y, z)$ at the axial position z from a hologram $u_h(x, y, z_m)$ obtained at an axial position z_m is described as

$$u(x, y, z) = F_t^{-1}\left(F_t\left[u_h\left(x, y, z_m\right)\right] \exp\left[-2\pi i\sqrt{\lambda^{-2} - v_x^2 - v_y^2}\,\left(z - z_m\right)\right]\right), \qquad (8.3)$$

where v_x and v_y are the 2D coordinates of the spatial frequency plane, F_t and F_t^{-1} are the 2D Fourier transform in the transverse directions and the inverse transform, respectively.

Figure 8.2 shows the flowchart of the procedure for 3D position measurement of the nanoparticle, performed on a computer. First, N holograms were captured at a fixed time interval Δt. The diffraction calculation of each hologram was performed on $z = z_t + n_z\Delta z$ ($n_z = -N_z/2$, $-N_z/2 + 1$, ... $N_z/2 - 1$), where z_t was an axial position where a template image was obtained and N_z was the number of diffraction calculations per hologram. $N_z = R_Z/\Delta z$, where R_Z was a measurement axial range. This number was closely related to the total calculation time, because the diffraction calculation was the biggest computational load in the calculation. $R_Z = 1.95\,\mu m$, which was larger than the movement of the trapped nanoparticle in each frame, was given.

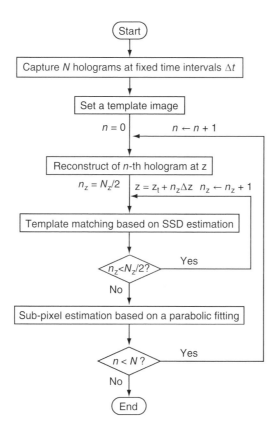

Figure 8.2 Flow chart

8.2.2.3 Template Matching

Template matching searches a 3D position that minimizes the sum of squared differences (SSD) between a diffraction image $I(x, y, z)$ and a template image $T(\xi, \zeta, z_0)$ that was previously set as the diffraction image where a nanoparticle was located at z_0. $I(x, y, z)$ is a diffraction image with $L \times L$ pixels, $T(\xi, \zeta, z_0)$ at the axial position z_0 is a template image with $M \times M$ pixels, and the SSD map is given by

$$SSD(x, y, z) = \sum_{\zeta=0}^{M-1} \sum_{\xi=0}^{M-1} \{I(x + \xi\Delta x, y + \zeta\Delta y, z) - T(\xi\Delta x, \zeta\Delta y, z_0)\}^2. \quad (8.4)$$

The minimum SSD value, $SSD_0 = SSD(x_{\min}, y_{\min}, z_{\min})$, was sought while changing x, y, and z from N_z diffraction images of the hologram obtained at a time t and the position $(x_{\min}, y_{\min}, z_{\min})$ was obtained. The lateral search region was restricted by $|x_n - x_{n-1}| < p\Delta x$ and $|y_n - y_{n-1}| < p\Delta y$, where $p = q = 25$, for reducing the calculation time.

Figure 8.3(a) shows an interference image $I(x, y, z)$ (a hologram of the 200 nm polystyrene particle). The image was clipped in a 51×51 pixel area around the position of the nanoparticle. Figures 8.3(b) and 8.3(c) show the diffraction images $I(x, y, z)$ at planes $z = -689$ nm and -1300 nm, respectively. Figure 8.3(d) shows the template image with 11×11 pixels $T(\xi, \zeta, z_0)$. The position that had the minimum amplitude was selected as the position of the nanoparticle from the diffraction image of the hologram obtained at $t = 0.00$ s, in this case, from Fig. 8.3(b).

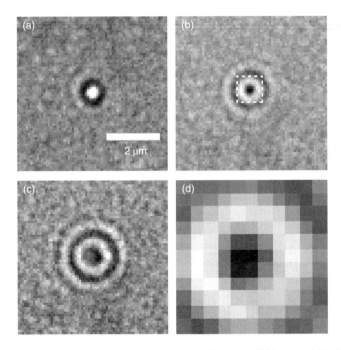

Figure 8.3 (a) Hologram of a polystyrene particle with a diameter of 200 nm and (b), (c) its diffraction images at distances $z = -689$ and -1300 nm. The square indicated in the dashed lines in (b) is clipped as the template in (d)

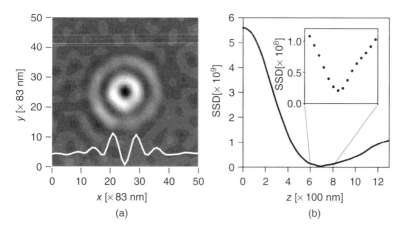

Figure 8.4 (a) The lateral SSD map and the profile through the center line including the minimum SSD. (b) The axial SSD profile. The inset shows a magnified view around the minimum SSD

This was selected experimentally, but optimization of the selection depending on the size and shape of the target object is important for improving the axial resolution. The size of the dark spot was 309 nm full width at half maximum (FWHM), which was slightly larger than the Airy disk diameter of 219 nm at FWHM. It means that the resolution of the holographic microscope was nearly equal to the theoretical resolution limit. The dashed square indicates the area clipped as $T(\xi, \zeta, z_0)$ from Fig. 8.3(b).

Figure 8.4(a) shows a lateral SSD map calculated from the diffraction image on the focus plane at $t = 1.17s$ and the template image (Fig. 8.3d). It contained the point that had SSD_0, and the profile indicated by a white curve was obtained on the line that included SSD_0. Figure 8.4(b) shows the axial SSD profile at the position (x_{min}, y_{min}) and the inset shows a magnified view around z_{min}.

8.2.2.4 Three-Dimensional Sub-Pixel Estimation

The sub-pixel estimation [30] was performed using parabolic fitting in the respective three-dimensional axes (see Fig. 8.4). From the SSD obtained in the template matching, the sub-pixel position x_{sub} was estimated from

$$x_{sub} = x_{min} + \Delta x \frac{SSD_{-1} - SSD_1}{2SSD_{-1} - 4SSD_0 - 2SSD_1}. \tag{8.5}$$

where $SSD_n = SSD(x_{min} + n\Delta x, y_{min}, z_{min})$. Same calculations were performed in the y and z directions to obtain the sub-pixel positions y_{sub} and z_{sub}, and the position of the nanoparticle was estimated as $(x_{sub}, y_{sub}, z_{sub})$. The movement of the nanoparticle was measured by performing the previous procedure for all holograms. In the 2D sub-pixel estimation, the calculations in Eq. (8.5) were performed only for the x and y directions [29], then the axial resolution was improved by minimizing Δz. The axial sub-pixel estimation drastically decreased the value of the diffraction calculation.

8.3   Experimental Results of 3D Position Measurement of Nanoparticles

8.3.1   A 200 nm Polystyrene Particle Fixed on a Glass Substrate

The accuracy of the 3D position measurement was estimated using a 200 nm polystyrene particle fixed on glass. Figure 8.5(a) shows the temporal changes for the measured positions of the fixed particle using the 3D sub-pixel estimation with $\Delta z = 13$ nm. The SDs of the temporal changes (the noise levels) for the x, y, and z directions were 3.5, 3.4, and 3.2 nm, respectively. These were much smaller than the diffraction limit and the Rayleigh length of the microscope and were smaller than the reported SDs measured with a high-speed camera and a QPD.

The system has three species of noises: optical noise, optoelectronic noise, and mechanical noise. The optical noise is an interference noise mainly caused by pollution on the optics. This was almost constant in this system. The optoelectronic noise mainly generated on the image sensor is temporally random. The main components of the mechanical noise were low frequency components because the vibration-isolation optical table eliminated the high frequency components that come from an external source. Therefore, the high frequency fluctuation in Fig. 8.5(a) was mainly caused by the optoelectronic noise. The thick gray curve

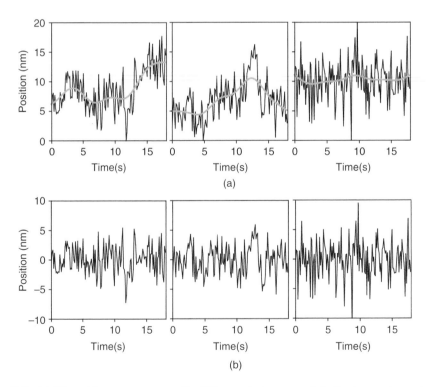

Figure 8.5   (a) 3D position measurement of a 200 nm polystyrene particle fixed on a glass substrate. The gray curves indicate the low frequency components of the temporal traces extracted by the low-pass filter based on the FFT. (b) The high frequency components that subtracted the low frequency components from the original temporal traces of the position measurements

indicates the low frequency components less than 0.3 Hz extracted by the low-pass filtering based on the FFT. Their SDs for the x, y, and z directions were 2.2, 2.0, and 0.4 nm, respectively. Figure 8.5(b) shows the high frequency components that subtracted the low frequency components from the original signals. The SDs for the x, y, and z directions were 2.3, 2.5, and 3.1 nm, respectively. The low frequency components had smaller values in the z-direction than in the x- and y-directions, and the high frequency components had larger values in the z-direction than in x- and y-directions.

8.3.2   Axial Step in 3D Sub-Pixel Estimation

To increase the axial step Δz in the 3D sub-pixel estimation leads to a decrease in the computational cost of the diffraction calculation that is the biggest load in 3D position measurement. Figure 8.6 shows the measurement error for the axial step in the sub-pixel estimation. The filled circles indicate the results of the 3D sub-pixel estimation. For comparison, the results of the 2D sub-pixel estimation without the axial sub-pixel estimation are indicated by the open circles. The circles on the solid curves indicate the SDs of the measured axial position, described as

$$SD_z = \sqrt{\frac{1}{N}\sum_{n=1}^{N}\left(z_{sub}(n) - \overline{z_{sub}}\right)^2}, \tag{8.6}$$

where $\overline{z_{sub}}$ is the mean value of $z_{sub}(n)$. This well-known equation is noted here for easily understanding comparison with the next equation. For 2D sub-pixel estimation, $z_{min}(n)$ and $\overline{z_{min}}$ were used instead of $z_{sub}(n)$ and $\overline{z_{sub}}$ in the calculations.

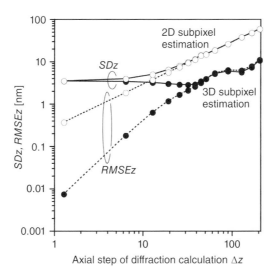

Figure 8.6 SD_z and $RMSE_z$ of the position measurements of the fixed particle for Δz. The open and the filled circles indicate the 2D and 3D sub-pixel estimations, respectively. The solid and dashed curves indicate SD_z and $RMSE_z$, respectively

The circles on the dashed curves indicate the root mean square error (RMSE) of the measured axial position $z_{\min}(n)|_{\Delta z=0.13\,nm}$, which is the most probable value of the true position when the minimum $\Delta z = 0.13$ nm was given and the 2D sub-pixel estimation was performed, given by

$$RMSE_z = \sqrt{\frac{1}{N}\sum_{n=1}^{N}\left(z_{sub}(n) - z_{\min}(n)|_{\Delta z=0.13\,nm}\right)^2}. \tag{8.7}$$

Similarly, $z_{\min}(n)$ was used instead of $z_{sub}(n)$ for 2D sub-pixel estimation.

First, the results of the 2D and 3D sub-pixel estimations were compared. The 3D sub-pixel estimation had smaller errors than the 2D sub-pixel estimation for any Δz. When $\Delta z < {\sim}40$ nm, SD_z of the 3D sub-pixel estimation reached the lower limit of ${\sim}3$ nm. SD_z represented the system resolution, which was determined by the SNR of the interference fringes, the SNR of image sensor noise, and mechanical stability of the optical system, as described before.

$RMSE_z$ was free from the optical, optoelectronic, and mechanical noises, which determined the system resolution. $RMSE_z$ depended only on the axial step Δz in the diffraction calculation; therefore, it decreased with a smaller Δz, which got near to the sufficiently small value of 0.13 in this study. In the 2D sub-pixel estimation, $RMSE_z$ had a linear dependence for Δz because it was caused by quantization error. In the 3D sub-pixel estimation, a quadratic dependence was observed when $\Delta z < 10$ nm because parabola fitting was used, and the axial profile of SSD was sufficiently fitted with the parabola function. When $\Delta z > 30$ nm, $RMSE_z$ showed a complex curve because the axial profile of SSD was different from the parabola function (see Fig. 8.4b) and the dependencies departed from the parabola function with increasing Δz.

8.3.3 Brownian Motion of a 200 nm Polystyrene Particle Held in Optical Tweezers

Figure 8.7 (Plate 14) shows the 3D positions of a 200 nm polystyrene particle exhibiting Brownian motion in the trapping volume of the optical tweezers. The traveling direction of the laser beam was from the bottom to the top of the graph and the gravity direction was the reverse. The holograms were recorded at 8.6 fps. The laser intensities were $I = 4.4$, 5.6, and 14.8 MW/cm^2, respectively. With the increase of laser intensity, the Brownian motion was reduced. Figure 8.8 shows the change in Brownian motion in response to changing the trapping laser power. The Brownian motion of the 200 nm particle changed drastically when $I < 14.8$ MW/cm^2. When $I < 4.4$ MW/cm^2, the Brownian motion was too strenuous and the particle was trapped for only a few seconds. When $I > 18.7$ MW/cm^2, the particle showed movement of less than two times greater than the noise level: that is, almost motionless. Similarly, the Brownian motion of a 500 nm particle became large when $I < 0.15$ MW/cm^2, and was almost completely motionless when $I > 0.97$ MW/cm^2. Here, we define the threshold power of the optical trapping as the smallest I where a particle was trapped over a long time. The threshold intensities for the 200 and 500 nm polystyrene particles were 4.4 and 0.53 MW/cm^2, respectively. This method is very useful for determining the threshold intensity of optical trapping and the 3D directional property.

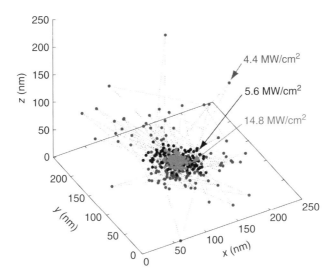

Figure 8.7 (Plate 14)   Movement of a 200 nm polystyrene particle in the trapping volume at different intensities: (blue) 4.4 MW/cm^2, (black) 5.6 MW/cm^2, and (red) 14.8 MW/cm^2. *See plate section for the color version*

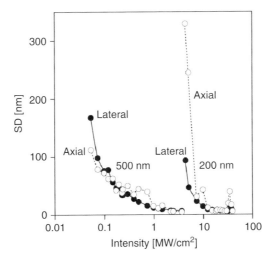

Figure 8.8   Change in Brownian motion with a change of the trapping laser intensity. The particles' diameters are 200 and 500 nm. The filled and open circles indicate the SDs of the lateral and axial movements, respectively

8.3.4 Brownian Motion of a 60 nm Gold Nanoparticle Held in Optical Tweezers

Figure 8.9 shows the measured 3D position of the optically trapped gold nanoparticle with a diameter of 60 nm. The detectable axial range roughly estimated from the 3D movement of gold nanoparticle was 1200 nm (from +900 to −300 nm from the trapping point) (see Fig. 8.10 (Plate 15)). When the laser intensity was 5.6 MW/cm^2 (Fig. 8.9a), the gold nanoparticle moved drastically according to the Brownian motion, especially in the axial direction because the axial force was lower than the lateral force. The axial movement was bigger toward the beam direction and the gold nanoparticle sometimes went off from the laser focus region to the axial upper direction (the beam direction) beyond the axial detectable region. The behavior was

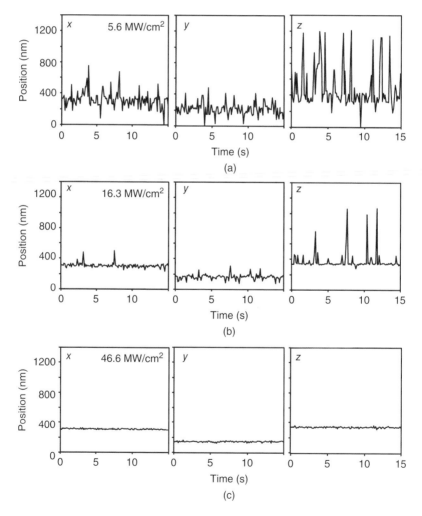

Figure 8.9 Three-dimensional position of a 60 nm gold nanoparticle trapped in optical tweezers with intensity of (a) I = 5.6 MW/cm^2, (b) 16.3 MW/cm^2, and (c) 46.6 MW/cm^2

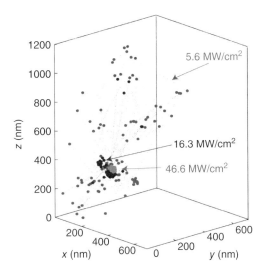

Figure 8.10 (Plate 15) Three-dimensional display of the gold nanoparticle movements. *See plate section for the color version*

different from that of a 200 nm polystyrene particle. It is dependent on the balance between the scattering force and the gradient force. The gradient force for a gold nanoparticle is much smaller than the scattering force. With an increase in laser intensity (Figs. 8.9b and 8.9c), the movement was reduced. When the laser intensity was 46.6 MW/cm², the gold nanoparticle was almost fixed and the lateral and axial variations were 6.1 and 7.1 nm, respectively. The spatial resolutions were a little lower than those of a 200 nm polystyrene particle.

Figure 8.11 shows the Brownian motion in response to changing the trapping laser intensity. The motion of the gold nanoparticle became bigger with the intensity smaller

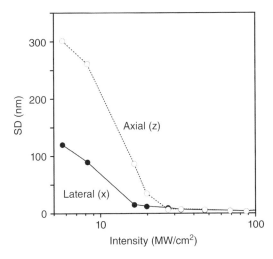

Figure 8.11 Brownian motion of a 60 nm gold nanoparticle dependent on the trapping laser intensity

than $I < 20\,\text{MW/cm}^2$. When $I = 5.6\,\text{MW/cm}^2$, the change was too strenuous, and the gold nanoparticle went off sometimes and returned back to the trapping region. When $I < 5.6\,\text{MW/cm}^2$, the gold nanoparticle went off for a long time and did not return. Here, we define the threshold intensity of the optical trapping of a 60 nm gold nanoparticle under this laser focusing conditions as the smallest I where a particle was trapped over a long time period. When $I > 30\,\text{MW/cm}^2$, the gold nanoparticle was almost completely motionless. The SDs of the measurement variations along the lateral and axial directions were 6.3 and 7.1 nm, respectively. When $I < 20\,\text{MW/cm}^2$, the larger variation for smaller nanoparticle depended on the Brownian motion. When $I > 20\,\text{MW/cm}^2$, the Brownian motion was suppressed and the random signals were caused only by noises in the measurement system.

8.4 Twilight Field Technique for Holographic Position Detection of Nanoparticles

8.4.1 Twilight Field Optical Microscope

When a nanoparticle of a much smaller size is used as a target, the quality of interference fringes is greatly decreased because of the ultra-weak scattered light intensity, and as a result, the accuracy of 3D measurement is decreased. A method for obtaining interference fringes, with higher signal-to-noise ratios (SNR) of amplitude in the 3D position measurement of a nanoparticle by using an in-line digital holographic microscope, is described in this section. The key component in the method is a low-frequency attenuation filter (LFAF) that only reduces the intensity of the reference light, which is non-scattered straight-through light, close to the intensity of the object light from one of the nanoparticles. When the LFAF has a transmittance of 1, the digital holographic microscope works as an ordinary bright field; conversely when the LFAF has a transmittance close to 0, it works like a dark field microscope. This method uses the intermediate transmittance, that is, the intermediate between bright field and dark field. Therefore, the proposed system is called a *twilight field optical microscope* [59].

When a nanoparticle with a radius r and a refractive index n_1 surrounded by a solution with a refractive index n_0 is in an in-line DHM, the angular distribution of the scattered light intensity at a distance d with respect to incident light intensity I_{inc} with wave-number k is given by the well-known Rayleigh scattering theory:

$$I_{sca}(\theta, d) = \frac{k^4 r^6}{d^2} \left(\frac{m^2 - 1}{m^2 + 1} \right) \left(\frac{1 + \cos^2\theta}{2} \right) I_{inc} = \alpha(\theta, d) I_{inc}, \tag{8.8}$$

where $m = n_1/n_0$ and α is a scattering coefficient depending on the angle θ and the distance d. This equation shows that I_{sca} decreases with the sixth power of r. The interference signal I between the scattered light and the straight-through reference light I_{ref} is described as

$$I = I_{ref} + I_{sca} + 2\sqrt{I_{ref}I_{sca}} \cos(\theta) + N, \tag{8.9}$$

where θ is a phase, N represents a magnitude of optical and optoelectronic noises. From Eq. (8.8), for a nanoparticle $I_{sca} \ll I_{inc}$: So $I_{ref} \approx I_{inc}$. SNR of the fringes, which is defined as the fringe amplitude $2(I_{ref}I_{sca})^{1/2}$ divided by N, is given by

$$SNR = \frac{2\sqrt{I_{ref}I_{sca}}}{N} \approx \frac{2I_{inc}\sqrt{\alpha}}{N} \approx \frac{2I_{ref}\sqrt{\alpha}}{N}. \tag{8.10}$$

From Eq. (8.10), it can be seen that the *SNR* can be increased by decreasing N and increasing α. N can be decreased by using an image sensor with small noise and by reducing undesired interference (speckle noise). α can be increased by using a light source with a large wave-number. Once these improvements have been realized for a certain application, the *SNR* can be increased only by increasing I_{inc}. However, the increase of I_{inc} is limited by the maximum detectable value of the image sensor, denoted as I_{max}. Consequently, the *SNR* has an upper limit according to the nanoparticle's physical properties, after selecting the system components.

Now it is assumed that I_{ref} at the plane after light scattering from the nanoparticle is decreased by the LFAF with a transmittance of T_R, and I_{inc} is increased by a factor of $g(>1)$. The interference signal is

$$I = gT_R I_{ref} + gI_{sca} + 2g\sqrt{T_R I_{ref} I_{sca}}\cos(\theta) + N. \tag{8.11}$$

If $T_R I_{ref} \gg I_{sca}$ and g is regulated to satisfy the condition $gT_R = 1$, technically speaking, the condition $I < I_{max}$ is always satisfied, the optical noise, which is directly proportional to the intensity of the light falling onto the image sensor is retained, and so the *SNR* increases in proportion to $T_R^{-1/2}$, as indicated in Eq. (8.12). This equation shows that it is possible to obtain an arbitrary value of the *SNR* by using an LFAF with appropriate design, that is, with an appropriate T_R, as described by:

$$SNR \approx \frac{2I_{inc}\sqrt{\alpha/T_R}}{N}. \tag{8.12}$$

From this equation, the highest *SNR* is obtained under $\alpha = T_R$, however, then a very strong light source satisfying $gT_R = 1$ is required. In practice, T_R uses the full dynamic range of the image sensor with dependence on the light source intensity and an acceptable intensity to the sample.

8.4.2   Low-Coherence, In-Line Digital Holographic Microscope with the LFAF

A low-coherence, in-line digital holographic microscope with the LFAF is shown in Fig. 8.12. The light source was an optical fiber coupled with a light emitting diode (LED), a center wavelength of 450 nm, and a spectral width of 25 nm. The light was focused near the sample to obtain strong illumination. The sample was a 100 nm-diameter polystyrene particle in water, sandwiched by a slide glass and a cover glass. The light scattered from the nanoparticle and the non-scattered reference light was magnified by a 60× objective lens (1.25NA) and lenses

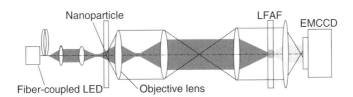

Figure 8.12   In-line digital holographic microscope with LFAF

through the LFAF. The LFAF had a simple structure, made of a thin transparent film with a semi-transparent circle of a 1 mm diameter. The interference fringes were detected by an electron-multiplying CCD (EMCCD) image sensor (DU-888, Andor) with a low noise and a frame interval of 29.6 ms. The captured interference images were reconstructed based on the angular spectrum method on a computer. The axial magnification M_a was $M_a = 5.1 \times 10^3$. The diffraction image was calculated according to the propagation distance z, which was scanned in the maximum movement range of the nanoparticle at intervals of $\Delta z = 0.1$ nm. Image processing was performed, including the pattern matching and the subpixel display, but only in the transverse directions, and was almost same as the method described in Section 8.2.2.

8.4.3 Improvement of Interference Fringes of a 100 nm Polystyrene Nanoparticle

Figure 8.13 shows the interference fringes of a 100 nm polystyrene nanoparticle fixed on the substrate glass, located at a position of $z = 1.5\,\mu m$ from the focal plane of the objective lens and captured (a) without the LFAF ($T_R = 1.0$), (b) with the LFAF ($T_R = 0.19$), and (c) with the LFAF ($T_R = 0.08$). These images show the increase in contrast of the interference fringes with the use of the denser LFAF. In the experiment, a standard environmental noise deviation of 10 counts/s was measured when the light source was blocked. The incident light intensity on the CCD image sensor was adjusted to an average intensity of 4.0×10^3 counts/s, and it was kept the same during the experiment in both cases, with and without the LFAF.

Figure 8.14(a) shows SNR of the fringes as a function of $T_R^{-1/2}$. The filled and open circles indicate the results when the nanoparticle positions were $z = 1.5$ and $3.0\,\mu m$ from the focal plane, respectively. The contrast of the interference fringes was calculated from the maximum and minimum intensities of the circular fringes, which were respectively measured from the center dark spot and the surrounding bright ring in the diffraction image of the hologram. SNR was improved by using the LFAF and increasing I_{inc}. As theoretically expected from Eq. (8.11), SNR was proportional to $T_R^{-1/2}$ because of $\alpha/T_R \ll 1$ in this experiment. The 3D positions of a 100 nm particle were obtained from 100 holograms, and the standard deviations σ_x, σ_y, and σ_z along the x, y, and z directions were calculated to evaluate the accuracy of the measurement. Figures 8.14(b) and 8.14(c) show σ_y and σ_z as functions of $T_R^{-1/2}$, respectively. Here, σ_x is not shown because it was similar to σ_y. The improvement in SNR that was realized by using the dense LFAF increased the accuracy of the 3D position measurement of the nanoparticle.

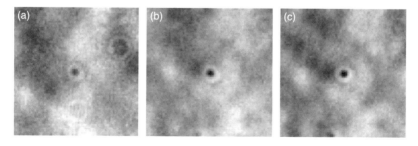

Figure 8.13 Holographic images of nanoparticle captured (a) without the LFAF ($T_R = 1.0$), (b) with the LFAF ($T_R = 0.19$), and (c) with the LFAF ($T_R = 0.08$)

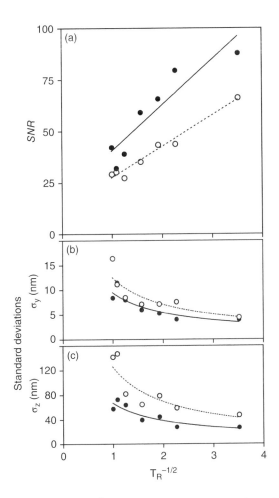

Figure 8.14   (a) *SNR* as a function of $T_R^{-1/2}$ under constant reference light intensity. Standard deviations along the (b) *y*- and (c) *z*-directions as a function of $T_R^{-1/2}$. The filled and open circles indicate the results when the nanoparticle positions were $z = 1.5$ and $3.0\,\mu m$ from the focal plane, respectively

8.5   Conclusion

This chapter described a position measurement of an optically-trapped nanoparticle using a low-coherence, in-line digital holographic microscope, with the pattern matching method based on the SSD estimation and the 3D sub-pixel estimation of the SSD map. The optical setup, the image processing procedure, and sample preparation were described. The behaviors of the trapped nanoparticle and the spatial resolution of the microscope were demonstrated.

The 3D position measurement of a 200 nm polystyrene particle held by optical tweezers in water was demonstrated. The optical tweezers performed the 3D positioning of the 200 nm polystyrene particle in a volume smaller than ~3 nm in SD, which corresponded to the noise level; that is, the spatial resolution of the holographic microscope. The lateral and axial resolutions were obtained from the position measurements of a 200 nm polystyrene particle fixed on the glass substrate. They were much smaller than the diffraction limit and the Rayleigh length of the microscope, respectively. The movement of the 200 nm polystyrene particle trapped by the optical tweezers in water was measured. When $I > 18.7 \, MW/cm^2$, the particle had the movement of less than two times larger than the noise level; that is, almost motionless. The movement was compared with that of the 500 nm polystyrene particle. The results revealed that the threshold laser intensity for trapping the 200 and 500 nm particles were 4.4 and $0.53 \, MW/cm^2$, respectively.

The 3D position of a 60 nm gold particle held by optical tweezers in water was also measured. The threshold trapping intensity was $I = 5.6 \, MW/cm^2$ and the gold nanoparticle was almost fixed with $I > 30 \, MW/cm^2$. From the comparison of movements between a gold nanoparticle fixed on a glass substrate and fixed by optical tweezers with a sufficient optical trapping laser intensity, we analyzed the three kinds of noise; optical, optoelectronic, and mechanical.

This holographic method was a very powerful tool for detailed study of the threshold of optical trapping, especially the ability of 3D measurements to measure a directional property. The threshold intensity depends on not only the particle's property but also a relationship between a particle and its surroundings. The further study of the threshold will be very interesting. This system will be useful for 3D mechanical measurement using optically trapped nanoparticles in fluid and biological systems.

Toward the observation of a smaller particle using an in-line digital holographic microscope, the twilight field method based on increased illumination and use of a light attenuation filter at a low spatial frequency was developed. The method can overcome the limitations in conventional holographic measurement of nanoparticles. The method improved the quality of the interference fringes and the accuracy of 3D position measurement of a nanoparticle.

Finally, in this section, an optical tweezers method for positioning a nanoparticle, an in-line digital holographic microscope with some image processing methods for measuring a nanoparticle's position, and a method for obtaining higher quality interference fringes of a nanoparticle were proposed. Consequently, these useful methods could well be required for improvements in advanced microscopy. Nanoparticle technology is very promising for a variety of scientific and engineering fields in future, and 3D position measurement of nanoparticles will be very effective for them.

References

[1] Ashkin, A., J. M. Dziedzic, J. E. Bjorkholm, and S. Chu, "Observation of a single-beam gradient force optical trap for dielectric particles," *Opt. Lett.* **11**, 288–290 (1986).

[2] Ashkin, A. and J. M. Dziedzic, "Optical trapping and manipulation of viruses and bacteria," *Science* **235**, 1517–1520 (1987).

[3] Tinoco, Jr., I., D. Collin, and P. T. X. Li, "Unfolding single RNA molecules: bridging the gap between equilibrium and non-equilibrium statistical thermodynamics," *Q. Rev. Biophys.* **38**, 291–301 (2005).

[4] Neuman, K. C. and A. Nagy, "Single-molecule force spectroscopy: optical tweezers, magnetic tweezers and atomic force microscopy," *Nature Methods* **5**, 491–505 (2008).

[5] Ichikawa, M., Y. Matsuzawa, Y. Koyama, and K. Yoshikawa, "Molecular fabrication: aligning DNA molecules as building blocks," *Langmuir* **19**, 5444–5447 (2003).

[6] Svoboda, K., C. F. Schmidt, B. J. Schnapp and S. M. Block, "Direct observation of 18 kinesin stepping by optical trapping interferometry," *Nature* **365**, 721–727 (1993).

[7] Greenleaf, W. J. and S. M. Block, "Single-molecule, motion-based DNA sequencing using RNA polymerase," *Science* **313**, 801–803 (2006).

[8] MacDonald, M. P., G. C. Spalding, and K. Dholakia, "Microfluidic sorting in an optical lattice," *Nature* **426**, 421–424 (2003).

[9] Ladavac, K., K. Kasza, and D. G. Grier, "Sorting mesoscopic objects with periodic potential landscapes: optical fractionation," *Phys. Rev. E* **70**, 010901(R) (2004).

[10] Miyazaki, M. and Y. Hayasaki, "Motion control of low-index microspheres in liquid based on optical repulsive force of a focused beam array," *Opt. Lett.* **34**, 821–823 (2009).

[11] Nakanishi, S., S. Shoji, S, Kawata, and H.-B. Sun, "Giant elasticity of photopolymer nanowires," *Appl. Phys. Lett.* **91**, 063112 (2007).

[12] Eom, S. I., Y. Takaya, and T. Hayashi, "Novel contact probing method using single fiber optical trapping probe," *Prec. Eng.* **33**, 235–242 (2009).

[13] Dufresne, E. R. and D. G. Grier, "Optical tweezer arrays and optical substrates created with diffractive optical elements," *Rev. Sci. Instrum.* **69**, 1974–1977 (1998).

[14] Hayasaki, Y., M. Itoh, T. Yatagai, and N. Nishida, "Nonmechanical optical manipulation of microparticles using spatial light modulator," *Opt. Rev.* **6**, 24–27 (1999).

[15] Reicherter, M., T. Haist, E. U. Wagemann, and H. J. Tiziani, "Optical particle trapping with computer-generated holograms written on a liquid-crystal display," *Opt. Lett.* **24**, 608–610 (1999).

[16] Jordan, P., H. Clare, L. Flendrig, J. Leach, J. Cooper, and M. Padgett, "Permanent 3D microstructures in a polymeric host created using holographic optical tweezers," *J. Mod. Opt.* **51**, 627–632 (2004).

[17] Roichman, Y. and D. G. Grier, "Holographic assembly of quasicrystalline photonic heterostructures," *Opt. Express* **13**, 5434–5439 (2005).

[18] Agarwal, R., K. Ladavac, Y. Roichman, G. Yu, C. M. Lieber, and D. G. Grier, "Manipulation and assembly of nanowires with holographic optical traps," *Opt. Express* **13**, 8906–8912 (2005).

[19] Gittes, F. and C. F. Schmidt, "Interference model for back-focal-plane displacement detection in optical tweezers," *Opt. Lett.* **23**, 7–9 (1998).

[20] Pralle, A., M. Prummer, E. L. Florin, E. H. K. Stelzer, and J. K. H. Hörber, "Three-dimensional high-resolution particle tracking for optical tweezers by forward scattered light," *Microsc. Res. Tech.* **44**, 378–386 (1999).

[21] Meiners, J. S. and S. Quake, "Direct measurement of hydrodynamic cross correlations between two particles in an external potential," *Phys. Rev. Lett.* **82**, 2211–2214 (1999).

[22] Huisstede, J. H. G., K. O. van der Werf, M. L. Bennink, and V. Subramaniam, "Force detection in optical tweezers using backscattered light," *Opt. Express* **13**, 1113–1123 (2005).

[23] Tolic-Norrelykke, S. F., E. Schäffer, J, Howard, F. S. Pavone, F. Jülicher, and H. Flyvbjerg, "Calibration of optical tweezers with positional detection in the back focal plane," *Rev. Sci. Instrum.* **77**, 103101 (2006).

[24] Gibson, G. M., J. Leach, S. Keen, A. J. Wright, and M. J. Padgett, "Measuring the accuracy of particle position and force in optical tweezers using high-speed video microscopy," *Opt. Express* **16**, 14561–14570 (2008).

[25] Otto, O., C. Gutsche, F. Kremer, and U. F. Keyser, "Optical tweezers with 2.5 kHz bandwidth video detection for single-colloid electrophoresis," *Rev. Sci. Instrum.* **79**, 023710 (2008).

[26] Lee, S. H. and D. G. Grier, "Holographic microscopy of holographically trapped three-dimensional structures," *Opt. Express* **15**, 1505–1512 (2007).

[27] Bowman, R., G. Gibson, and M. Padgett, "Particle tracking stereomicroscopy in optical tweezers: control of trap shape," *Opt. Express* **18**, 11785–11790 (2010).

[28] Dixon, L., F. C. Cheong, and D. G. Grier, "Holographic deconvolution microscopy for high-resolution particle tracking," *Opt. Express* **19**, 16410–16417 (2011).

[29] Higuchi, T., Q. D. Pham, S. Hasegawa, and Y. Hayasaki, "Three-dimensional positioning of optically-trapped nanoparticles," *Appl. Opt.* **50**, H183–H188 (2011).

[30] Sato, A., Q. D. Pham, S. Hasegawa, and Y. Hayasaki, "Three-dimensional subpixel estimation in holographic position measurement of an optically trapped nanoparticle," *Appl. Opt.* **52**, A216–A222 (2013).

[31] Onural, L. and P. D. Scott, "Digital recording of in-line holograms," *Opt. Eng.* **26**, 1124–1132 (1987).

[32] Schnars, U. and W. Jüptner, "Direct recording of holograms by a CCD target and numerical reconstruction," *Appl. Opt.* **33**, 179–181 (1994).

[33] Skarman, B., K. Wozniac, and J. Becker, "Simultaneous 3D-PIV and temperature measurement using a new CCD based holographic interferometer," *Flow Meas. Instrum.* **7**, 1–6 (1996).

[34] Garcia-Sucerquia, J., W. Xu, S. K. Jericho, P. Klages, M. H. Jericho, and H. J. Kreuzer, "Digital in-line holographic microscopy," *Appl. Opt.* **45**, 836–850 (2006).

[35] Yu, L. and L. Cai, "Iterative algorithm with a constraint condition for numerical reconstruction of a three-dimensional object from its hologram," *J. Opt. Soc. Am. A* **18**, 1033–1045 (2001).

[36] Ma, L., H. Wang, Y. Li, and H. Jin, "Numerical reconstruction of digital holograms for three-dimensional shape measurement," *J. Opt. A* **6**, 396–400 (2004).

[37] Yang, Y., B. S. Kang, and Y. J. Choo, "Application of the correlation coefficient method for determination of the focus plane to digital particle holography," *Appl. Opt.* **47**, 817–824 (2008).

[38] Liebling, M. and M. Unser, "Autofocus for digital Fresnel holograms by use of a Fresnelet-sparsity criterion," *J. Opt. Soc. Am. A* **21**, 2424–2430 (2004).

[39] Dubois, F., C. Schockaert, N. Callens, and C. Yourassowski, "Focus plane detection criteria in digital holography microscopy by amplitude analysis," *Opt. Express* **14**, 5895–5908 (2006).

[40] Mallahi, E. and F. Dubois, "Dependency and precision of the refocusing criterion based on amplitude analysis in digital holographic microscopy," *Opt. Express* **19**, 6684–6698 (2011).

[41] Dubois, F., N. Callens, C. Yourassowsky, M. Hoyos, P. Kurowski, and O. Monnom, "Digital holographic microscopy with reduced spatial coherence for three dimensional particle flow analysis," *Appl. Opt.* **45**, 864–871 (2006).

[42] Cheong, F. C., B. J. Krishnatreya, and D. G. Grier, "Strategies for three-dimensional particle tracking with holographic video microscopy," *Opt. Express* **18**, 13563–13573 (2010).

[43] Pedrini, G. and H. J. Tiziani, "Short-coherence digital microscopy by use of a lensless holographic imaging system," *Appl. Opt.* **41**, 4489–4496 (2002).

[44] Tamano, S., Y. Hayasaki, and N. Nishida, "Phase-shifting digital holography with a low-coherence light source for reconstruction of a digital relief object hidden behind a light-scattering medium," *Appl. Opt.* **45**, 953–959 (2006).

[45] Sugiura, T., T. Okada, Y. Inouye, O. Nakamura, and S. Kawata, "Gold-bead scanning near-field optical microscope with laser-force position control," *Opt. Lett.* **22**, 1663–1665 (1997).

[46] Kalkbrenner, T., M. Ramstein, J. Mlynek, and V. Sandoghdar, "A single gold particle as a probe for apertureless scanning near-field optical microscopy," *J. Microscopy* **202**, 72–76 (2001).

[47] Ukita, H., H. Uemi, and A. HIrata, "Near field observation of a refractive index grating and a topologocal grating by an optically-trapped gold particle," *Opt. Rev.* **11**, 365–369 (2004).

[48] Seol, Y., A. E. Carpenter, and T. T. Perkins, "Gold nanoparticles: enhanced optical trapping and sensitivity coupled with significant heating," *Opt. Lett.* **31**, 2429–2431 (2006).

[49] Hansen, P. M., V. K. Bhatia, N. Harrit, and L. Oddershede, "Expanding the optical trapping range of gold nanoparticles," *Nano Lett.* **5**, 1937–1942 (2005).

[50] Bosanac L., T. Aabo, P. M. Bendix, and L. B. Oddershede, "Efficient optical trapping and visualization of silver nanoparticles," *Nano Lett.* **8**, 1486–1491 (2008).

[51] Selhuber-Unkel, C., I. Zins, O. Schubert, and C. Sönnichsen, L. B. Oddershede, "Quantitative optical trapping of single gold nanorods," *Nano Lett.* **8**, 2998–3003 (2008).

[52] Furukawa, H. and I. Yamaguchi, "Optical trapping of metallic particles by a fixed Gaussian beam," *Opt. Lett.* **23**, 216–218 (1998).

[53] Gu, M. and D. Morrish, "Three-dimensional trapping of Mie metallic particles by the use of obstructed laser beams," *J. Appl. Phys.* **91**, 1606–1612 (2002).

[54] Saija, R., P. Denti, F. Borghese, O. M. Maragò, and M. A. Iatì, "Optical trapping calculations for metal nanoparticles. Comparison with experimental data for Au and Ag spheres," *Opt. Express* **17**, 10231–10241 (2009).

[55] Hajizadeh, F. and S. N. S. Reihani, "Optimized optical trapping of gold nanoparticles," *Opt. Express* **18**, 551–559 (2010).

[56] Atlan, M., M. Gross, P. Desbiolles, É. Absil, G. Tessier, and M. Coppey-Moisan, "Heterodyne holographic microscopy of gold particles," *Opt. Lett.* **33**, 500–502 (2008).

[57] Absil, E., G. Tessier, M. Gross, M. Atlan, N. Warnasooriya, S. Suck, *et al.*, "Photothermal heterodyne holography of gold nanoparticles," *Opt. Express* **18**, 780–786 (2010).

[58] Verpillat, F., F. Joud, P. Desbiolles, and M. Gross, "Dark-field digital holographic microscopy for 3D-tracking of gold nanoparticles," *Opt. Express* **19**, 26044–26055 (2011).

[59] Pham, Q. D., Y. Kusumi, S. Hasegawa, and Y. Hayasaki, "Digital holographic microscope with low-frequency attenuation filter for position measurement of nanoparticle," *Opt. Lett.* **37**, 4119–4121 (2012).

[60] Kerker, M., *The Scattering of Light and Other Electromagnetic Radiation*. (Academic Press, New York, 1969), Ch. 3.

[61] Bohren C. F. and D. R. Huffman, *Absorption and Scattering of Light by Small Particles*. (John Wiley & Sons, Inc., New York, 1983), Ch. 4.

[62] Goodman, J. W. *Introduction to Fourier Optics*, 2nd edn (McGraw-Hill, New York, 1996), Ch. 3.10.

9

Digital Holographic Microscopy: A New Imaging Technique to Quantitatively Explore Cell Dynamics with Nanometer Sensitivity

Pierre Marquet[1,2] and Christian Depeursinge[3]

[1]*Centre de Neurosciences Psychiatriques, Centre Hospitalier Universitaire Vaudois, Département de Psychiatrie, Switzerland*
[2]*Brain Mind Institute, Institute of Microengineering, École Polytechnique Fédérale de Lausanne, Switzerland*
[3]*Institute of Microengineering, École Polytechnique Fédérale de Lausanne, Switzerland*

9.1 Chapter Overview

In the first part of this chapter, we summarize how the new concept of digital optics, applied to the field of holographic microscopy, has allowed the development of a reliable and flexible digital holographic quantitative phase microscopy (DH-QPM) technique at the nanoscale, particularly suitable for cell imaging. In the second part, particular emphasis is placed on the original biological information provided by the quantitative phase signal. We present the most relevant DH-QPM applications in the field of cell biology, including automated cell counts, recognition, classification, three-dimensional tracking, and discrimination between physiological and pathophysiological states. In addition, we present how the phase signal can be used to specifically calculate some important biophysical cell parameters including dry mass, protein content and production, membrane fluctuations at the nanoscale, absolute volume, transmembrane water permeability, and how these different biophysical parameters can be used to perform a non-invasive multiple-site optical recording of neuronal activity.

Multi-dimensional Imaging, First Edition. Edited by Bahram Javidi, Enrique Tajahuerce and Pedro Andrés.
© 2014 John Wiley & Sons, Ltd. Published 2014 by John Wiley & Sons, Ltd.

9.2   Introduction

Historically, optical microscopy has been one of the most productive scientific instruments in technology and medicine. Cells and micro-organisms were identified for the first time in the nineteenth century. These observations can be considered the beginning of developments in modern biology and medicine. On the other hand, some limitations of optical microscopy soon appeared due in particular to the lack of resolution formulated by the well-known Abbe's law, as well as the lack of quantitative information due to the analogical nature of conventional optical microscopes.

Overcoming these limitations is particularly crucial in the field of cell biology, in which to be able to quantitatively appreciate cell structure and dynamics in ever-increasing detail is essential to elucidate the mechanisms underlying physiological or pathological cell processes. In addition, considering that most biological cells differ only slightly from their surroundings in terms of optical properties (including absorbance, reflectance etc.) obtaining a high resolution and quantitative visualization of cell structure and dynamics remains a difficult challenge.

Consequently, several modes of contrast generation have been developed to overcome these limitations. Among the many contrast-generating modes, those based on wavefront phase information, representing an intrinsic contrast of transparent specimens, have demonstrated their relevance for noninvasive visualization of cell structure, in particular the Zernicke's invention of *phase contrast* (PhC) in the mid-twentieth century [1]. Currently, PhC, as well as Normarski's differential interference contrast (DIC), are widely used contrast-generating techniques available for high-resolution light microscopy. In contrast to fluorescence techniques, PhC and DIC allow the visualization of transparent specimens, making visible, in particular, the fine subcellular structural organization without using any staining contrast agent. Basically, these two noninvasive contrast-generating techniques, PhC and DIC, result from their capacity to transform, in detectable modulation intensity, the minute relative phase shift that a transparent microscopic object, differing from the surroundings only by a slight difference of refractive index, induces between the transmitted wave light and the undeviated background wave (PhC) and between two orthogonally polarized transmitted waves (DIC). However, PhC and DIC do not allow the direct and quantitative measurement of phase shift or optical path length. Consequently, DIC or PhC signal variations are only qualitative and remain difficult to interpret in terms of quantitative modification of specific biophysical cell parameters.

In contrast, interference microscopy has the capacity to provide a direct measurement of the optical path length based on interference between the light waves passing the specimen, called the *object wave*, and *reference wave*. Although quantitative phase measurements with interference microscopy applied to cell imaging were already known in the 1950s, since the seminal work of Barer [2] only a few attempts have been reported to dynamically image live cells in biology [3]. Indeed phase shifts are very sensitive to experimental artifacts, including lens defects, and noise originating from vibrations or thermal drift. Temporal phase shifting interferometry therefore requires demanding and costly opto-mechanical designs preventing wider applications in biology.

In parallel, holography techniques were developed by Gabor in 1948 [4] who demonstrated its lensless imaging capabilities thanks to the reconstruction of an exact replica of the full wavefront (amplitude and phase) emanating from the observed specimen (object wave). Due to costly opto-mechanical designs and the non-availability of long coherence sources in optics, few applications were developed at that time.

Among the many contrast-generating mode, fluorescence microscopy in a confocal config-uration (confocal fluorescent lasers canning microscopy, CLSM) and its extension to multi-photon fluorescent excitation for a few decades has been a powerful and widely used cellular imaging technique in biology [5,6].

Nowadays, the chance to bring quantitative and specific data in optical microscopy on one hand, and to reach the nanometer scale on the other, appears more and more clearly to be an incentive to invent and develop new concepts in optical microscopy. Most of them call for the new means offered by informatics and algorithmics. Indeed, optical data are now increas-ingly easily and rapidly processed, bringing an important push to new applications, improved performance, and reproducibility of scientific results. Practically, the development of quan-titative microscopy helps much in extracting meaningful data from images. More generally, the quantitative evaluation of an increasing number of parameters extracted from microscopic images including phase, fluorescence, or nonlinear images such as second (SHD) [7] and third harmonics (THD), or even *Coherent Anti-Stokes* Raman Scattering (CARS) [8], are at the basis of a new field of investigation, sometime called *bio-image informatics*. A clear trend towards super-resolution microscopy is another strand in the revival of optical microscopy [9]. It is the fruit of better knowledge about the physical principles underlying microscopy. The use of coherence properties of light waves has, henceforth, altogether permitted break-ing the diffraction limit and providing a full 3D image of microscopic or nanometric objects. Super-resolution is a commonly accepted term to designate in optical microscopy allowing the imaging of objects beyond the diffraction limit (defined by Abbe's law).

Practically, interferences created by partially coherent waves allow structuring of light beams: fringes are generated for SIM (*Structured Illumination Microscopy*) whereas a combination of focal spots is used for STED (*STimulated Emission Depletion microscopy*) and GSD (*Ground State Depletion microscopy*). These beam structuring techniques are used to enlarge the bandpass of the image beyond the bandpass of the *microscope objective* itself. These new super-resolution are all based on fluorescence imaging of natural or selected fluorophores. Most of them (except SIM) exploit the nonlinear response of fluorescent dyes or proteins to extend further the spectral domain of the images, thereby permitting the achievement of super-resolved images.

The use of statistical treatment of optical signals is a complimentary approach to improve the localization of single fluorescent molecules and thereby push the resolution beyond the diffrac-tion limit [10]. Palm, Storm, and derived methods are examples of that approach. Resulting from fluorescent tag availability, these methods are priceless in biology. However, they do not provide any information about the dielectric properties of specimens, which also give access to important biophysical parameters. Practically, these dielectric properties can be efficiently detected using new digital interferometric and/or holographic approaches.

Indeed, in the field of holography and interferometry, scientific advances lowering the cost of lasers and data acquisition equipment, as well as the development of computing facili-ties, the large spread of PCs and digital signal processors, have completely changed perspec-tives. Practically, this led to the development of various quantitative phase microscopy (QPM) approaches related to holography [11–41], interferometry [42–78] being considerably simpler to implement than classical interference microscopy, while providing a reliable and quantita-tive phase mapping of the observed specimen. It may be noted that QPM techniques based on other approaches, including transport-intensity equations [79,80], quadriwave lateral shearing interferometer [81] or phase-retrieval algorithmic [82–85], have also been developed. Finally, we note some attempts to use DIC and PhC as quantitative imaging techniques [86–92].

In the first part of this chapter, we present the principle of classical holography as well as the current state-of-the-art in digital holography microscopy. Emphasis is given to the key benefits provided by digital means to develop a reliable and flexible quantitative phase microscopy technique with a nanometric axial sensitivity. In a second part, the most relevant applications in the field of cell biology provided by *digital holographic quantitative phase microscopy* (DH-QPM) are presented. Particular attention is paid to how specific biophysical cell parameters can be determined from the quantitative phase signal and how such parameters can be used to address important biological questions, including the optical monitoring of neuronal activity.

9.3 Holographic Techniques

9.3.1 Classical Holography

Holography techniques were developed by Gabor in 1948 with the aim of improving the detection of spatial resolution in the X-ray wavelength by exploiting its lensless imaging capabilities [93]. This resulted in the possibility of generating, during the illumination of the recorded hologram (reconstruction process), an exact replica with a specific magnification of the full object wavefront created by the observed specimen [94,95]. However, as already identified by Gabor, the imaging possibilities of holography are greatly reduced in quality owing to the presence of different diffraction orders in the propagation of the diffracted wavefront by the hologram when illuminated during the reconstruction process. This was resolved by Leith and Upatneik, who proposed the use of a reference wave from a slightly different propagation direction than the object wave [96,97]. This method, referred to as *off-axis geometry*, has been analyzed from a computational point of view [98] with a formalism based on diffraction, and was the first to use quantitative phase measurements [99]. Practically, the first developments in off-axis configuration were performed through a common-path configuration due to the coherence limitation of the light sources. Moreover, the emergence of laser light sources, enabling very long coherence lengths with high power, takes advantage of the versatility of "interferometric configuration". Note that short coherence length sources have been investigated recently in various cases [73,100,101]. Shorter coherence lengths have the capability to improve the lateral resolution as well as to decrease the coherence noise, which could limit the quality of the reconstructed image, particularly the phase sensitivity. However, these low coherence implementations require more complicated arrangements where the coherence zone (in both spatial and temporal domains) must be adapted to ensure optimal interference [102].

9.3.2 From Classical to Digital Holography

The use of digital means in holography gradually occurred at the end of the 1960s when Goodman used a vidicon detector to encode a hologram, which could be reconstructed on a computer [103]. However, the interest in digital holography rose with the availability of cheaper digital detectors and charge-coupled device (CCD) cameras. The use of CCD cameras for holographic applications was validated in the mid-1990s in the case of reflection macroscopic holograms [104], and microscopic holograms in endoscopic applications [105]. Another approach to hologram reconstruction was developed by taking advantages of the capability of digital detectors to rapidly record multiple frames, through the use of a phase-shifting technique [14] as developed first for interferometry [106,107]. Up to this point, holography was essentially considered an imaging technique, enabling the possibility of lensless imaging or the capability of focusing images recorded out of focus through the recovery of the full wavefront. The digital treatment, contrary to the classical approach, considered the wavefront to be a combination of amplitude and phase, which led to the development of quantitative phase imaging through holography [13].

9.3.3   Digital Holography Methods

The two main approaches to recovering the object wave are namely *temporal decoding*, that is, phase shifting, and *spatial decoding*, that is, off-axis methods. Phase shifting reconstruction methods are based on the combination of several frames, which enables the suppression of the zero order and one of the cross terms through temporal sampling [108,109]. The most well-known phase-shifting algorithm, proposed by Yamaguchi [14], is based on the recording of four frames separated by a phase shift of a quarter of a wavelength. Various combinations of frames derived from interferometry have been considered [107,108] and many different approaches have been developed to produce the phase shift, including high precision piezo-electric transducers that move a mirror in the reference wave, or acousto-optics modulators using the light frequency shift, and so on. One of the main issues in the phase shifting method is the requirement of several frames for reconstruction in interferometric setups, which are commonly very sensitive to vibrations, so it could be difficult to ensure stable phase shifts and an invariant sample state during acquisition. In addition, the requirements on the accuracy of the phase-shifts are rather high with regard to displacements in the magnitude of hundreds of nanometers, implying the use of high-precision transducers. Consequently, several attempts were made to either reduce the required number of frames for reconstruction, which led to two-frame reconstruction [110,111], or to enable the recoding of the various phase shift frames simultaneously by employing, for example, multiplexing methods [112]. On the other hand, more refined algorithms were developed in order to loosen the accuracy requirements of phase shift methods [113–115].

   The second main approach to recovering the object wave is based on an off-axis configuration, so that the different diffraction terms encoded in the hologram (zero-order wave, real image, and virtual image) propagate in different directions, enabling their separation for reconstruction. This configuration was the one employed for the first demonstration of a fully numerical recording and reconstruction holography [105,116].

   In practice, reconstruction methods based on off-axis configuration usually rely on Fourier methods to filter one of the diffraction terms. This concept was first proposed by Takeda *et al.* [117] in the context of interferometric topography. The method was later extended for smooth topographic measurements for phase recovery [118], and generalized for use in digital holographic microscopy with amplitude and phase recovery [13]. As discussed in the following paragraph, the main characteristic of this approach is its capability of recovering the complex object wave through only one acquisition, thus greatly reducing the influence of vibrations. However, as the diffraction terms are spatially encoded in the hologram, this one-shot capability potentially comes at the cost of usable bandwidth. In addition, frequency modulation, induced by the angle between the reference and the object wave, has to guarantee the separability of the information contained in the different diffraction terms that are encoded in the hologram, while carrying a frequency compatible with the sampling capacity of digital detectors.

9.3.4   Digital Holographic Microscopy

9.3.4.1   State-of-the-Art

The applications of digital holography to microscopy are characterized by the fact that the wavefront diffracted by the microscopic specimen is acquired and reconstructed in a numerical form. The innovative aspects of digital holographic microscopy reside in the fact that there is no need for focusing before image acquisition. Compared to the acquisition of the formed image, hologram acquisition with a digital camera appears more flexible and "information rich". The result of this holographic approach is a new microscope that we have called the *Digital Holographic Microscope* (DHM) [13,119]. It comprises a microscope objective to adapt the sampling capacity of the camera to the information content of the hologram, enabling reconstruction of the complex 3D wavefront scattered by the microscopic specimen (see Fig. 9.1b later). DHM delivers the data describing the complex wavefront that can be extracted directly from the digitalized hologram. This set of complex numbers is then propagated to the image plane of the specimen, thereby restoring the true magnified image of the object. The complex data provided by the numerical reconstruction of the wavefront opens the way to the possibility of fully simulating light wave propagation and conditioning by numerical methods, correcting aberrations, and distortions, thereby avoiding the complexity of an optical setup to achieve the necessary beam manipulations [120–123]. In many ways, the introduction of numerical procedures to mimic complex optical systems represents a breakthrough in modern optics [124]. This modality allows for subsequent reconstruction and focusing of the intensity or quantitative phase image. Practically, the propagated wavefront delivers the distribution of the image in depth and therefore allows for extension of the depth of focus without resorting to complex optical realizations [125]. Moreover the high-sensitivity phase measurement provided by the reconstructed wavefront, resulting from an interferometric detection, permits the evaluation of optical path lengths with ultrahigh resolution, in practice down to the subnanometer scale, depending on wavelength and other parameters that include the integration time. Such a wavefront reconstruction can be achieved in real time with a personal computer, the hologram being recorded with a digital camera.

Optical Set-up: A digital holographic microscope [126] consists of both an optical setup devoted to hologram formation and software specially developed to process numerically the digitized hologram. The hologram results from the interference of the object beam with a reference wave that can be kept separate from the object beam. The reference beam is controllable both in intensity and polarization in order to optimize the contrast and signal. The goal is to estimate precisely the propagated wavefront corresponding to either the virtual or real image of the specimen magnified by the microscope objective. Computing the wavefront at various positions along the beam by progressively increasing the reconstruction distance will yield the three-dimensional distribution of the wavefront. Different optical setups have been proposed to perform microscopy with holography [127–129]. We have given preference to optical setups making use of a microscope objective offering the largest numerical apertures [13]. Typical

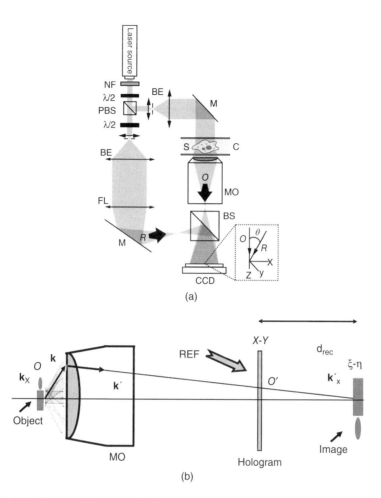

Figure 9.1 Optical setup: (a) for transmission DHM, (b) role of the microscope objective (MO): Magnifying the image of the specimen so that the hologram can be sampled by an electronic camera according to Shannon's rules. O: object beam, R: reference beam, BS: Beam splitter, PBS: polarizing beam splitter, M: Mirrors, CCD: camera, MO: Microscope objective, NF: neutral filter; $\lambda/2$: half wave plate, FL: field lens; S: specimen; C: specimen chamber. *Insets*: details showing off-axis geometry at the incidence on the CCD. *Source*: Marquet, P., Depeursinge, C., Magistretti, P.J., 2013, Neural cell dynamics explored with digital holographic microscopy, Annual Review of Biomedical Engineering, (15), 407–431

arrangements developed and used by us for exploring cell structure and dynamics is depicted in Fig. 9.1(a): DHM allows the recording of a digital hologram corresponding to the wave scattered by the specimen in a transmission configuration by means of a standard CCD (or CMOS) camera inserted at the exit of a Mach–Zehnder interferometer.

Other configurations are possible, depending on the targeted application, but will not be reviewed in detail here. An important issue is the need for a reference beam, which should be controllable both in intensity and polarization. Contrast and signal could be accordingly improved. The holographic principle also permits other valuable concepts to be built on: In particular the possibility of superimposing several holograms. Holograms with several reference waves corresponding to several polarization states can be generated in order to analyze the birefringence properties of specimens: strained dielectrics or biological molecules [130,131]. Reference waves corresponding to different wavelengths can also be generated, permitting the use of synthetic wavelengths from a single hologram [132].

Reconstruction: As in other developments in digital holography, the reconstruction method is based on the theory of diffraction. Intercepting the wavefront provided by the microscope objective (MO) at finite distance from the specimen image plan gives rise to holograms in the Fresnel zone. Huyghens–Fresnel expression of diffraction will therefore be considered as valid for the calculation of the propagation of the reconstructed wave.

The (MO) allows us to adapt the object wavefield to the sampling capacity of the camera: the lateral components of the wavevector $k_{x \, or \, y}$ can be divided by the magnification factor M of the MO, therefore permitting an adequate sampling of the off-axis hologram interference fringes by the electronic camera (Fig. 9.1b).

The hologram taken after the MO results from the superposition of the object wave O' generated by the MO from the wavefield O, emitted by the specimen and a reference wave R, having some non-cancelling mutual coherence with O'. Practically, the hologram intensity is given by expression (9.1):

$$I_H(x, y) = (R + O')^* \cdot (R + O') = |R|^2 + |O'|^2 + R^*O' + RO'^* \tag{9.1}$$

As is usual in holography, the reconstruction of the wavefront can be achieved in the plane of the hologram by illuminating it with a wave matching the reference wave. Taking advantage of a full digital approach, the reconstruction of the object wavefront can be fully achieved by a computer calculating the diffraction process of a simulated reference wave R_{num} by the hologram intensity distribution. Then the distribution of the object wave front in space is obtained by calculating the propagation of the object wavefront reconstructed in the plane of the hologram. As is well known by holographers, the reconstruction of the wavefront gives rise to several propagated beams, the zero order beam and higher order beams, mainly in the order of $+1$ and-1. These two orders provide virtual and real images that appear as twin images, which are the reflection of one another about the hologram plane. If the reference wave is "on axis", that is, propagating in parallel to the object wave O', propagated real and virtual images appear as superposed on any plane where the reconstruction is performed. The specimen image appears therefore as "blurred" by the presence of the unfocused twin image. In order to eliminate this blur, filtration must be performed to eliminate all scattered beam except one: In those conditions focused virtual (O'^*) and/or real (O') images can be reconstructed exactly.

Concretely, the reconstruction method is based on the restoration of the object wavefield O'_{rec}. The reconstruction or restoration of the object wavefield O'_{rec} is usually achieved in the hologram plane x-y by forming the product of the hologram intensity with the reference wave R_{num} generated in the computer and further adjusted to match the original reference wave:

$$O'_{rec}(x, y) = I_{H \; filtered}(x, y) \cdot R_{num}(x, y) / |R_{num}|^2 \tag{9.2}$$

Practically, hologram $I_H(x,y)$ filtering allows us to isolate one of the cross terms RO'^* or R^*O' appearing in Eq. (9.1). Filtering can be achieved either in the time domain or in the space domain.

In the time domain several phase shifted holograms (at least three, but usually four) are taken successively in order to eliminate the square part in Eq. (9.1) corresponding to the zero-order term and one of the cross terms corresponding to the virtual image. Time domain filtering is similar to the so-called "phase-shifting interferometry" and has been proposed by Yamaguchi [14,133] for application in holography. Although this method has the advantage of preserving the full spatial bandwidth, a major inconvenience is that multiple holograms must be taken; a fact that renders the reconstruction of movement blurred and instantaneous images difficult to take.

In the space domain, the filtering of cross terms can be achieved by taking a digitalized hologram in a slightly off-axis geometry [126]. Off-axis geometry introduces a spatial carrier frequency and demodulation restores the full spatial frequency content of the wavefront. The main advantage of this approach is that all the information for reconstructing the complex wavefield comes from a single hologram [13]. In microscopy the full bandwidth of the beam delivered by the MO can be acquired without limitation.

Finally, selecting the signal corresponding to the third term in Eq. (9.1) in the Fourier domain of the hologram [134] allows for the full restoration of the wavefront O'. Reconstructing the wavefront in 3D is therefore simply done by propagating the wavefront generated in the hologram plane x-y to the object plane $\xi-\eta$, which is situated at the distance d_{rec}. This can be simply achieved by computing the Fresnel transform of the wavefield. The mathematical expression Eq. (9.3) used for that computation is:

$$O'(\xi, \eta) = -i \cdot \exp(ikd_{rec}) \cdot F^\sigma_{Fresnel}[O'(x, y)], \tag{9.3}$$

which, in the paraxial approximation, can be put in the following form that is computed after discretization.

$$F^\sigma_{Fresnel}[O'(x, y)] = \frac{1}{\sigma^2} \exp\left[\frac{i\pi}{\sigma^2}\left(\xi^2 + \eta^2\right)\right] \cdot F_{Fouirer}\left\{O'(x, y) \cdot \exp\left[\frac{i\pi}{\sigma^2}\left(x^2 + y^2\right)\right]\right\} \tag{9.4}$$

with

$$\sigma = \sqrt{\lambda d_{rec}} = \sqrt{2\pi \frac{d_{rec}}{k}}. \tag{9.5}$$

When the reconstruction distance d_{rec} approaches infinity, the parameter σ also tends to infinity and the Fresnel transform becomes identical to Fourier transform.

In our DHM implementation, there is no time spent heterodyning or moving mirrors. The microscope design is therefore simple and robust. DHM brings quantitative data derived simultaneously from the amplitude and the quantitative phase of the complex reconstructed object wavefront.

Our approach requires the adjustment of several reconstruction parameters [13], which can done using a computer-aided method developed by our group. Some image processing is also needed to improve the accuracy of the phase [135]. Using a high numerical aperture, submicron transverse resolution has been achieved: to 300 nm lateral resolution, which corresponds to diffraction limited resolution. Accuracies of approximately 0.1° have been estimated for phase measurements. In reflection geometry, this corresponds to a vertical resolution less than 1 nm at a wavelength of 632 nm. In the transmission geometry, the resolution is limited to a few nanometers as far as living cells are concerned.

Characterization of the effect of noise on the formation of the hologram and the reconstruction of the image results in an improved signal-to-noise ratio by the coherent detection of low level of scattered light [136,137]. This improvement is often described as "coherent amplification" of the signal.

9.4   Cell Imaging with Digital Holographic Quantitative Phase Microscopy

Biological specimens such as living cells and tissues are usually phase objects; that is, they are transparent and made visible most often by PhC as explained earlier. The phase signal originates in the RI difference generated by the presence of organic molecules in cells, including proteins, DNA, organelles, and nuclei. Consequently, DH-QPM visualizes cells by quantitatively providing the phase retardation that they induce on the transmitted wavefront [15]. This quantitative phase signal is given by the following expression:

$$\Phi = \frac{2\pi}{\lambda}(\bar{n}_c - n_m)d, \qquad (9.6)$$

where d is the cellular thickness, $\bar{n}_c$ is the intracellular RI averaged over the OPL of optical rays crossing the specimen, and n_m is the RI of the surrounding medium.

DH-QPM allows for the precise determination of the phase or OPL, directly proportional to the (RI) integrated over the propagation of the light beam. Point-to-point OPL determinations yield an absolute PhC image in microscopy, and very high accuracies, comparable to those provided by high-quality interferometers, can be achieved. However, DH-QPM offers much-improved flexibility as well as the capability to adjust the reference plane with a computer (i.e., without positioning the beam or the object). A quantitative phase image, obtained for living neurons in culture, is presented in Fig. 9.2(b) (Plate 16).

9.4.1   Cell Counting, Recognition, Classification, and Analysis

Several original applications made with DH-QPM, in combination with the unique possibilities presented by digital optics (real-time imaging, extended depth of focus, etc.), in the field of cell imaging are presented next.

The DHM-QP modality offers a quantitative alternative to classical PhC and DIC. Practically, the quantitative phase signal is particularly well suited to the development of algorithms, based on biophysical cell parameters, allowing automated cell counting [16,17,138,139], recognition, and classification [18,19,21,140–143]. Considering the fact that the quantitative phase signal also provides some information about the intracellular content, which contributes to modifying the intracellular RI, interesting applications allowing discrimination between physiological and pathophysiological states have been achieved, particularly in the fields of assisted reproduction [20,22,144] and cancer research [23,24,61]. For example, proposing that quantitative evaluation of the phase shift recorded by DHM-QP could provide new information on the structure and composition of the sperm head or cell nucleus has been useful to clinical practice.

The ability of digital propagation to apply autofocusing [25–27] and extended depth of focus [28–30,125] has opened up the possibility of efficiently tracking particles [31,145,146] including those capable of second harmonic generation [147,148]. In addition, applications to the study of cell migration in 3D have been made [18,25,32,33,35,149–151] offering an alternative to the shallow depth of field of conventional microscopy, which hampers any fast 3D tracking of cells in their environment.

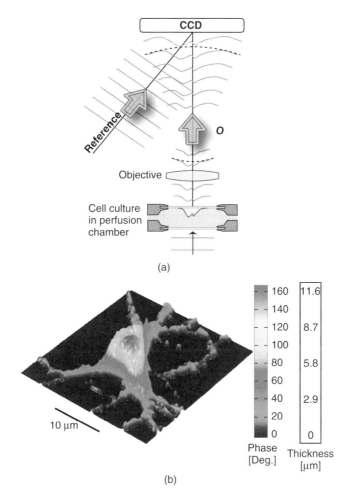

(a)

(b)

Figure 9.2 (Plate 16) Digital holographic microscopy (DHM) of living mouse cortical neurons in culture. (a) Schematic representation of cultured cells mounted in a closed perfusion chamber and trans-illuminated (b) 3D perspective image in false colors of a living neuron in culture. Each pixel represents a quantitative measurement of the phase retardation or cellular optical path length (OPL) induced by the cell with a sensitivity corresponding to a few tens of nanometers. By using the measured mean value of the neuronal cell body refractive index, resulting from the decoupling procedure, scales (*right*), which relate OPL (°) to morphology in the *z*-axis (μm), can be constructed. *See plate section for the color version*

9.4.2 Dry Mass, Cell Growth, and Cell Cycle

As described earlier, the measured quantitative phase shift induced by an observed cell on the transmitted light wavefront is proportional to the intracellular RI, which mainly depends on protein content. Therefore, this measure can be used to directly monitor protein production, owing to a relation established more than 50 years ago by Barer [2,152]. Within the framework of interference microscopy, the phase shift induced by a cell is related to its dry mass (DM) by the following equation (converted to the International System of Units):

$$DM = \frac{10\lambda}{2\pi\alpha} \int_{S_c} \Delta\varphi ds = \frac{10\lambda}{2\pi\alpha} \Delta\overline{\varphi} S_c, \tag{9.7}$$

where $\Delta\overline{\varphi}$ is the mean phase shift induced by the whole cell, λ is the wavelength of the illuminating light source, S_c is the projected cell surface, and α is a constant known as the specific refraction increment (in cubic meters per kilogram) and related to the intracellular content. α is approximated by $1.8–2.1 \times 10^{-3}$ m^3 kg^{-1} when considering a mixture of all the components of a typical cell.

Recently, several groups using various QPM techniques have begun to exploit this phase/DM relationship to study the dynamics of cell growth and the cell cycle [36,153–156]. The relationship has also been explicitly and implicitly used to calculate hemoglobin content in red blood cells (RBCs) [157–161].

9.4.3   Cell Membrane Fluctuations and Biomechanical Properties

RBCs are squeezed as they pass capillaries often smaller than their cell diameter. This ability can be attributed to the remarkable elastic properties of the membrane structure. This structure exhibits a high resistance to stretching ensuring that no leakage through the lipid bilayer occurs, whereas its low resistance to bending and shearing allows the cell to easily undergo morphological changes when passing through small capillaries. As a consequence of these elastic properties, RBCs show spontaneous cell membrane fluctuations (CMFs) at the nanometric scale, often called flickering. Owing to their high sensitivity, and allowing us to quantitatively measure RBC membrane fluctuations over the whole cell surface, different QPM techniques have shed new light on these CMFs by providing quantitative information about the biomechanical properties of RBC membranes [37,52,68,69,161,1621]. We also should mention the integration of DHM with optical tweezers, which is a very promising tool, especially with respect to monitoring trapped objects along the axial direction as well as manipulating and testing biomechanical properties of cells [18,38,163–165].

9.4.4   Absolute Cell Volume and Transmembrane Water Movements

The various applications presented previously highlight the wealth of information brought by the phase signal regarding cell dynamics. However, as indicated by Eq. (9.1), information concerning the intracellular content related to $\overline{n}_c$ is intrinsically mixed with morphological information relating to a thickness d. As a result of this dual dependence, the phase signal remains difficult to interpret. As an illustration, a simple hypotonic shock induces an *a priori* surprising phase signal decrease [39] that is difficult to interpret as cellular swelling, resulting coherently, however, from a decrease of $\overline{n}_c$ due to dilution of the intracellular content by an osmotic water influx. Accordingly, some strategies have been developed to separately measure cell morphology and RI. Some authors [166,167] measured the intracellular RI by trapping cells between two cover slips separated by a known distance. However, this approach, which prevents cell

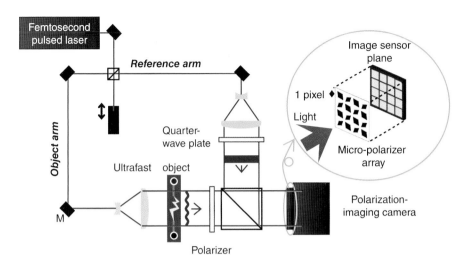

Plate 1 (Figure 1.13) Schematic diagram of parallel phase-shifting digital holography system using a femtosecond pulsed laser

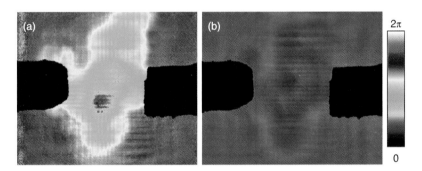

Plate 2 (Figure 1.15) Reconstructed images. (a) By parallel phase-shifting digital holography, (b) by a diffraction integral alone

Multi-dimensional Imaging, First Edition. Edited by Bahram Javidi, Enrique Tajahuerce and Pedro Andrés.
© 2014 John Wiley & Sons, Ltd. Published 2014 by John Wiley & Sons, Ltd.

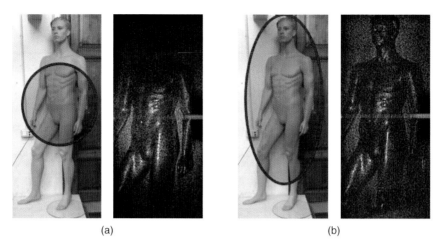

(a)   (b)

Plate 3 (Figure 2.9)   (a) Spherical lens configuration. Mannequin image with irradiated area in red; mannequin hologram amplitude reconstruction. (b) Cylindrical lens configuration. Mannequin image with irradiated area in red; mannequin hologram amplitude reconstruction

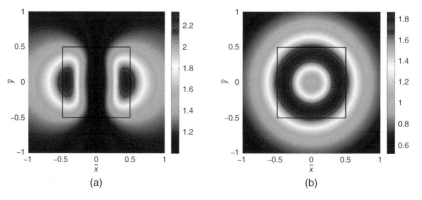

Plate 4 (Figure 2.10)   (From left to right) Mannequin image with irradiated area in red; mannequin hologram amplitude reconstruction at different scanning time and superposition of the most significant frames

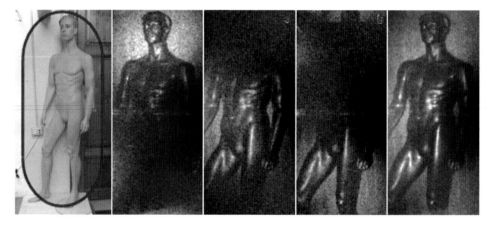

(a)   (b)

Plate 5 (Figure 3.10)   Single point resolution in a transversal plane (from Fournier *et al.* 2010): (a) x-resolution map normalized by the value of x-resolution on the optical axis; (b) normalized z-resolution map. The squares in the center of figures (a) and (b) represent the sensor boundaries. *Source*: Fournier C., Denis L., and Fournel T., 2010. Reproduced with permission from the Optical Society

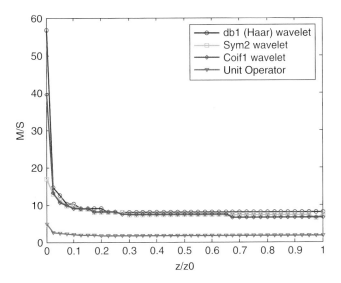

Plate 6 (Figure 4.6) Simulation results showing the normalized compressive sampling ratio for different sparsifying bases [26]. *Source*: Y. Rivenson, A. Stern, and B. Javidi 2013. Reproduced with permission from The Optical Society

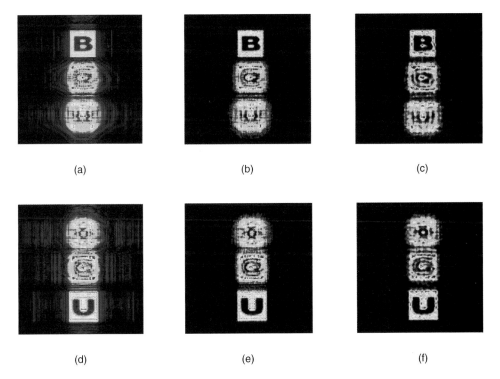

Plate 7 (Figure 4.9) Reconstruction examples of the B (forward) and U (backward) planes. (a) Reconstruction of the B plane form 100% of the projections. (b) CS reconstruction of the B plane forms 6% of the projections. (c) CS reconstruction of the B plane forms 2.5% of the projections. (d) Reconstruction of the U plane forms 100% of the projections. (e) CS reconstruction of the U plane forms 6% of the projections. (f) CS reconstruction of the U plane forms 2.5% of the projections

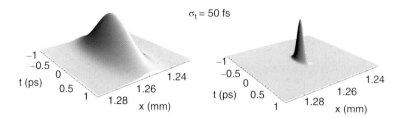

Plate 8 (Figure 5.6) Spatiotemporal profiles of the fifth diffraction order of a 100 lines/inch diffractive grating for an input pulse width of $\sigma_t = 50\,\text{fs}$. The right part of the figure is obtained after focusing with an achromatic lens doublet and the left part by focusing with the DOE-based system. Note that the time origin is chosen arbitrarily. *Source*: Mínguez-Vega, G., Tajahuerce, E., Fernández-Alonso, M., Climent, V., Lancis, J., Caraquitena, J., Andrés, P., (2007). Figure 6. Reproduced with permission from The Optical Society

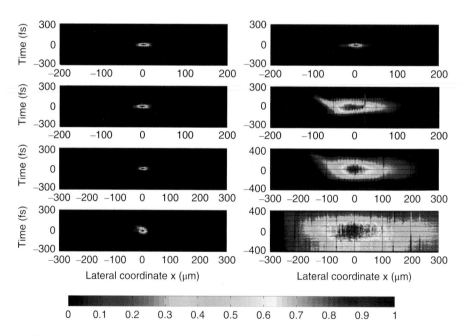

Plate 9 (Figure 5.9) Normalized spatiotemporal light intensity after low NA focusing of the beamlets coming from a diffractive grating (0th, +1st, +2nd, and +3rd diffraction orders from *top to bottom*) with (*left column*) and without (*right column*) DCM. Measurements were captured using STARFISH [18]. The maximum frequency component, for the third diffraction order, is 35.4 lp/mm. *Source*: Martínez-Cuenca, R., Mendoza-Yero, O., Alonso, B., Sola, Í. J., Mínguez-Vega, G., Lancis, J. (2012). Figure 2. Reproduced with permission from The Optical Society

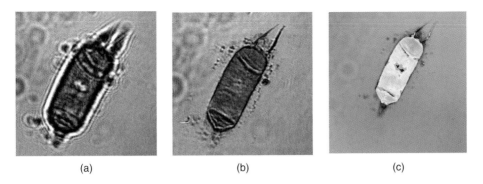

(a)   (b)   (c)

Plate 10 (Figure 6.6) (a) Out of focus color intensity of the alga *Odontella* sp., (b) refocused intensity, (c) composite phase image of the RGB channels

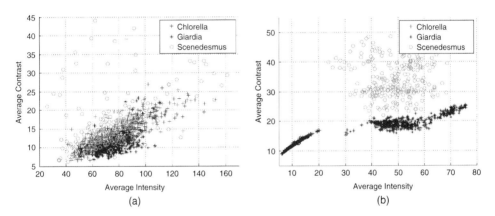

(a)   (b)

Plate 11 (Figure 6.13) Feature space representation using (a) intensity information of the detected particles of the three species, and (b) compensated phase of detected particles of the three species

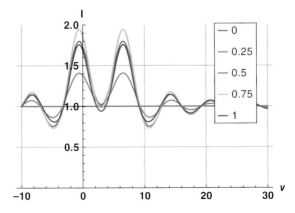

Plate 12 (Figure 7.14) Imaging of two neighboring point-like phase defects using VDIC with different parameters

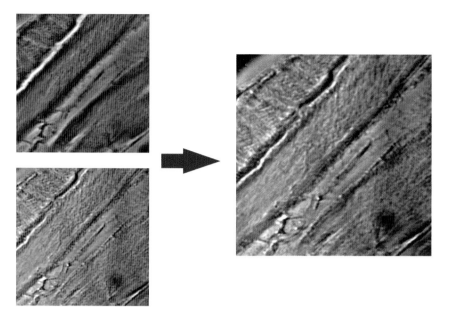

Plate 13 (Figure 7.15) Combination of an image obtained using Zernike phase contrast (hue) and one obtained using DIC (intensity) in order to visualize a phase structure

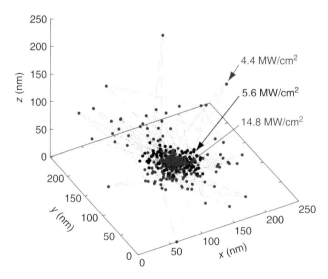

Plate 14 (Figure 8.7) Movement of a 200 nm polystyrene particle in the trapping volume at different intensities: (blue) 4.4 MW/cm^2, (black) 5.6 MW/cm^2, and (red) 14.8 MW/cm^2

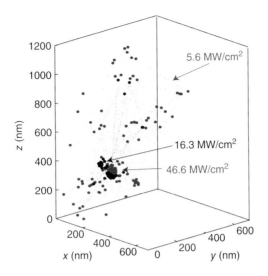

Plate 15 (Figure 8.10)   Three-dimensional display of the gold nanoparticle movements

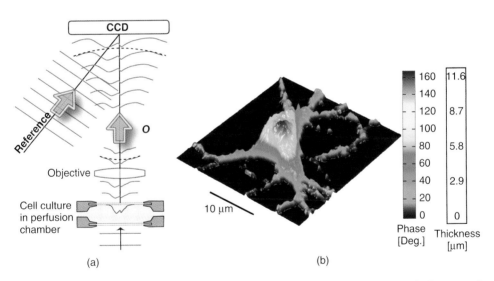

(a)   (b)

Plate 16 (Figure 9.2)   Digital holographic microscopy (DHM) of living mouse cortical neurons in culture. (a) Schematic representation of cultured cells mounted in a closed perfusion chamber and trans-illuminated (b) 3D perspective image in false colors of a living neuron in culture. Each pixel represents a quantitative measurement of the phase retardation or cellular optical path length (OPL) induced by the cell with a sensitivity corresponding to a few tens of nanometers. By using the measured mean value of the neuronal cell body refractive index, resulting from the decoupling procedure, scales (*right*), which relate OPL (°) to morphology in the z-axis (µm), can be constructed

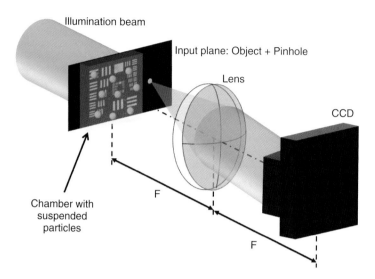

Plate 17 (Figure 10.4) Digital holography super-resolution setup. *Source*: Zalevsky Z., Gur E., Garcia J., Micó V., Javidi B. 2012. Figure 1. Reproduced with permission from The Optical Society

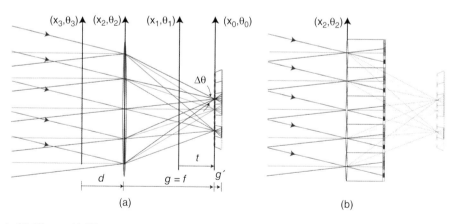

Plate 18 (Figure 11.17) (a) Scheme for the calculation of the plenoptic function in planes parallel to the MLA; (b) the plenoptic field as evaluated in the plane of the camera lens is equivalent to the one captured with an IP setup

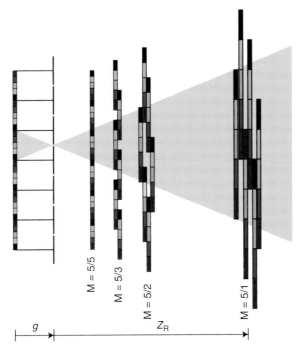

Plate 19 (Figure 11.20) Scheme of the conventional reconstruction algorithm. In this figure, the number of pixels per microlens is $N = 5$

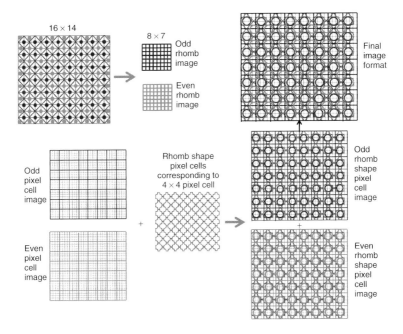

Plate 20 (Figure 12.12) The image format of the rhombus cell corresponding to a pixel cell with 4×4 pixels

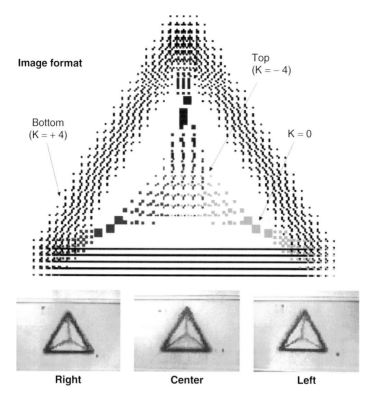

Plate 21 (Figure 12.16) Image format of a triangular pyramid

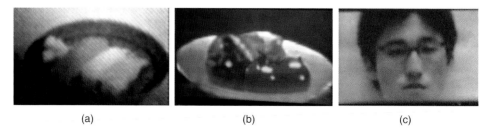

(b)

Plate 22 (Figure 13.8) Reconstruction of FP-HS. (b) Example of the reconstructed image from full-color FP-HS hologram [20]. *Source*: T. Utsugi and M. Yamaguchi 2013. Reproduced with permission from The Optical Society

(a)   (b)   (c)

Plate 23 (Figure 13.23) Reconstructed images from the FP HS recorded from the captured light-field data. (a) Sushi, (b) vegetables, and (c) human face

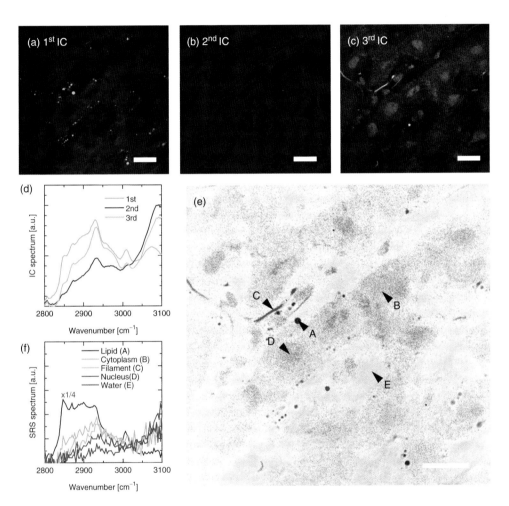

Plate 24 (Figure 15.12) Spectral imaging of a rat liver tissue [28]. 91 images at wavenumbers from 2800–3100 cm⁻¹ were taken and averaged over 10 times. The total acquisition time was <30 s. The spectral images were analyzed by using 5 ICs. (a) First IC image reflecting the distribution of lipid-rich region. (b) Second IC image reflecting the distribution of water-rich regions. (c) Third IC image reflecting the distribution of protein-rich region. (d) IC spectra. (e) Multicolor image produced by combining images (a)–(c) and inverting the contrast. (a)–(e) are explained in the text. (f) SRS spectra in locations indicated by arrows in (e). Scale bar: 20 μm. *Source*: Y. Ozeki, W. Umemura, Y. Otsuka, S. Satoh, H. Hashimoto, K. Sumimura, N. Nishizawa, K. Fukui, and K. Itoh 2012. Reproduced with permission from Nature Publishing

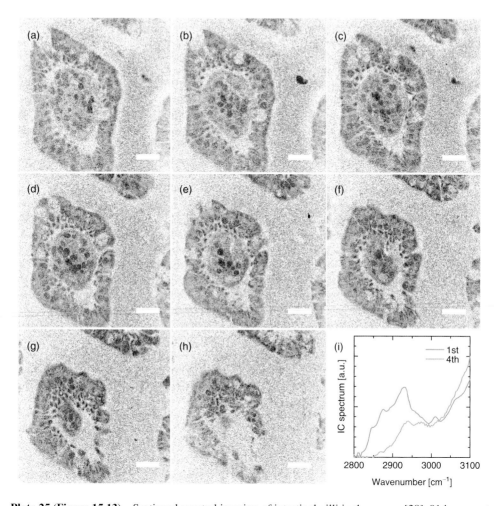

Plate 25 (Figure 15.13)   Sectioned spectral imaging of intestinal villi in the mouse [28]. 91 images at wavenumbers from 2800–3100 cm^{-1} were taken by changing the z position by 5.6 μm. The total acquisition time was 24 s. The spectral images were analyzed by using 4 ICs. The first IC (cytoplasm) and the fourth IC (nuclei) images were colored cyan and yellow, respectively, and then combined and the contrast was inverted. (a–h). Sectioned multicolor images. (f). Spectra of the first and fourth ICs. Scale bar: 20 μm. *Source*: Y. Ozeki, W. Umemura, Y. Otsuka, S. Satoh, H. Hashimoto, K. Sumimura, N. Nishizawa, K. Fukui, and K. Itoh 2012. Reproduced with permission from Nature Publishing

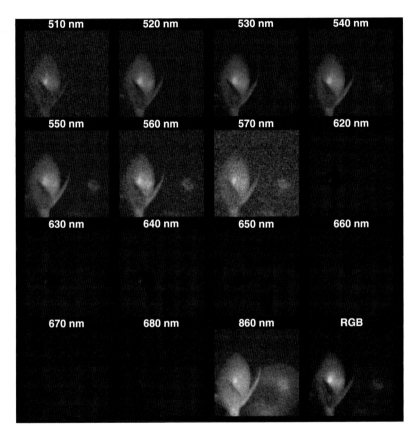

Plate 26 (Figure 16.5)   Multispectral data cube reconstructed using CS. In the VIS band, the reflectance for each spectral channel is represented by means of a 256 × 256 pseudo-color image. In the NIR band we show a gray-scale representation. A colorful image of the scene made up from the conventional RGB channels is also included. *Source*: F. Soldevila, E. Irles, V. Durán, P. Clemente, M. Fernández-Alonso, E. Tajahuerce, and J. Lancis 2013, Figure 4. Reproduced with permission from Springer

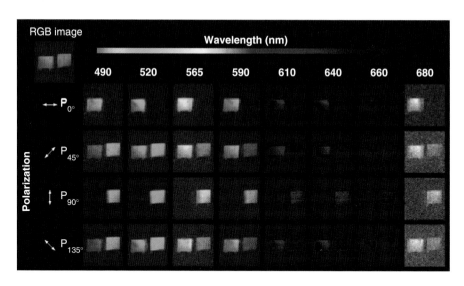

Plate 27 (Figure 16.8) Multispectral image cube reconstructed by CS algorithm for four different configurations of the polarization analyzer. The RGB image of the object is also included. In the VIS spectrum all channels are represented by pseudo-color images and a gray-scale representation is used for the wavelength closer to the NIR spectrum. *Source*: F. Soldevila, E. Irles, V. Durán, P. Clemente, M. Fernández-Alonso, E. Tajahuerce, and J. Lancis 2013, Figure 5. Reproduced with permission from Springer

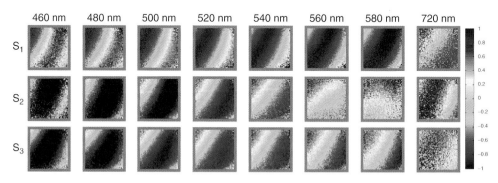

Plate 28 (Figure 16.11) Spatial distribution of the Stokes parameters of the polystyrene piece. Each distribution is represented by a pseudo-colored 128×128 pixels picture. The values range from -1 (blue) to 1 (red)

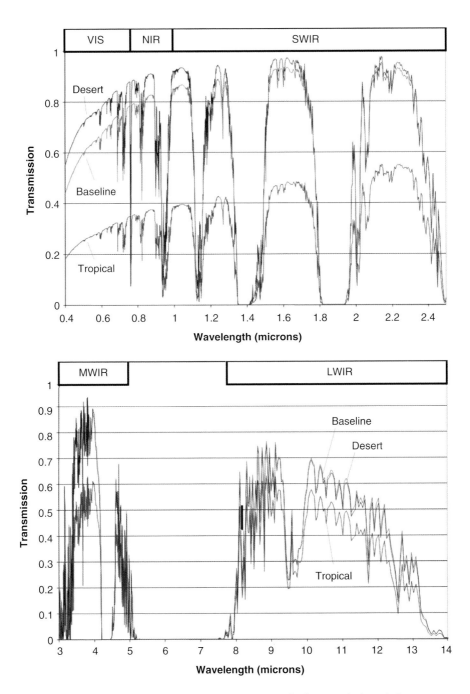

Plate 29 (Figure 17.11)   Ground to space atmospheric transmission windows

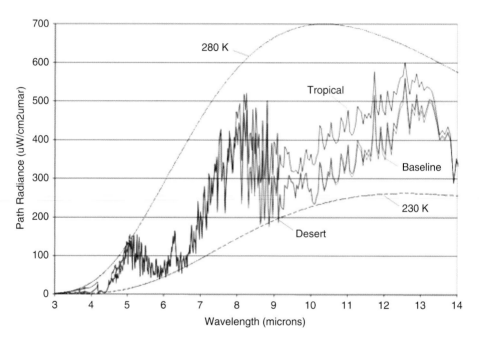

Plate 30 (Figure 17.15) Examples of path emission along a ground to space line of sight as a function of range and atmospheric conditions

movement, precludes the possibility of exploring dynamic cellular processes. Recently, spectroscopy phase microscopy approaches [134–136] have addressed this deficiency, at least as far as cells with high intrinsic dispersion properties are considered, including RBCs (owing to presence of the hemoglobin pigment). However, such spectroscopy approaches are applicable only to a very limited variety of cell types, most of which have intrinsic dispersions almost identical to that of water. We have developed another approach, called the decoupling procedure, to separately measure the parameters $\bar{n}_c$ and d from the phase signal Φ, based on a modification of the extracellular RI n_m. Basically, this method consists of performing a slight alteration of the extracellular RI n_m and recording two holograms that correspond to the two different values of n_m, thereby allowing reconstruction of two quantitative phase images (Φ_1 and Φ_2) described, for each pixel i, by the following system of equations:

$$\Phi_{1,i} = \frac{2\pi}{\lambda_1}(\bar{n}_{c,i} - n_{m,1})d_i \tag{9.8}$$

$$\Phi_{2,i} = \frac{2\pi}{\lambda_2}(\bar{n}_{c,i} - n_{m,2})d_i, \tag{9.9}$$

where λ_1 and λ_2 are the wavelengths of the light source.

By solving this system of equations, we obtain $\bar{n}_{c,i}$ and d_i for each pixel i. We have considered two different approaches to modify n_m: The first approach requires sequentially perfusing a standard cell perfusion solution and a second solution with a different RI but with the same osmolarity (to avoid cell volume variation) to record the two corresponding holograms at a single wavelength ($\lambda_1 = \lambda_2$) [39]. Practically, this procedure has permitted us to quantitatively measure some highly relevant RBC parameters, including mean corpuscular volume and mean corpuscular hemoglobin concentration [157]. However, owing to the solution exchange time, this approach precludes the possibility to monitor dynamic changes of cell morphology and RI that occur during fast biological processes. To overcome these drawbacks, we have developed a second approach, dual-wavelength ($\lambda_1 \neq \lambda_2$) DHM (DW-DHM) [132], which exploits the dispersion of the extracellular medium that is enhanced by the use of an extracellular dye ($n_{m,1} = n_m(\lambda_1) \neq n_m(\lambda_2) = n_{m,2}$) to achieve separate measurements of the intracellular RI and the absolute cell volume in real time [168].

This approach has been successfully applied to study the osmotic water membrane permeability P_f – representing the water volume flux per unit of time per unit of membrane surface for a given applied osmotic gradient – by monitoring cell volume changes while retaining the cell functionality [169]. Table 9.1 contains P_f measurements based on the monitoring of absolute cell volume in a variety of cell types.

Water crosses membrane though several routes (simple diffusion through the lipid bilayer, transmembrane proteins, specialized water channels, aquaporins, AQP, etc.). As regards the results of Table 9.1, the high membrane water permeability of human RBCs is mainly determined by the endogenously expressed water channel AQP1. However, the RBC provides an example where the water transport mechanism can easily be altered by pharmacological inhibition of AQP1 with mercury chloride $HgCl2$, a potent, rapid inhibitor of AQP1. The high water permeability in astrocytes is usually attributed to the endogenously expressed water channel aquaporin AQP4 and it has been shown that astrocytes from AQP4 deficient mice show a seven-fold reduced osmotic water permeability. The quite high water permeability measured in primary cultures of neurons could result from an increased expression of specific solute transporters that exhibit an intrinsic water permeability.

Table 9.1 Osmotic challenge: Measured parameters in different cell types. CHO cells (n = 46), HEK cells (n = 14), neurons (n = 11) astrocytes (n = 19) and red blood cells (RBCs) (n = 22). P_f: osmotic water membrane permeability , V_0 : initial cell volume, A_0: initial cell surface, Π_i / Π_0 : The respective ratio of the initial osmolarity to the hypoosmotic osmolarity. Adapted from [170]

	V_0 [μm^3]	Π_i / Π_0	A_0 [μm^2]	P_f [10^{-3} cm s^{-1}]
CHOs	1160 ± 669	1.452	747 ± 340	165 ± 0.318
HEKs	1996 ± 562	1.667	676 ± 153	3.04 ± 0.87
Neurons	1671 ± 1162	1.452	475 ± 226	4.69 ± 2.89
Astrocytes	861 ± 324	1.452	475 ± 137	7.64 ± 3.54
RBCs	95 ± 48	1.553	129 ± 44	5.2 ± 2.9
RBCs/HgCl$_2$	101 ± 49	1.553	135 ± 45	1.5 ± 1.2

Otherwise, resulting from the linear relationship between the intracellular RI and the dry mass, as well as from a dry mass balance equation, it has been possible to determine the RI of the transmembrane flux n_f [170]. The high precision with which n_f can be determined provides us whether the transmembrane water flux is accompanied by solute transport and therefore leads to characterization of the involved transmembrane transport process. A sustained application of glutamate on astrocytes has permitted us to detect a dry mass accumulation in good agreement with the expected uptake of glutamate through a specific co-transport.

Consequently, P_f, n_f, as well as absolute cell volume, represent indices providing some highly relevant information about the different mechanisms involved in the transmembrane water movements and in the complex question of cell volume regulation.

9.4.5 Exploration of Neuronal Cell Dynamics

A distinct feature of the nervous tissue is the intricate network of synaptic connections among neurons of diverse phenotypes. Although initial connections are formed largely through molecular mechanisms there is little doubt that electrical activities, influencing neuronal function and connectivity on multiple time scales, as well as on multiple levels of specificity, play a pivotal role in the development and integrative functions of neural networks/circuits. Nevertheless, how electrical activity affects the structure and function of neuronal networks is very limited. Consequently, the development of techniques that allow the noninvasive resolution of local neuronal network activity is required.

With respect to the study of neuronal activity electrophysiological approaches, in particular, voltage clamp and patch-clamp techniques, have permitted a major breakthrough by setting a voltage across the neuronal membrane and directly measuring the current flowing through a single ion channel. The patch clamp is the gold standard for assessing ion channel function, allowing discrimination of ionic currents in the femtoampere (10^{-15} A) range and with microsecond time resolution.

However, the patch clamp is still a highly invasive, laborious process requiring precise micromanipulations and a high degree of operational skill, which generally restricts voltage recording to a limited number of cells that form a neuronal network. Optical techniques seem to be an ideal solution for measuring membrane potentials because these techniques are relatively noninvasive and work at both low and high magnification. For instance, a calcium indicator used in combination with high-resolution 2P microscopy allows measurement of the spiking activity of hundreds, or even several thousand, neurons in mammalian circuits while still keeping track of the activity of each neuron individually [171,172]. However, calcium imaging has its shortcomings and cannot substitute for voltage imaging [172].

Practically, voltage imaging methods have lagged behind calcium imaging as a result of important challenges related to physical constraints of the measurement themselves, including an electrical field closely located to the thin membrane region, an essentially two-dimensional plasma membrane that cannot contain an arbitrary number of voltage chromophores without disrupting its properties, and a plasma membrane representing only a small proportion of the total membrane surface in the neuron on which chromophores are attached. Finally, the relatively high speed of the electrical responses of mammalian neurons is also a serious challenge for voltage measurements.

Consequently, despite some promising features, the different voltage imaging methods suffer from poor signal-to-noise ratios and secondary side effects, and they fall short of providing single-cell resolution when imaging the activity of neuronal populations [172].

9.4.5.1 Measurement of Transmembrane Water Movements to Resolve Neuronal Network Activity

It is well known that neuronal activity induces modifications of intrinsic optical properties at the subcellular [173–176] cellular [177,178] and tissue levels [179–181]. Thus, we have studied the early stage of neuronal responses, induced by glutamate (the main excitatory neurotransmitter in the brain that is released in 80% of synapses), with multimodality microscopes, combining either DH-QPM with epifluorescence microscopy [182] or DH-QPM with electrophysiology [41], thereby allowing simultaneous monitoring of the DHM quantitative phase signal and the dynamics of intracellular ionic homeostasis or transmembrane ionic currents. Practically, experiments combining electrophysiology and DH-QPM, performed in a well-established biological model, allowed us to accurately study the relationship between transmembrane ionic currents and water movements. Concretely a mathematical relationship, involving some cell morphology parameters and a parameter ε (ml C^{-1}) representing the volume of the water movement associated with the net charge transported through the cell membrane, has been established between net transmembrane currents and the phase signals, thus opening the possibility of performing simultaneous multiple-site optical recording of transmembrane currents with DH-QPM [41].

Glutamate is known to induce an early neuronal swelling and to activate specific ionotropic receptors, namely the *N*-methyl-D-aspartate (NMDA), the 2-amino-3-(3-hydroxy-5-methyl-isoxazol-4-yl) propionate (AMPA), and the kainate receptors, which induce the opening of their associated ion channels, allowing influxes of Ca^{2+} and Na^+ down their electrochemical gradient. Concretely, following various applications of glutamate in primary cultures of mouse cortical neurons, the multimodality DH-QPM electrophysiology setup recorded both a strong inward current and a decrease in phase signal, the amplitude of which was proportional to the concentration of glutamate and to the duration of the application. This inward current is consistent with glutamate-mediated activation of ionotropic receptors, and the phase decrease results from the dilution of the intracellular RI due to the osmotic water entry accompanying the net ionic influx, leading to the expected neuronal swelling. The receptor-specific

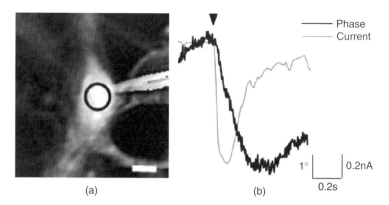

Figure 9.3 *Phase shift associated with glutamate-mediated neuronal activity.* (a) Quantative phase image of a patched mouse cortical neuron in primary culture recorded by DHM. The full circles in the middle of the cell correspond to the region of interest where the phase signal is recorded (scale bar: 10 μm). (B) Local application of glutamate (500 μM, 200 ms; arrow head) on the neuron triggered both a strong transient decrease of the phase signal associated to an inward current. Phase is expressed in degrees. *Source*: Marquet, P., Depeursinge, C., Magistretti, P.J., 2013, Neural cell dynamics explored with digital holographic microscopy, Annual Review of Biomedical Engineering, (15), 407–431

nature of these phase signals has been demonstrated by their suppression in the presence of MK801 and CNQX, two specific antagonists of NMDA and AMPA/kainate ionotropic receptors, respectively. Consequently these experiments have revealed that transmembrane water movements are one of the mechanisms inducing activity-related modifications of the intrinsic optical properties of neuronal cells. Actually, DH-QPM showed that glutamate produces mainly three distinct optical responses in neurons, which are predominantly mediated by NMDA receptors: biphasic, reversible decrease (RD), and irreversible decrease (ID) responses. The shape and amplitude of the optical signal were not associated with a particular neuronal phenotype but reflected the physiopathological status of neurons linked to the degree of NMDA activity [40]. These different phase responses can be separated into two components: a rapid one accompanying glutamate-mediated current (I_{GLUT}), the phase decrease depicted in Fig. 9.3, and a slow one corresponding generally to a phase recovery or a phase plateau (ID responses) when $I_{GLUT} = 0$. The phase recovery, which is much slower than the fast component, is likely to correspond to a nonelectrogenic neuronal volume regulation involving several mechanisms. Actually, furosemide and bumetanide, two inhibitors of sodium-coupled and/or potassium-coupled chloride movements, strongly modified the phase recovery, suggesting an involvement of two neuronal cotransporters, NKCC1 (Na-K-Cl) and KCC2 (K-Cl), in the optical response. This observation is of particular interest because it shows that DH-QPM is the first imaging technique able to monitor dynamically and in situ the activity of these nonelectrogenic co-transporters under physiological and/or pathological neuronal conditions [40]. Interestingly, the time course presented in Fig. 9.3 shows that the water movements are not significantly delayed, at the tenth-of-a-second scale at least, relative to the recorded current.

The measurement of I_{GLUT} as well as the concomitant rapid phase decrease corresponding to the early neuronal swelling allowed the estimation of the parameter ε_{GLUT} (ml C^{-1}). Practically, values of ε_{GLUT} lie within the range of 60–120 μm^3 nC^{-1}, equivalent to 340–620 water molecules transported per ion having crossed the membrane. The typical intracelluar RI change induced by a glutamate pulse (500 uM, 0.2 s) is around 0.002–0.003 corresponding to a substantial variation (7–10%) of the scattering potential proportional to the parameter $(\bar{n}_c^2 - n_m^2)$ and from which the scattering coefficient can be evaluated. The associated neuronal swelling is around 100 fL for a typical neuronal cell body of 1500 fL, corresponding to a 6–7% cell volume variation. It should be mentioned that these orders of magnitude correspond to an exogenous glutamate application during a few tenths of a second. Physiological release of endogenous glutamate could possibly induce smaller water movements and intracellular RI changes. Moreover, the mechanisms underlying these early significant and rapid fluxes of water, in contrast to those occurring during the phase recovery, remain to be clarified.

9.5   Future Issues

Holography has found an attractive field of application in microscopy. Because of its flexibility, DHM could easily be adapted to the need of microscopists and biologists. Complex wavefront reconstruction yields quantitative phase and accuracies in the nanometer, or even in the subnanometer range, with some statistical treatment, can be obtained. A major advantage of the holographic approach is that in-depth reconstruction and imaging is possible from a single hologram that can be acquired in a very short time interval.

As illustrated through these various applications, DH-QPM is well adapted to quantitatively study cell structure and dynamics. In practice, low energy levels and short exposure times are required to prevent photodamage, even during long experiments. In addition, DHM, as a label-free technique, does not require any solution change or insertion of dye, providing efficient conditions for high-throughput screening. However, although the quantitative phase signal provides unique information with high sensitivity about cell morphology and content, its interpretation in terms of biophysical parameters for analyzing specific biological mechanisms remains a major issue. Consequently, as has been illustrated with the decoupling procedures, any developments allowing us to separately measure the information corresponding to cell morphology and to cell content from the phase signal will represent an important step toward addressing some relevant biological questions. Within this framework, the development of a real time DHM-based optical diffraction tomography, by enabling both a direct access to the 3D RI map and to synthesize an enlarged numerical aperture (NA) providing super-resolved phase images represents a very promising way forward [183–188]. Indeed, future developments allowing high-resolution 3D mapping of the intracellular RI, could provide invaluable information about cytoarchitecture and compartmentalization of cytoplasm, which plays a critical role in several fundamental cell mechanisms, including protein synthesis.

On the other hand, integrating DH-QPM into a multimodal microscope is able to provide various different types of information on the cell state, also represent a promising way to obtain a comprehensive understanding of cell structure and dynamics. For instance multimodality approaches, combining DH-QPM with epifluorescence microscopy and/or with electrophysiology recordings have permitted us to efficiently explore the mechanisms underlying the complex processes of cell volume regulation, ion homeostasis, and transmembrane water movements, as well as their involvements in neuronal activity.

Acknowledgments

This article is the result of a close collaboration between the Microvision and Microdiagnosis Group (SCI/STI/CHD group) of EPFL with Florian Charrière, Jonas Kühn, Nicolas Pavillon, Etienne Shafer, and Yann Cotte; the Laboratory of Neuroenergetics and Cellular Dynamics at Brain and Mind Institute of EPFL with Benjamin Rappaz and Pascal Jourdain; the Center for Psychiatric Neuroscience at CHUV with Daniel Boss; and Lyncée Tec SA (www.lynceetec.com) with Etienne Cuche, Tristan Colomb, Frederic Montfort, Nicolas Aspert, and Yves Emery. We thank the Swiss National Science Foundation (SNSF grant CR3213_132993) for its support.

References

[1] Zernike F. 1942. Phase contrast, a new method for the microscopic observation of transparent objects. *Physica* **9**:686–6898.

[2] Barer R. 1952. Interference microscopy and mass determination. *Nature (London)* **169**:366–367.

[3] Dunn GA, Zicha D. 1993. Phase-shifting interference microscopy applied to the analysis of cell behavior. *Sym Soc Exp Biol* **47**:91–106.

[4] Gabor D. 1948. A new microscopic principle. *Nature* **161**:777–778.

[5] Conchello JA, Lichtman JW. 2005. Optical sectioning microscopy. *Nat Methods* **2**:920–931.

[6] Giepmans BN, Adams SR, Ellisman MH, Tsien RY. 2006. The fluorescent toolbox for assessing protein location and function. *Science* **312**:217–224.

[7] Campagnola P. 2011. Second harmonic generation imaging microscopy: applications to diseases diagnostics. *Anal Chem* **83**:3224–3231.

[8] Fujita K, Smith NI. 2008. Label-free molecular imaging of living cells. *Molecules and cells* **26**:530-5.

[9] Schermelleh L, Heintzmann R, Leonhardt H. 2010 A guide to super-resolution fluorescence microscopy. *J Cell Biol* **190**:165–175.

[10] Gould TJ, Hess ST, Bewersdorf J. 2012 Optical nanoscopy: from acquisition to analysis. *Annu Rev Biomed Eng* **14**:231–254.

[11] Indebetouw G, Klysubun P. 2001. Spatiotemporal digital microholography. *J Opt Soc Am A* **18**:319–325.

[12] Klysubun P, Indebetouw G. 2001. A posteriori processing of spatiotemporal digital microholograms. *J Opt Soc Am A* **18**:326–331.

[13] Cuche E, Marquet P, Depeursinge C. 1999. Simultaneous amplitude-contrast and quantitative phase-contrast microscopy by numerical reconstruction of Fresnel off-axis holograms. *Appl Optics* **38**:6994–7001.

[14] Yamaguchi I, Zhang T. 1997. Phase-shifting digital holography. *Opt Lett* **22**:1268–1270.

[15] Marquet P, Rappaz B, Magistretti PJ, Cuche E, Emery Y, *et al.* 2005. Digital holographic microscopy: a noninvasive contrast imaging technique allowing quantitative visualization of living cells with subwavelength axial accuracy. *Opt Lett* **30**:468–470.

[16] Mihailescu M, Scarlat M, Gheorghiu A, Costescu J, Kusko M, *et al.* 2011. Automated imaging, identification, and counting of similar cells from digital hologram reconstructions. *Appl Optics* **50**:3589–3597.

[17] Molder A, Sebesta M, Gustafsson M, Gisselson L, Wingren AG, Alm K. 2008. Non-invasive, label-free cell counting and quantitative analysis of adherent cells using digital holography. *Journal of Microscopy* **232**:240–247.

[18] DaneshPanah M, Zwick S, Schaal F, Warber M, Javidi B, Osten W. 2010. 3D holographic imaging and trapping for non-invasive cell identification and tracking. *J Disp Technol* **6**:490–499.

[19] Moon I, Javidi B, Yi F, Boss D, Marquet P. 2012. Automated statistical quantification of three-dimensional morphology and mean corpuscular hemoglobin of multiple red blood cells. *Opt Express* **20**:10295–10309.

[20] Crha I, Zakova J, Huser M, Ventruba P, Lousova E, Pohanka M. 2011. Digital holographic microscopy in human sperm imaging. *J Assist Reprod Gen* **28**:725–729.

[21] Liu R, Dey DK, Boss D, Marquet P, Javidi B. 2011. Recognition and classification of red blood cells using digital holographic microscopy and data clustering with discriminant analysis. *J Opt Soc Am A* **28**:1204–1210.

[22] Memmolo P, Di Caprio G, Distante C, Paturzo M, Puglisi R, *et al.* 2011. Identification of bovine sperm head for morphometry analysis in quantitative phase-contrast holographic microscopy. *Opt Express* **19**:23215-26.

[23] Janeckova H, Vesely P, Chmelik R. 2009. Proving tumour cells by acute nutritional/energy deprivation as a survival threat: a task for microscopy. *Anticancer Res* **29**:2339–2345.

[24] Mann CJ, Yu LF, Lo CM, Kim MK. 2005. High-resolution quantitative phase-contrast microscopy by digital holography. *Opt Express* **13**:8693–8698.

[25] Choi YS, Lee SJ. 2009. Three-dimensional volumetric measurement of red blood cell motion using digital holographic microscopy. *Appl Optics* **48**:2983–2990.

[26] Langehanenberg P, Ivanova L, Bernhardt I, Ketelhut S, Vollmer A, *et al.* 2009. Automated three-dimensional tracking of living cells by digital holographic microscopy. *J Biomed Opt* **14**:014018.

[27] Toy MF, Richard S, Kuhn J, Franco-Obregon A, Egli M, Depeursinge C. 2012. Digital holographic microscopy for the cytomorphological imaging of cells under zero gravity. *Three-Dimensional and Multidimensional Microscopy: Image Acquisition and Processing XIX*, p. 8227.

[28] Antkowiak M, Callens N, Yourassowsky C, Dubois F. 2008. Extended focused imaging of a microparticle field with digital holographic microscopy. *Opt Lett* **33**:1626–1628.

[29] Colomb T, Pavillon N, Kuhn J, Cuche E, Depeursinge C, Emery Y. 2010. Extended depth-of-focus by digital holographic microscopy. *Opt Lett* **35**:1840–1842.

[30] McElhinney C, Bryan Hennelly, Naughton T. 2008. Extended focused imaging for digital holograms of macroscopic three-dimensional objects. *Appl Optics* **47**:D71–D79.

[31] Warnasooriya N, Joud F, Bun P, Tessier G, Coppey-Moisan M, *et al.* 2010. Imaging gold nanoparticles in living cell environments using heterodyne digital holographic microscopy. *Opt Express* **18**:3264–3273.

[32] Bohm M, Mastrofrancesco A, Weiss N, Kemper B, von Bally G, *et al.* 2011. PACE4, a member of the prohormone convertase family, mediates increased proliferation, migration and invasiveness of melanoma cells in vitro and enhanced subcutaneous tumor growth in vivo. *J Invest Dermatol* **131**:S108.

[33] Dubois F, Yourassowsky C, Monnom O, Legros JC, Debeir O, *et al.* 2006. Digital holographic microscopy for the three-dimensional dynamic analysis of in vitro cancer cell migration. *J Biomed Opt* **11**:054032 (5 pages).

[34] Sun H, Song B, Dong H, Reid B, Player MA, *et al.* 2008. Visualization of fast-moving cells in vivo using digital holographic video microscopy. *J Biomed Opt* **13**:014007–014009.

[35] DaneshPanah M, Javidi B. 2007. Tracking biological microorganisms in sequence of 3D holographic microscopy images. *Opt Express* **15**:10761–10766.

[36] Rappaz B, Cano E, Colomb T, Kuhn J, Depeursinge C, *et al.* 2009. Noninvasive characterization of the fission yeast cell cycle by monitoring dry mass with digital holographic microscopy. *J Biomed Opt* **14**:034049.

[37] Rappaz B, Barbul A, Hoffmann A, Boss D, Korenstein R, *et al.* 2009. Spatial analysis of erythrocyte membrane fluctuations by digital holographic microscopy. *Blood Cell Mol Dis* **42**:228–232.

[38] Cardenas N, Yu LF, Mohanty SK. 2011. Stretching of red blood cells by optical tweezers quantified by digital holographic microscopy. *Optical Interactions with Tissue and Cells XXII* p. 7897.

[39] Rappaz B, Marquet P, Cuche E, Emery Y, Depeursinge C, Magistretti PJ. 2005. Measurement of the integral refractive index and dynamic cell morphometry of living cells with digital holographic microscopy. *Opt Express* **13**:9361–9373.

[40] Jourdain P, Pavillon N, Moratal C, Boss D, Rappaz B, *et al.* 2011. Determination of transmembrane water fluxes in neurons elicited by glutamate ionotropic receptors and by the cotransporters KCC2 and NKCC1: A digital holographic microscopy study. *J Neurosci* **31**:11846–11854.

[41] Jourdain P, Boss D, Rappaz B, Moratal C, Hernandez MC, *et al.* 2012. Simultaneous optical recording in multiple cells by digital holographic microscopy of chloride current associated to activation of the ligand-gated chloride channel GABA(A) receptor. *PloS One* **7**:e51041.

[42] Popescu G, Ikeda T, Best CA, Badizadegan K, Dasari RR, Feld MS. 2005. Erythrocyte structure and dynamics quantified by Hilbert phase microscopy. *J Biomed Opt* **10**:060503.

[43] Veselov O, Lekki J, Polak W, Strivay D, Stachura Z, *et al.* 2005. The recognition of biological cells utilizing quantitative phase microscopy system. *Nucl Instrum Meth B* **231**:212–217.

[44] Indebetouw G, Tada Y, Leacock J. 2006. Quantitative phase imaging with scanning holographic microscopy: an experimental assessment. *Biomed Eng Online* **5**.

[45] Amin MS, Park Y, Lue N, Dasari RR, Badizadegan K, *et al.* 2007. Microrheology of red blood cell membranes using dynamic scattering microscopy. *Opt Express* **15**:17001–17009.

[46] Fang-Yen C, Oh S, Park Y, Choi W, Song S, *et al.* 2007. Imaging voltage-dependent cell motions with heterodyne Mach–Zehnder phase microscopy. *Opt Lett* **32**:1572–1574.

[47] Park Y, Popescu G, Badizadegan K, Dasari RR, Feld MS. 2007. Fresnel particle tracing in three dimensions using diffraction phase microscopy. *Opt Lett* **32**:811–813.

[48] Whelan MP, Lakestani F, Rembges D, Sacco MG. 2007. Heterodyne interference microscopy for non-invasive cell morphometry – art. no. 66310E. *Novel Optical Instrumentation for Biomedical Applications III* **6631**:E6310.

[49] Brazhe AR, Brazhe NA, Rodionova NN, Yusipovich AI, Ignatyev PS, *et al.* 2008. Non-invasive study of nerve fibres using laser interference microscopy. *Philos T R Soc A* **366**:3463–3481.

[50] Lue N, Choi W, Badizadegan K, Dasari RR, Feld MS, Popescu G. 2008. Confocal diffraction phase microscopy of live cells. *Opt Lett* **33**:2074–2076.

[51] Pavani SRP, Libertun AR, King SV, Cogswell CJ. 2008. Quantitative structured-illumination phase microscopy. *Appl Optics* **47**:15–24.

[52] Popescu G, Park Y, Choi W, Dasari RR, Feld MS, Badizadegan K. 2008. Imaging red blood cell dynamics by quantitative phase microscopy. *Blood Cell Mol Dis* **41**:10–16.

[53] Tychinsky VP, Kretushev AV, Klemyashov IV, Vyshenskaya TV, Filippova NA, *et al.* 2008. Quantitative real-time analysis of nucleolar stress by coherent phase microscopy. *J Biomed Opt* **13**:064032.

[54] Warger WC, DiMarzio CA. 2008. Modeling of optical quadrature microscopy for imaging mouse embryos – art. no. 68610T. In *Three-Dimensional and Multidimensional Microscopy: Image Acquisition and Processing XV*, eds JA Conchello, CJ Cogswell, T Wilson, TG Brown, **6861**:T8610-T.

[55] Lee S, Lee JY, Yang W, Kim DY. 2009. The measurement of red blood cell volume change induced by Ca(2+) based on full field quantitative phase microscopy. *Imaging, Manipulation, and Analysis of Biomolecules, Cells, and Tissues VII*, p. 7182.

[56] Park Y, Yamauchi T, Choi W, Dasari R, Feld MS. 2009. Spectroscopic phase microscopy for quantifying hemoglobin concentrations in intact red blood cells. *Opt Lett* **34**:3668–3670.

[57] Shaked NT, Zhu YZ, Rinehart MT, Wax A. 2009. Two-step-only phase-shifting interferometry with optimized detector bandwidth for microscopy of live cells. *Opt Express* **17**:15585–15591.

[58] Moradi AR, Ali MK, Daneshpanah M, Anand A, Javidi B. 2010. Detection of calcium-induced morphological changes of living cells using optical traps. *IEEE Photonics J* **2**:775–783.

[59] Shaked NT, Finan JD, Guilak F, Wax A. 2010. Quantitative phase microscopy of articular chon-drocyte dynamics by wide-field digital interferometry. *J Biomed Opt* **15**(1):010505.

[60] Tychinsky VP, Tikhonov AN. 2010. Interference microscopy in cell biophysics. 2. Visualization of individual cells and energy-transducing organelles. *Cell Biochem Biophys* **58**:117–128.

[61] Wang P, Bista RK, Khalbuss WE, Qiu W, Uttam S, *et al*. 2010. Nanoscale nuclear architec-ture for cancer diagnosis beyond pathology via spatial-domain low-coherence quantitative phase microscopy. *J Biomed Opt* **15**:066028.

[62] Xue L, Lai JC, Li ZH. 2010. Quantitative phase microscopy of red blood cells with slightly-off-axis interference. *Optics in Health Care and Biomedical Optics IV* **7845**:784505–784508.

[63] Yamauchi T, Sugiyama N, Sakurai T, Iwai H, Yamashita Y. 2010. Label-free classification of cell types by imaging of Cell Membrane Fluctuations Using Low-Coherent Full-Field Quantitative Phase Microscopy. *Three-Dimensional and Multidimensional Microscopy: Image Acquisition and Processing XVII* **7570**:75700X-X-8.

[64] Yang W, Lee S, Lee J, Bae Y, Kim D. 2010. Silver nanoparticle-induced degranulation observed with quantitative phase microscopy. *J Biomed Opt* **15**:045005.

[65] Gonzalez-Laprea J, Marquez A, Noris-Suarez K, Escalona R. 2011. Study of bone cells by quan-titative phase microscopy using a Mirau interferometer. *Rev Mex Fis* **57**:435–440.

[66] Lee S, Kim YR, Lee JY, Rhee JH, Park CS, Kim DY. 2011. Dynamic analysis of pathogen-infected host cells using quantitative phase microscopy. *J Biomed Opt* **16**:036004.

[67] Kim M, Choi Y, Fang-Yen C, Sung YJ, Dasari RR, *et al*. 2011. High-speed synthetic aperture microscopy for live cell imaging. *Opt Lett* **36**:148–150.

[68] Lee S, Lee JY, Park CS, Kim DY. 2011. Detrended fluctuation analysis of membrane flickering in discocyte and spherocyte red blood cells using quantitative phase microscopy. *J Biomed Opt* **16**:076009.

[69] Park YK, Best CA, Auth T, Gov NS, Safran SA, *et al*. 2010. Metabolic remodeling of the human red blood cell membrane. *P Natl Acad Sci USA* **107**:1289–1294.

[70] Wang P, Bista R, Bhargava R, Brand RE, Liu Y. 2011. Spatial-domain Low-coherence Quanti-tative Phase Microscopy for Cancer Diagnosis. *Optical Coherence Tomography and Coherence Domain Optical Methods in Biomedicine XV* **7889**.

[71] Wang R, Ding HF, Mir M, Tangella K, Popescu G. 2011. Effective 3D viscoelasticity of red blood cells measured by diffraction phase microscopy. *Biomed Opt Express* **2**:485–490.

[72] Yamauchi T, Iwai H, Yamashita Y. 2011. Label-free imaging of intracellular motility by low-coherent quantitative phase microscopy. *Opt Express* **19**:5536–5550.

[73] Yaqoob Z, Yamauchi T, Choi W, Fu D, Dasari RR, Feld MS. 2011. Single-shot full-field reflection phase microscopy. *Opt Express* **19**:7587–7595.

[74] Ansari R, Aherrahrou R, Aherrahrou Z, Erdmann J, Huttmann G, Schweikard A. 2012. Quan-titative analysis of cardiomyocyte dynamics with optical coherence phase Microscopy. *Opti-cal Coherence Tomography and Coherence Domain Optical Methods in Biomedicine Xvi* **8213**:821338.

[75] Cardenas N, Mohanty SK. 2012. Investigation of shape memory of red blood cells using optical tweezers and quantitative phase microscopy. *Imaging, Manipulation, and Analysis of Biomolecules, Cells, and Tissues X* **8225**:82252B.

[76] Pan F, Liu S, Wang Z, Shang P, Xiao W. 2012. Dynamic and quantitative phase-contrast imaging of living cells under simulated zero gravity by digital holographic microscopy and superconduct-ing magnet. *Laser Phys* **22**:1435–1438.

[77] Tychinsky VP, Kretushev AV, Vyshenskaya TV, Shtil AA. 2012. Dissecting eukaryotic cells by coherent phase microscopy: quantitative analysis of quiescent and activated T lymphocytes. *J Biomed Opt* **17**:076020.

[78] Yamauchi T, Sakurai T, Iwai H, Yamashita Y. 2012. Long-term measurement of spontaneous membrane fluctuations over a wide dynamic range in the living cell by low-coherent quantitative phase microscopy. *Imaging, Manipulation, and Analysis of Biomolecules, Cells, and Tissues X* **8225**:82250A.

[79] Curl CL, Bellair CJ, Harris PJ, Allman BE, Roberts A, *et al*. 2004. Quantitative phase microscopy: a new tool for investigating the structure and function of unstained live cells. *Clin Exp Pharmacol P* **31**:896–901.

[80] Almoro PF, Waller L, Agour M, Falldorf C, Pedrini G, *et al*. 2012. Enhanced deterministic phase retrieval using a partially developed speckle field. *Opt Lett* **37**:2088–2090.

[81] Bon P, Maucort G, Wattellier B, Monneret S. 2009. Quadriwave lateral shearing interferometry for quantitative phase microscopy of living cells. *Opt Express* **17**:13080–13094.

[82] Almoro PF, Pedrini G, Gundu PN, Osten W, Hanson SG. 2010. Phase microscopy of technical and biological samples through random phase modulation with a diffuser. *Opt Lett* **35**:1028–1030.

[83] Zhang Y, Pedrini G, Osten W, Tiziani HJ. 2004. Phase retrieval microscopy for quantitative phase-contrast imaging. *Optik* **115**:94–96.

[84] Chhaniwal VK, Anand A, Faridian A, Pedrini G, Osten W, Javidi B. 2011. Single beam quantitative phase contrast 3D microscopy of cells. *Proc Spie* **8092**:80920D-D-8.

[85] Bao P, Situ G, Pedrini G, Osten W. 2012. Lensless phase microscopy using phase retrieval with multiple illumination wavelengths. *Appl Optics* **51**:5486–5494.

[86] Cogswell CJ, Smith NI, Larkin KG, Hariharan P. 1997. Quantitative DIC microscopy using a geometric phase shifter. *Three-Dimensional Microscopy: Image Acquisition and Processing IV, Proceedings of* **2984**:72–81.

[87] Ishiwata H, Yatagai T, Itoh M, Tsukada A. 1999. Quantitative phase analysis in retardation modulated differential interference contrast (RM-DIC) microscope. *Interferometry '99: Techniques and Technologies* **3744**:183–187.

[88] Totzeck M, Kerwien N, Tavrov A, Rosenthal E, Tiziani HJ. 2002. Quantitative Zernike phase-contrast microscopy by use of structured birefringent pupil-filters and phase-shift evaluation. *P Soc Photo-Opt Ins* **4777**:1–11.

[89] King SV, Libertun A, Piestun R, Cogswell CJ, Preza C. 2008. Quantitative phase microscopy through differential interference imaging. *J Biomed Opt* **13**:024020.

[90] Kou SS, Waller L, Barbastathis G, Sheppard CJR. 2010. Transport-of-intensity approach to differential interference contrast (TI-DIC) microscopy for quantitative phase imaging. *Opt Lett* **35**:447–449.

[91] Fu D, Oh S, Choi W, Yamauchi T, Dorn A, *et al*. 2010. Quantitative DIC microscopy using an off-axis self-interference approach. *Opt Lett* **35**:2370–2302.

[92] Gao P, Yao BL, Harder I, Lindlein N, Torcal-Milla FJ. 2011. Phase-shifting Zernike phase contrast microscopy for quantitative phase measurement. *Opt Lett* **36**:4305–4307.

[93] Pavillon N. 2011. *Cellular dynamics and three-dimensional refractive index distribution studied with quantitative phase imaging* (Doctoral dissertation).

[94] Gabor D. 1948. A new microscopic principle. *Nature* **161**:777.

[95] Gabor D. 1949. Microscopy by reconstructed wave-fronts. *Proceedings of the Royal Society of London. Serie A, Mathematical and Physical Sciences* **197**:454–487.

[96] Leith EN, Upatnieks J. 1964. Wavefront reconstruction with diffused illumination and three-dimensional objects. *J Opt Soc Am* **54**:1295–1301.

[97] Leith EN, Upatniek.J. 1962. Reconstructed wavefronts and communication theory. *J Opt Soc Am* **52**:1123.

[98] Wolf E, Shewell JR. 1970. Diffraction theory of holography. *J Math Phys* **11**:2254.

[99] Carter WH. 1970. Computational reconstruction of scattering objects from holograms. *J Opt Soc Am* **60**:306–314.

[100] Dubois F, Callens N, Yourassowsky C, Hoyos M, Kurowski P, Monnom O. 2006. Digital holographic microscopy with reduced spatial coherence for three-dimensional particle flow analysis. *Appl Optics* **45**:864–871.

[101] Kemper B, Stürwald S, Remmersmann C, Langehanenberg P, von Bally G. 2008. Characterisation of light emitting diodes (LEDs) for application in digital holographic microscopy for inspection of micro and nanostructured surfaces. *Opt Laser Eng* **46**:499–507.

[102] Ansari Z, Gu Y, Tziraki M, Jones R, French PMW, *et al.* 2001. Elimination of beam walk-off in low-coherence off-axis photorefractive holography. *Opt. Lett.* **26**:334–336.

[103] Goodman JW, Lawrence RW. 1967. Digital image formation from electronically detected holograms. *Applied Physics Letters* **11**:77–79.

[104] Schnars U, Jüptner W. 1994. Direct recording of holograms by a CCD target and numerical reconstruction. *Appl Optics* **33**:179–181.

[105] Coquoz O, Conde R, Taleblou F, Depeursinge C. 1995. Performances of endoscopic holography with a multicore optical-fiber. *Appl Optics* **34**:7186–7193.

[106] Bruning JH, Herriott DR, Gallaghe JE, Rosenfel DP, White AD, Brangacc DJ. 1974. Digital wavefront measuring interferometer for testing optical surfaces and lenses. *Appl Optics* **13**:2693–2703.

[107] Carré P. 1966. Installation et utilisation du comparateur photoélectrique et interférentiel du Bureau International des Poids et Mesures. *Metrologia* **2**:13–33.

[108] Kreis T. 2005. *Handbook of Holographic Interferometry: Optical and Digital Methods*. Weinheim, FRG: Wiley-VCH Verlag GmbH & Co. KGaA.

[109] Rastogi P. 1994. *Holographic Interferometry: Principles and Methods*. NY: Springer-Verlag.

[110] Guo P, Devaney AJ. 2004. Digital microscopy using phase-shifting digital holography with two reference waves. *Opt Lett* **29**:857–859.

[111] Liu JP, Poon TC. 2009. Two-step-only quadrature phase-shifting digital holography. *Opt Lett* **34**:250–252.

[112] Awatsuji Y, Sasada M, Kubota T. 2004. Parallel quasi-phase-shifting digital holography. *Applied Physics Letters* **85**:1069–1071.

[113] Guo CS, Zhang L, Wang HT, Liao J, Zhu YY. 2002. Phase-shifting error and its elimination in phase-shifting digital holography. *Opt Lett* **27**:1687–1689.

[114] Xu XF, Cai LZ, Wang YR, Meng XF, Sun WJ, *et al.* 2008. Simple direct extraction of unknown phase shift and wavefront reconstruction in generalized phase-shifting interferometry: algorithm and experiments. *Opt. Lett.* **33**:776–778.

[115] Wang Z, Han B. 2004. Advanced iterative algorithm for phase extraction of randomly phase-shifted interferograms. *Opt Lett* **29**:1671–1673.

[116] Schnars U, Juptner W. 1994. Direct recording of holograms by a CCD target and numerical reconstruction. *Appl Optics* **33**:179–181.

[117] Takeda M, Ina H, Kobayashi S. 1982. Fourier-transform method of fringe-pattern analysis for computer-based topography and interferometry. *J Opt Soc Am* **72**:156–160.

[118] Kreis T. 1986. Digital holographic interference-phase measurement using the Fourier-transform method. *Journal of the Optical Society of America A. Optics and Image Science* **3**:847–855.

[119] Cuche E, Poscio P, Depeursinge C. 1997. Optical tomography by means of a numerical low-coherence holographic technique. *Journal of Optics-Nouvelle Revue D Optique* **28**:260–264.

[120] Montfort F, Charrière F, Colomb T, Cuche E, Marquet P, Depeursinge C. 2006. Purely numerical compensation for microscope objective phase curvature in digital holographic microscopy: influence of digital phase mask position. *J Opt Soc Am A* **23**:2944–2953.

[121] Colomb T, Montfort F, Kühn J, Aspert N, Cuche E, *et al.* 2006. Numerical parametric lens for shifting, magnification and complete aberration compensation in digital holographic microscopy. *J Opt Soc Am A* **23**:3177–190.

[122] Colomb T, Kühn J, Charrière F, Depeursinge C, Marquet P, Aspert N. 2006. Total aberrations compensation in digital holographic microscopy with a reference conjugated hologram. *Opt Express* **14**:4300–4306.

[123] Colomb T, Cuche E, Charrière F, Kühn J, Aspert N, *et al*. 2006. Automatic procedure for aberration compensation in digital holographic microscopy and applications to specimen shape compensation. *Appl Optics* **45**:851–863.

[124] Colomb T, Charriere F, Kuhn J, Marquet P, Depeursinge C. 2008. Advantages of digital holographic microscopy for real-time full field absolute phase imaging – art. no. 686109. *Three-Dimensional and Multidimensional Microscopy: Image Acquisition and Processing XV* **6861**:86109.

[125] Ferraro P, Grilli S, Alfieri D, Nicola SD, Finizio A, *et al*. 2005. Extended focused image in microscopy by digital Holography. *Opt Express* **13**:6738–6749.

[126] Cuche E, Bevilacqua F, Depeursinge C. 1999. Digital holography for quantitative phase-contrast imaging. *Opt Lett* **24**:291–293.

[127] Haddad WS, Cullen D, Solem JC, Longworth JW, McPherson A, *et al*. 1992. Fourier-transform holographic microscope. *Appl Optics* **31**:4973–4978.

[128] Takaki Y, Kawai H, Ohzu H. 1999. Hybrid holographic microscopy free of conjugate and zero-order images. *Appl Optics* **38**:4990–4996.

[129] Xu W, Jericho MH, Meinertzhagendagger IA, Kreuzer HJ. 2001. Digital in-line holography for biological applications. *PNAS* **98**:11301–11305.

[130] Colomb T, Dahlgren P, Beghuin D, Cuche E, Marquet P, Depeursinge C. 2002. Polarization imaging by use of digital holography. *Appl Optics* **41**:27–37.

[131] Colomb T, Dürr F, Cuche E, Marquet P, Limberger H, *et al*. 2005. Polarization microscopy by use of digital holography: application to optical fiber birefringence measurements. *Appl Optics* **44**:4461–4469.

[132] Kuhn J, Colomb T, Montfort F, Charriere F, Emery Y, *et al*. 2007. Real-time dual-wavelength digital holographic microscopy with a single hologram acquisition. *Opt Express* **15**:7231–7242.

[133] Zhang T, Yamaguchi I. 1998. Three-dimensional microscopy with phase-shifting digital holography. *Opt Lett* **23**:1221–1223.

[134] Cuche E, Marquet P, Depeursinge C. 2000. Spatial filtering for zero-order and twin-image elimination in digital off-axis holography. *Appl Optics* **39**:4070–4075.

[135] Cuche E, Marquet P, Depeursinge C. 2000. Aperture apodization using cubic spline interpolation: application in digital holographic microscopy. *Opt Commun* **182**:59–69.

[136] Charriere F, Colomb T, Montfort F, Cuche E, Marquet P, Depeursinge C. 2006. Shot-noise influence on the reconstructed phase image signal-to-noise ratio in digital holographic microscopy. *Appl Optics* **45**:7667–7673.

[137] Charrière F, Rappaz B, Kühn J, Colomb T, Marquet P, Depeursinge C. 2007. Influence of shot noise on phase measurement accuracy in digital holographic microscopy. *Opt Express* **15**:8818–8831.

[138] Seo S, Isikman SO, Sencan I, Mudanyali O, Su TW, *et al*. 2010. High-throughput lens-free blood analysis on a chip. *Analytical Chemistry* **82**:4621–4627.

[139] Milgram JH, Li WC. 2002. Computational reconstruction of images from holograms. *Appl Optics* **41**:853–864.

[140] Yi F, Lee CG, Moon IK. 2012. Statistical analysis of 3D volume of red blood cells with different shapes via digital holographic microscopy. *J Opt Soc Korea* **16**:115–120.

[141] Moon I, Yi F, Javidi B. 2010. Automated three-dimensional microbial sensing and recognition using digital holography and statistical sampling. *Sensors-Basel* **10**:8437–8451.

[142] Javidi B, Daneshpanah M, Moon I. 2010. Three-dimensional holographic imaging for identification of biological micro/nanoorganisms. *IEEE Photonics J* **2**:256–259.

[143] Anand A, Chhaniwal VK, Javidi B. 2011. imaging embryonic stem cell dynamics using quantitative 3-D digital holographic microscopy. *IEEE Photonics J* **3**:546–554.

[144] Miccio L, Finizio A, Memmolo P, Paturzo M, Merola F, *et al*. 2011. Detection and visualization improvement of spermatozoa cells by digital holography. In *Molecular Imaging III*, eds CP Lin, V Ntziachristos, p. 8089.

[145] Antkowiak M, Callens N, Schockaert C, Yourassowsky C, Dubois F. 2008. Accurate three-dimensional detection of micro-particles by means of digital holographic microscopy – art. no. 699514. In *Optical Micro- and Nanometrology in Microsystems Technology II*, eds C Gorecki, AK Asundi, W Osten, **6995**:99514.

[146] Bae Y, Lee S, Yang W, Kim DY. 2010. Three-dimensional Single Particle Tracking using Off-axis Digital Holographic Microscopy. *Nanoscale Imaging, Sensing, and Actuation for Biomedical Applications VII* **7574**:757408.

[147] Hsieh CL, Grange R, Pu Y, Psaltis D. 2009. Three-dimensional harmonic holographic microcopy using nanoparticles as probes for cell imaging. *Opt Express* **17**:2880–2891.

[148] Shaffer E, Depeursinge C. 2010. Digital holography for second harmonic microscopy: application to 3D-tracking of nanoparticles. *Biophotonics: Photonic Solutions for Better Health Care II* **7715**.

[149] Mann CJ, Yu LF, Kim MK. 2006. Movies of cellular and sub-cellular motion by digital holographic microscopy. *Biomed Eng Online* **5**:21.

[150] Sun HY, Song B, Dong HP, Reid B, Player MA, *et al*. 2008. Visualization of fast-moving cells in vivo using digital holographic video microscopy. *J Biomed Opt* **13**:014007.

[151] Javidi B, Moon I, Daneshpanaha M. 2010. Detection, identification and tracking of biological micro/nano organisms by computational 3D optical imaging. *Biosensing III* **7759**:77590R-R-6.

[152] Barer R. 1953. Determination of dry mass, thickness, solid and water concentration in living cells. *Nature (London)* **172**:1097–1098.

[153] Popescu G, Park Y, Lue N, Best-Popescu C, Deflores L, *et al*. 2008. Optical imaging of cell mass and growth dynamics. *Am J Physiol-Cell Ph* **295**:C538–C544.

[154] Kemper B, Bauwens A, Vollmer A, Ketelhut S, Langehanenberg P, *et al*. 2010. Label-free quantitative cell division monitoring of endothelial cells by digital holographic microscopy. *J Biomed Opt* **15**:036009.

[155] Mir M, Wang Z, Shen Z, Bednarz M, Bashir R, *et al*. 2011. Optical measurement of cycle-dependent cell growth. *P Natl Acad Sci USA* **108**:13124–13129.

[156] Zicha D, Dunn GA. 1995. An image-processing system for cell behavior studies in subconfluent cultures. *J Microsc-Oxford* **179**:11–21.

[157] Rappaz B, Barbul A, Emery Y, Korenstein R, Depeursinge C, *et al*. 2008. Comparative study of human erythrocytes by digital holographic microscopy, confocal microscopy, and impedance volume analyzer. *Cytom Part A* **73A**:895–903.

[158] Yusipovich AI, Parshina EY, Brysgalova NY, Brazhe AR, Brazhe NA, *et al*. 2009. Laser interference microscopy in erythrocyte study. *J Appl Phys* **105**:102037.

[159] Pham H, Bhaduri B, Ding HF, Popescu G. 2012. Spectroscopic diffraction phase microscopy. *Opt Lett* **37**:3438–3440.

[160] Rinehart M, Zhu YZ, Wax A. 2012. Quantitative phase spectroscopy. *Biomed Opt Express* **3**:958–965.

[161] Jang Y, Jang J, Park Y. 2012. Dynamic spectroscopic phase microscopy for quantifying hemoglobin concentration and dynamic membrane fluctuation in red blood cells. *Opt Express* **20**:9673–9681.

[162] Boss D, Hoffmann A, Rappaz B, Depeursinge C, Magistretti PJ, *et al*. 2012. Spatially-resolved eigenmode decomposition of red blood cells membrane fluctuations questions the role of ATP in flickering. *Plos One* **7**:e40667.

[163] Mauritz JMA, Esposito A, Tiffert T, Skepper JN, Warley A, *et al*. 2010. Biophotonic techniques for the study of malaria-infected red blood cells. *Med Biol Eng Comput* **48**:1055–1063.

[164] Cardenas N, Yu LF, Mohanty SK. 2011. Probing orientation and rotation of red blood cells in optical tweezers by digital holographic microscopy. *Optical Diagnostics and Sensing Xi: Toward Point-of-Care Diagnostics and Design and Performance Validation of Phantoms Used in Conjunction with Optical Measurement of Tissue Iii* **7906**:790613–790619.

[165] Esseling M, Kemper B, Antkowiak M, Stevenson DJ, Chaudet L, *et al.* 2012. Multimodal biophotonic workstation for live cell analysis. *J Biophotonics* **5**:9–13.

[166] Lue N, Popescu G, Ikeda T, Dasari RR, Badizadegan K, Feld MS. 2006. Live cell refractometry using microfluidic devices. *Opt Lett* **31**:2759–2761.

[167] Kemper B, Kosmeier S, Langehanenberg P, von Bally G, Bredebusch I, *et al.* 2007. Integral refractive index determination of living suspension cells by multifocus digital holographic phase contrast microscopy. *J Biomed Opt* **12**:054009 (5 pages).

[168] Rappaz B, Charrière F, Depeursinge C, Magistretti PJ, Marquet P. 2008. Simultaneous cell morphometry and refractive index measurement with dual-wavelength digital holographic microscopy and dye-enhanced dispersion of perfusion medium. *Opt Lett* **33**:744–746.

[169] Boss D, Kuhn J, Jourdain P, Depeursinge C, Magistretti PJ, Marquet P. Measurement of absolute cell volume, osmotic membrane water permeability, and refractive index of transmembrane water and solute flux by digital holographic microscopy. *J Biomed Opt* **18**:036007.

[170] Boss D, Kuehn J, Jourdain P, Depeursinge C, Magistretti PJ, Marquet P. 2013. Measurement of absolute cell volume and osmotic water membrane permeability by real time dual wavelength holographic microscopy. *J Biomed Opt* in press.

[171] Cossart R, Aronov D, Yuste R. 2003. Attractor dynamics of network UP states in the neocortex. *Nature* **423**:283–288.

[172] Peterka DS, Takahashi H, Yuste R. 2011. Imaging voltage in neurons. *Neuron* **69**:9–21.

[173] Carter KM, George JS, Rector DM. 2004. Simultaneous birefringence and scattered light measurements reveal anatomical features in isolated crustacean nerve. *J Neurosci Methods* **135**:9–16.

[174] Cohen LB. 1973. Changes in neuron structure during action potential propagation and synaptic transmission. *Physiol Rev* **53**:373–418.

[175] Tasaki I, Watanabe A, Sandlin R, Carnay L. 1968. Changes in fluorescence, turbidity, and birefringence associated with nerve excitation. *Proc Natl Acad Sci U S A* **61**:883–888.

[176] Hill DK, Keynes RD. 1949. Opacity changes in stimulated nerve. *J Physiol* **108**:278–281.

[177] Tasaki I, Byrne PM. 1992. Rapid structural changes in nerve fibers evoked by electric current pulses. *Biochem Biophys Res Commun* **188**:559–564.

[178] Stepnoski RA, LaPorta A, Raccuia-Behling F, Blonder GE, Slusher RE, Kleinfeld D. 1991. Noninvasive detection of changes in membrane potential in cultured neurons by light scattering. *Proc Natl Acad Sci U S A* **88**:9382–9386.

[179] Holthoff K, Witte OW. 1996. Intrinsic optical signals in rat neocortical slices measured with near-infrared dark-field microscopy reveal changes in extracellular space. *J Neurosci* **16**:2740–2749.

[180] Andrew RD, Adams JR, Polischuk TM. 1996. Imaging NMDA- and kainate-induced intrinsic optical signals from the hippocampal slice. *J Neurophysiol* **76**:2707–2717.

[181] MacVicar BA, Hochman D. 1991. Imaging of synaptically evoked intrinsic optical signals in hippocampal slices. *J Neurosci* **11**:1458–1469.

[182] Pavillon N, Benke A, Boss D, Moratal C, Kuhn J, *et al.* 2010. Cell morphology and intracellular ionic homeostasis explored with a multimodal approach combining epifluorescence and digital holographic microscopy. *J Biophotonics* **3**:432–436.

[183] Charrière F, Marian A, Montfort F, Kühn J, Colomb T, *et al.* 2006. Cell refractive index tomography by digital holographic microscopy. *Opt Lett* **31**:178–810.

[184] Debailleul M, Georges V, Simon B, Morin R, Haeberle O. 2009. High-resolution three-dimensional tomographic diffractive microscopy of transparent inorganic and biological samples. *Opt Lett* **34**:79–81.

[185] Choi W, Fang-Yen C, Badizadegan K, Oh S, Lue N, *et al.* 2007. Tomographic phase microscopy. *Nat Methods* **4**:717–719.

[186] Sheppard CJR, Kou SS. 2010. 3D imaging with holographic tomography. *International Conference on Advanced Phase Measurement Methods in Optics an Imaging* **1236**:65–69.

[187] Park Y, Diez-Silva M, Fu D, Popescu G, Choi W, *et al.* 2010. Static and dynamic light scattering of healthy and malaria-parasite invaded red blood cells. *J Biomed Opt* **15**:020506.

[188] Cotte Y, Toy FM, Jourdain P, Pavillon N, Boss D, *et al.* 2013. Marker-free phase nanoscopy. *Nature Photonics* **7**, 113–117.

10

Super Resolved Holographic Configurations

Amihai Meiri[1], Eran Gur[2], Javier Garcia[3], Vicente Micó[3], Bahram Javidi[4] and Zeev Zalevsky[1]

[1] *Faculty of Engineering, Bar-Ilan University, Israel*
[2] *Department of Electrical Engineering and Electronics, Azrieli – College of Engineering, Israel*
[3] *Departamento de Óptica, Universitat Valencia, Spain*
[4] *Department of Electrical and Computer Engineering, University of Connecticut, USA*

10.1 Introduction

Optical imaging suffers from a drawback inherent to the process of recording: the recording media (either photographic film or a digital camera) captures only the intensity of the incident light, sacrificing the three-dimensional data in the process, which lies in the phase of the electric field. In 1948 Dennis Gabor invented a technique which circumvents the loss of phase by adding a reference field to the recorded object [1]. This technique is dubbed *holography*, where the interference patterns between the object and the reference field can be recorded. These interference fringes depend on the phase of the object, therefore maintaining this information. If we write the recorded object in terms of amplitude and phase, $a = |a|e^{j\phi}$ add a reference field A and record the obtained intensity we have

$$||a|e^{j\phi} + A|^2 = |a|^2 + |A|^2 + aA \cos \phi \qquad (10.1)$$

The first term is the intensity of the recorded field as in a conventional imaging system and the second term is the intensity of the reference field. The last term reveals that the phase information is maintained and recorded on the imaging medium as well.

As in every imaging system, holography is limited in both resolution and field of view (FOV). Another drawback in this technique is the excess information: the intensity of the reference beam (the second term in Eq. (10.1)) and the so-called twin image problem which is explained in the next section. Both factors deteriorate the quality of the hologram.

In this chapter we show how metal nanoparticles come to the aid of holography in improving the resolution, FOV, and eliminating the twin image and reference field.

10.2 Digital Holography

The original holography scheme, suggested by Dennis Gabor, is presented in Fig. 10.1. In the Gabor hologram the object is assumed to be highly transmissive, with transmission function

$$T(x, t) = T + \Delta T(x, y), \tag{10.2}$$

where T is the average transmission and $\Delta T(x, y)$ is the variation in space. The high transmissivity of the object means that $\Delta T << T$. In this configuration, T serves as the reference beam. Looking as before at the scattered object beam a and the reference beam A we can write the recorded intensity as

$$I = |A + a|^2 = |A|^2 + |a|^2 + A^*a + Aa^*. \tag{10.3}$$

If we illuminate the developed recording medium I by a reference beam A', the result is $I' = A'I$. Note that, in writing both the recorded intensity and the reconstructed intensity, we assumed that the amplitude transmittance of the developed hologram is proportional to the exposure, and for simplicity of argument it is assumed that this proportionality constant is 1. The generalization of the theory in that case is straightforward [2]. Looking at the four terms of Eq. (10.3) we can neglect the second term due to our assumption about the transmissivity of the object. We are left with three terms: the first term of Eq. (10.3) that is the reference beam, the third term of Eq. (10.3) that is our desired field multiplied by a constant, and the last term of Eq. (10.3) that is the complex conjugate of the field, termed the *twin image*. The reference beam and the twin image deteriorate the quality of the hologram and are a major drawback of the on-axis hologram.

To solve the DC term (reference wave) and twin image problem Leith and Hupetnieks [3] suggested using an off-axis setup, where the reference is incident on the recording medium at

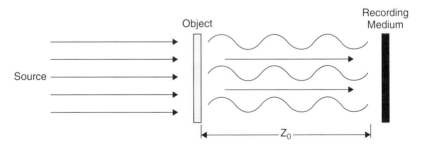

Figure 10.1 Gabor hologram. *Source*: Amihai Meiri, Eran Gur, Javier Garcia, Vicente Micó, Bahram Javidi, Zeev Zalevsky 2013. Reproduced with permission from SPIE

an angle: $Ae^{-j2\pi \sin 2\theta x/\lambda}$, where θ is the angle between the reference beam and the optical axis. The recorded hologram in this case is

$$I = \left| a + Ae^{-j2\pi \sin 2\theta x/\lambda} \right|^2 = |A|^2 + |a|^2 + A^* e^{j2\pi \sin 2\theta x/\lambda} a + Ae^{-j2\pi \sin 2\theta x/\lambda} a^*. \qquad (10.4)$$

The phase factors of the last two terms translate into spatial shift as a result of the free space propagation. Now we have two on axis fields – the intensity of the object and the reference beam, and two off axis – the last two terms, one above the optical axis and one below it. This enables us to spatially filter out the unwanted terms. The information bandwidth captured by this solution is smaller than the on-axis hologram.

Since all the information is present in the recorded hologram, a digitally recorded hologram does not need a reconstruction method, but the information can be digitally analyzed. As an example to such a procedure we can look at the focusing of the hologram. In presenting the basic theory of holography we neglected the axial position of the hologram. Looking at a point source in the reference beam location, the object source location, and the reconstruction source location, we can show that the object beam results in a virtual image positioned at a distance of Z_0 before the recording medium and a twin image being a real image formed at a distance of Z_0, behind the recording medium [2]. Assuming that the distance Z_0 is relatively large, we look at the unfocused recorded hologram. By digitally implementing a Fresnel diffraction integral we can propagate the recorded hologram by a distance of $-Z_0$ and obtain the focused virtual image. Now the real image (the twin image) is highly unfocused and its influence on the quality of the recorded hologram is relatively small. In the case of microscopy, where the distances are small, such a solution is inadequate and the twin image must be eliminated by other means, such as phase retrieval [4], deconvolution [5], and recording of multiple holograms in different axial locations [6].

10.3   Metal Nanoparticles

Owing to their various size, shapes and materials, metal nanoparticles became a major interest, especially in the fields of imaging and biomedical engineering. Their attractiveness stems from the sensitivity of their absorption and scattering spectrum to these attributes [7,8]. This sensitivity is explained by the oscillation of electrons in the conduction bend as a result of an incident electric field. When these electrons oscillate coherently the absorption and scattering are at their peak and a distinct resonance can be observed [9]. This phenomenon is known as *surface plasmon resonance* (SPR).

The simplest of shapes for these nanoparticles is a sphere (see Fig. 10.2). Looking at gold (Au) spheres, their apparent color is red, where the exact shade depends on the size of the nanoparticles. An example of SPR can be seen in Fig. 10.3, where the absorption spectra of variously sized nanospheres are presented.

The sensitivity of the extinction spectrum of metal nanoparticles is not limited to their size. Their shape is also an important factor. Nanorods, cigar shaped nanoparticles, exhibit different spectra to nanospheres where the peak is red-shifted. In addition, their aspect ratio, the ratio between their length and diameter, also determines the location of the extinction peak [10]. Other factors that determine their spectrum are the surrounding medium's refractive index [11], the material from which the nanoparticles are made [7], and their density due to inter-particle coupling [12,13].

Figure 10.2 Metal nanoparticles are fabricated in various shapes. *Left* – a nanorod with an aspect ratio of L/R. *Right* – a nanoshpere. *Source*: Amihai Meiri, Eran Gur, Javier Garcia, Vicente Micó, Bahram Javidi, Zeev Zalevsky 2013. Reproduced with permission from SPIE

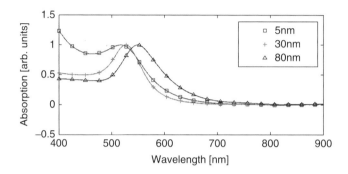

Figure 10.3 Measured spectra of various gold nanospheres. *Source*: Amihai Meiri, Eran Gur, Javier Garcia, Vicente Micó, Bahram Javidi, Zeev Zalevsky 2013. Reproduced with permission from SPIE

Metal nanoparticles are attractive not only due to SPR. They also exhibit a high scattering cross section, which reduces the image acquisition time and high stability [14], all in comparison to fluorescent markers, which are in common use in various imaging systems [15,16].

Usually metal nanoparticles are suspended in liquid and due to their small size they move in Brownian motion. The Brownian motion was used by Gur *et al.* [17] as a random encoding mask for super-resolved microscopy, a technique that is the foundation of the methods presented in this chapter. This technique is presented briefly here.

When metal nanoparticles are suspended in liquid and are placed close to a high resolution object, their location constantly changes. Each metal nanoparticle couples the non-propagating near field to a propagating field that can be recorded by a camera. Since the near field is not diffraction limited, its resolution is not limited, but by the size of the nanoparticles in a manner that resembles apertureless *Near field Scanning Optical Microscope* (aNSOM) [18]. If we denote the high resolution object by $s(x)$, the time varying random nanoparticles mask by $g(x, t)$, and the point spread function (PSF) of the optical system by $p(x)$, each frame that is captured by the camera can be written as

$$I(x, t) = \int s(x')g(x', t)p(x - x')dx'. \tag{10.5}$$

When multiple frames are recorded, over time, the nanoparticles cover the entire area of the object to be imaged. Due to their small size, each nanoparticle can be considered as a point source and therefore, on the recording medium it has a shape of a PSF. When these nanoparticles are sparse enough (a distance of at least half a wavelength is required), each of these nanoparticles can be localized and the mask can be computed for each frame. The center of the PSF has a Gaussian like shape; therefore, it can be localized precisely by simple Gaussian fitting. This localization can be shown to have an error that depends on the number of detected photons and reach accuracy of a few nanometers [19–22].

Multiplying each frame by the computed nanoparticles mask and applying time averaging, results in

$$R(x) = \int \left[\int s\left(x'\right) g(x',t)p(x-x')dx' \right] \widetilde{g}(x,t)dt, \tag{10.6}$$

where $\widetilde{g}(x,t)$ is the computed nanoparticle mask. Due to the random nature of the nanoparticle mask and the high localization accuracy we can assume that

$$\int g(x',t)\widetilde{g}(x,t)dt = \kappa + \delta(x'-x). \tag{10.7}$$

Using this in Eq. (10.6) results in

$$R(x) = s(x)p(0) + \kappa \int s(x')p(x-x')dx'. \tag{10.8}$$

The first term here is the high resolution object multiplied by a constant (the value of the PSF at the center) and the second term is the low resolution object obtained by the optical system. Since this term is the conventionally captured image, it can be subtracted from $R(x)$ to obtain $s(x)$.

10.4 Resolution Enhancement in Digital Holography

The principle presented in the previous Section 10.3 can be adapted to digital holography (DH) configuration [23]. The setup is presented in Fig. 10.4 (Plate 17), in which a Fourier plane hologram is recorded. In this setup the object is placed close to a pinhole, which serves as the reference beam that is required for the holography recording process.

We denote the high resolution object we would like to resolve by $s(x_1)$ and the random decoding pattern by $g(x_1,t)$. The pinhole is positioned in the Fourier domain hologram at a distance of Δx from the object $s(x_1)$. The detector captures the Fourier transform of the composed input field. Due to the nature of the CCD the result is multiplied by a rectangular function with a width of D. At the CCD plane we write the expression obtained at the CCD as:

$$I(x_2,t) = \left| \int \left(\delta\left(x-\Delta x\right) + s(x_1)g(x_1,t) \right) e^{-2\pi i x_1 x_2/\lambda F} dx_1 \right|^2 rect\left(\frac{x_2}{D}\right). \tag{10.9}$$

This expression can be expended into three terms $I(x_2,t) = T_1(x_2,t) + T_2(x_2,t) + T_3(x_2,t)$ where

$$T_1(x_2,t) = \left(1 + \int s\left(x_1\right) g(x_1,t)e^{-2\pi i x_1 x_2/\lambda F} dx_1 \right) rect\left(\frac{x_2}{D}\right).$$

$$T_2(x_2,t) = \left(e^{-2\pi i \Delta x x_2/\lambda F} \int s\left(x_1\right) g(x_1,t)e^{-2\pi i x_1 x_2/\lambda F} dx_1 \right) rect\left(\frac{x_2}{D}\right)$$

$$T_3(x_2,t) = \left(e^{2\pi i \Delta x x_2/\lambda F} \int s^*\left(x_1\right) g^*(x_1,t)e^{2\pi i x_1 x_2/\lambda F} dx_1 \right) rect\left(\frac{x_2}{D}\right) \tag{10.10}$$

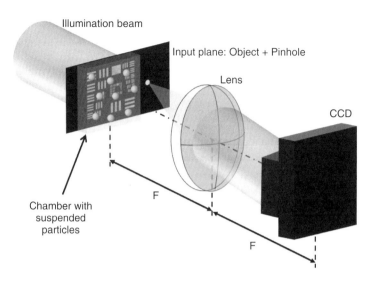

Figure 10.4 (Plate 17) Digital holography super-resolution setup. *Source*: Zalevsky Z., Gur E., Garcia J., Micó V., Javidi B. 2012. Figure 1. Reproduced with permission from The Optical Society. *See plate section for the color version*

The first step in the decoding is the inverse Fourier transform of $I(x_2, t)$, which results in a separation of the three terms in space. The T_1 term will appear on the optical axis and the T_2, T_3 terms will appear on the +1 and −1 orders due to the exponential term $e^{\pm 2\pi i \Delta x x_2 / \lambda F}$. This separation allows us to take only the term we are interested in, that is, T_2. The inverse Fourier transform of T_2 can be written as:

$$I.F.T. \left\{ T_2 \left(x_2, t \right) \right\} = s \left(x_3 - \Delta x \right) g \left(x_3 - \Delta x, t \right) \otimes \left(D \sin c \left(\frac{x_3}{\lambda F} D \right) \right), \tag{10.11}$$

where $\otimes$ denotes convolution. The first two terms stem from the inverse Fourier transform of the Fourier transform of the object multiplied by the mask and the shift due to the delta function. The delta function is a result of the inverse Fourier transform of the exponential term before the integral in T_2. The last term is the Fourier transform of the *rect* function.

In the same manner as that presented in Section 10.3 we can calculate the decoding mask. We now multiply the resulting term by the decoding pattern and perform time averaging to obtain

$$R \left(x_3 \right) = \int I.F.T \left\{ T_2 \left(x_3, t \right) \right\} g \left(x_3 - \Delta x, t \right) dt. \tag{10.12}$$

By substituting (10.11) into (10.12) we have

$$R(x_3) = \int s(x_3 - \Delta x) g(x_3 - \Delta x, t) \otimes \left(D \sin c \left(\frac{x_3}{\lambda F} D \right) \right) g(x_3 - \Delta x, t) dt. \tag{10.13}$$

We now look at the time-dependent terms in Eq. (10.13), which is the nanoparticles mask. Due to their small size of the nanoparticles and their random distribution we can write

$$\int g(x' - \Delta x, t) g(x_3 - \Delta x, t) dt = \kappa + \delta(x' - x_3). \tag{10.14}$$

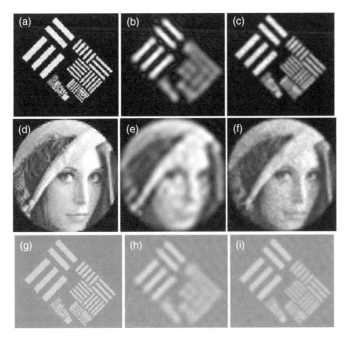

Figure 10.5 Numerical simulations for a binary amplitude resolution target (*top row*), Lena image (*middle row*), and phase resolution target (*bottom row*). The left column is the original high resolution object. The low resolution DH is in the center column and the super resolved object using metal nanoparticles is in the right column. *Source*: Zalevsky Z., Gur E., Garcia J., Micó V., Javidi B. 2012. Figure 2. Reproduced with permission from The Optical Society

We use the last equation to obtain the reconstruction

$$R(x_3) = D\kappa \int s(x' - \Delta x) \sin c \left(\frac{x_3 - x'}{\lambda F} D \right) dx' + \Delta x s(x_3 - \Delta x). \tag{10.15}$$

The first term here is the convolution from Eq. (10.13) and it corresponds to the low resolution image obtained by the convolution between the high resolution object and the *sinc* function caused by the finite size of the CCD. The second term is the term of interest to us and it is the high resolution reconstruction obtained by the proposed technique. Simulation results for the original object, the low resolution object and the super-resolution reconstruction are shown in Fig. 10.5.

10.5 Field of View Enhancement in Digital Holography

The resolution increase due to the placement of the nanoparticles in the object plane can be explained by considering the uncertainty principle of the Fourier transform. This principle states that

$$\Delta x \Delta f_x = const \tag{10.16}$$

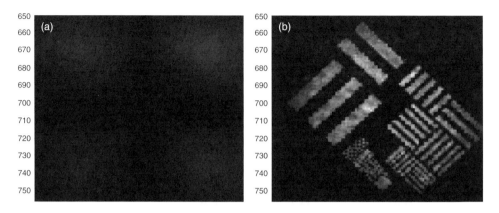

Figure 10.6 Numerical simulations for the FOV DH reconstruction. (a) The FOV reduced by a factor of 8. (b) Successful recovery of the original FOV with the nanoparticles' mask method. *Source*: Zalevsky Z., Gur E., Garcia J., Micó V., Javidi B. 2012. Figure 3. Reproduced with permission from The Optical Society

where x is the spatial coordinate and f_x is the spatial frequency. Δ designates the uncertainty or the range of values for each one of those two parameters. By using the metal nanoparticles in the object plane we limit our sampling of the object in each time frame to smaller Δx which now has the size of a metal nanoparticle; therefore, Δf_x increases; that is, more spatial frequencies can be recorded by the camera. Since, in the optical Fourier transform the frequency corresponds to real spatial coordinates, it means that in the Fourier plane we effectively need to increase the size of the CCD. Following this discussion, and remembering that up to multiplicative factors, the Fourier transform and Inverse Fourier transform are the same, if we reduce the sampling point in the CCD plane (the Fourier plane), we can obtain a larger FOV in the object plane. The finite size of the pixel on the CCD limits our ability to obtain large FOV, but using metal nanoparticles we can obtain sub-sampling in the CCD plane and thus larger FOV in the object plane.

In this case the term we are interested in, T_2 equals:

$$T_2(x_2) = \left[\left(e^{-2\pi i \Delta x x_2 / \lambda F} \int s\left(x_1\right) e^{-2\pi i x_1 x_2 / \lambda F} dx_1 \right) g(x_2, t) \right] \otimes p(x_2). \qquad (10.17)$$

Here $p(x_2)$ is the PSF that reduces the resolution of the recorded DH (the outcome of the finite sized pixels of the CCD). The decoding will include multiplication by high resolution pattern $g(x_2, t)$, inverse Fourier transform and time averaging:

$$R\left(x_3\right) = \int I.F.T\left\{T_2\left(x_2\right) g\left(x_3, t\right)\right\} dt, \qquad (10.18)$$

thus

$$R\left(x_3\right) = \int \left[\left(s\left(x_3 - \Delta x\right) \otimes G\left(\frac{x_3}{\lambda F}, t\right) \right) P\left(\frac{x_3}{\lambda F}, t\right) \right] \otimes G\left(\frac{x_3}{\lambda F}, t\right) dt, \qquad (10.19)$$

where G and P are the Fourier transforms of g and p respectively. Since

$$\int G\left(\frac{x'}{\lambda F}, t\right) G\left(\frac{x_3 - x_3'}{\lambda F}, t\right) dt = \delta\left(x' - x_3 - x_3'\right) + \kappa, \tag{10.20}$$

the result is

$$R\left(x_3\right) = s\left(x_3 - \Delta x\right) \int P\left(\frac{x_3'}{\lambda F}, t\right) dx_3' + \kappa \eta_s \eta_P, \tag{10.21}$$

where η_s, η_P are the averages of s and P respectively. The simulation results of this method are shown in Fig. 10.6.

10.6   Eliminating the DC Term and the Twin Images

A similar technique can be implemented in on-axis holographic scheme in order to eliminate the DC term and the twin image in the recorded hologram. Looking at the hologram with metal nanoparticles that are placed in proximity to the object, the time varying recorded frame can be written as

$$I(x, t) = |a(x) g(x, t) + A|^2, \tag{10.22}$$

where the reference is assumed to be a plane wave. Again, as before we capture multiple frames, localize the nanoparticles, and compute a decoding mask for each frame $\bar{g}(x, t) \approx g(x, t)$. Each frame is then digitally multiplied by the complex conjugate of the nanoparticles mask and time averaging of all frames is performed:

$$O(x) = \int I(x, t) \bar{g}^*(x, t) dt, \tag{10.23}$$

which can be written as

$$O(x) = \int |A|^2 \bar{g}^*(x, t) dt + \int |a(x) g(x, t)|^2 \bar{g}^*(x, t) dt + \int A^* a(x) g(x, t) \bar{g}^*(x, t) dt$$

$$+ \int A a^*(x) g^*(x, t) \bar{g}^*(x, t) dt \tag{10.24}$$

We now wish to eliminate all the terms but the third one which contains the actual imaged object. For this the nanoparticles should be chosen such that

$$g(x, t) = \exp(i\phi(x, y)), \tag{10.25}$$

where $\phi(x, t)$ can have one of four values: $\phi(x, t) = n\pi/2, n = 0, 1, 2, 3$. These four values can come from four different nanoparticle species. Since in the computed decoding mask we have to know not only the location of the nanoparticle, but also its phase, we can determine it by identifying to which of the four species the localized nanoparticle belongs. This can be accomplished by taking nanoparticles in different colors, shapes, scattering cross sections and so on. By considering Eq. (10.25), we observe that the third term in (10.24) equals

$$\int A^* a(x) g(x, t) \bar{g}^*(x, t) dt = \int A^* a(x) dt = \Delta T A^* a, \tag{10.26}$$

where ΔT is the integration time. Due to the random nature of the nanoparticles mask, and the phase values associated with each nanoparticle, the averaging of the other terms is zero, that is

$$\int \overline{g}^*(x,t)\,dt = \int g^*(x,t)\,\overline{g}^*(x,t)\,dt = \int |g(x,t)|^2\overline{g}^*(x,t)\,dt = 0. \qquad (10.27)$$

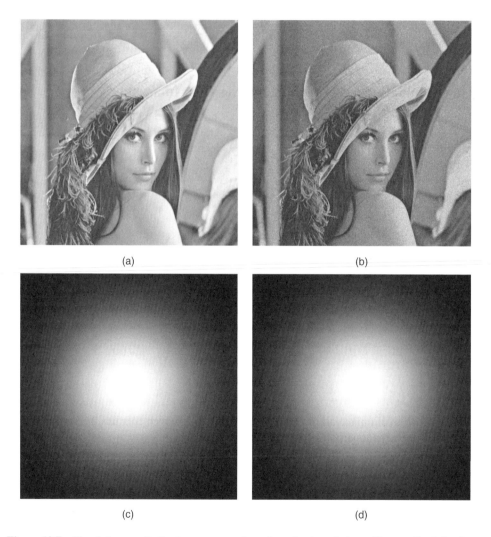

(a)   (b)

(c)   (d)

Figure 10.7   Simulation results for the reconstruction of amplitude and phase. The amplitude is a Lena image and the phase has a Gaussian profile. (a) Original field amplitude. (b) Reconstructed field amplitude. (c) Original field phase. (d) Reconstructed field phase. *Source*: Amihai Meiri, Eran Gur, Javier Garcia, Vicente Micó, Bahram Javidi, Zeev Zalevsky 2013. Reproduced with permission from SPIE

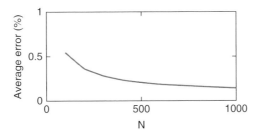

Figure 10.8   Phase error as a function of number of frames. *Source*: Amihai Meiri, Eran Gur, Javier Garcia, Vicente Micó, Bahram Javidi, Zeev Zalevsky 2013. Reproduced with permission from SPIE

Therefore we are left with

$$O(x) = \Delta T A^* a, \tag{10.28}$$

which is our imaged object, where the amplitude and phase are maintained. The simulated results for such a reconstruction are shown in Fig. 10.7 for 1000 recorded frames. The noise pattern of the reconstructed amplitude image in Fig. 10.7(a) is a result of this reconstruction and depends on the number of frames.

The relative error in phase reconstruction was calculated from the results of the simulations and shown on Fig. 10.8. This error was averaged over all pixels of the image and the results indicate that, even for a 100 frames, the error is as small as 0.5% and reduces rapidly with an increase in number of frames. This shows that using metal nanoparticles results in a very accurate reconstruction of the object phase.

10.7   Additional Applications

The random nanoparticle mask encoding can be used in other applications as well. As an example, we show the joint transform correlator (JTC) [24]. The JTC is a system that can be used to calculate the convolution of correlation between two functions in an all-optical setup. The setup for the JTC is presented in Fig. 10.9. A collimated beam is incident on two objects, h_1, h_2, which is located at the focal plane of lens L_1. At the back focal plane of the lens the recording medium is placed: see Fig. 10.9(a). In order to obtain the desired output (correlation of convolution between h_1 and h_2) we illuminate the recorded transparency by a collimated beam. The transparency is located at the focal plane of lens L_2 and the output is obtained at the back focal plane of the lens, see Fig. 10.9(b).

Using the metal nanoparticles random mask is depicted in Fig. 10.9(c). Here the nanoparticles are placed close to object h_1 and have the different phases as in Eq. (10.25). We now look at the intensity recorded by the transparency. The field at plane x_1 can be written as

$$U_1(x_1, t) = h_1(x_1 - X/2) g(x_1 - X/2, t) + h_2(x_1 + X/2), \tag{10.29}$$

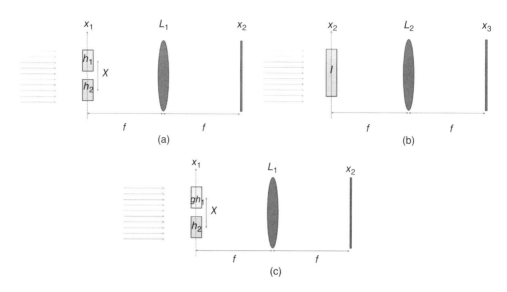

Figure 10.9 Joint transform correlator. (a) Recording the filter. (b) Obtaining the output

where g is the time varying nanoparticles phase encoding. In the rear focal plane of the lens L_1 we find the Fourier transform of the field

$$U_2(x_2,t) = \frac{1}{\lambda f}H_1\left(\frac{x_2}{\lambda f}\right) \otimes G\left(\frac{x_2}{\lambda f},t\right)e^{-j\pi x_2 X/\lambda f} + \frac{1}{\lambda f}H_2\left(\frac{x_2}{\lambda f}\right)e^{j\pi x_2 X/\lambda f}, \qquad (10.30)$$

where $\otimes$ denotes convolution. Therefore, the intensity recorded by the transparency is:

$$I(x_2,t) = \frac{1}{\lambda^2 f^2}\left[\left|H_1\left(\frac{x_2}{\lambda f}\right) \otimes G\left(\frac{x_2}{\lambda f},t\right)\right|^2 + \left|H_2\left(\frac{x_2}{\lambda f}\right)\right|^2\right]$$

$$+ \frac{1}{\lambda^2 f^2}\left[H_1\left(\frac{x_2}{\lambda f}\right) \otimes G\left(\frac{x_2}{\lambda f},t\right) H_2^*\left(\frac{x_2}{\lambda f}\right)e^{-j2\pi x_2 X/\lambda f}\right]$$

$$+ \frac{1}{\lambda^2 f^2}\left[H_1^*\left(\frac{x_2}{\lambda f}\right) \otimes G^*\left(\frac{x_2}{\lambda f},t\right) H_2\left(\frac{x_2}{\lambda f}\right)e^{j2\pi x_2 X/\lambda f}\right]. \qquad (10.31)$$

We now compute the Fourier transform of the nanoparticle mask, multiply each frame by the corresponding computed mask, and perform time averaging:

$$\int I(x_2,t) G\left(\frac{x_2}{\lambda f},t\right) dt \qquad (10.32)$$

The result is

$$I\left(x_2,t\right) = \frac{1}{\lambda^2 f^2} \int \left[\left|H_1\left(\frac{x_2}{\lambda f}\right) \otimes G\left(\frac{x_2}{\lambda f},t\right)\right|^2 + \left|H_2\left(\frac{x_2}{\lambda f}\right)\right|^2\right] G\left(\frac{x_2}{\lambda f},t\right) dt$$

$$+ \frac{1}{\lambda^2 f^2}\left[\int H_1\left(\frac{x_2}{\lambda f}\right) \otimes G\left(\frac{x_2}{\lambda f},t\right) H_2^*\left(\frac{x_2}{\lambda f}\right) e^{-j2\pi x_2 X/\lambda f} G\left(\frac{x_2}{\lambda f},t\right) dt\right]$$

$$+ \frac{1}{\lambda^2 f^2}\left[\int H_1^*\left(\frac{x_2}{\lambda f}\right) \otimes G^*\left(\frac{x_2}{\lambda f},t\right) H_2\left(\frac{x_2}{\lambda f}\right) e^{j2\pi x_2 X/\lambda f} G\left(\frac{x_2}{\lambda f},t\right) dt\right] \quad (10.33)$$

We look at the time integrals and obtain that:

$$\int \left|H_1\left(\frac{x_2}{\lambda f}\right) \otimes G\left(\frac{x_2}{\lambda f},t\right)\right|^2 G\left(\frac{x_2}{\lambda f},t\right) dt = 0$$

$$\int \left|H_2\left(\frac{x_2}{\lambda f}\right)\right|^2 G\left(\frac{x_2}{\lambda f},t\right) dt = 0$$

$$\int G\left(\frac{x_2}{\lambda f},t\right) G\left(\frac{x_2}{\lambda f},t\right) dt = 0$$

$$\int G^*\left(\frac{x_2}{\lambda f},t\right) G\left(\frac{x_2}{\lambda f},t\right) dt = const \quad (10.34)$$

as in the first three integrals of (10.34) there is a remaining time varying phase, which is averaged to zero while only in forth integral the time varying phase term is cancelled. The calculation of the nanoparticle mask is performed in the same way as in Section 10.6: we have four nanoparticle species, with different phases $\phi(x,t) = n\pi/2, n = 0, 1, 2, 3$. The nanoparticle mask equals

$$g\left(x,t\right) = \exp\left(i\phi\left(x,y\right)\right). \quad (10.35)$$

We can now localize each nanoparticle and identify its associated phase by identifying the species. Therefore

$$\int I\left(x_2,t\right) G\left(\frac{x_2}{\lambda f},t\right) dt = \frac{1}{\lambda^2 f^2} H_1^*\left(\frac{x_2}{\lambda f}\right) H_2\left(\frac{x_2}{\lambda f}\right) e^{j2\pi x_2 X/\lambda f}. \quad (10.36)$$

When illuminating the transparency as in Fig. 10.9(b) the field at the output is the Fourier transform of Eq. (10.36) and can now be expressed as

$$U_3\left(x_3\right) = \frac{1}{\lambda f}\left[h_1^*\left(-x_3\right) \otimes h_2\left(x_3\right) \otimes \delta\left(x_3 + X\right)\right]. \quad (10.37)$$

The result is the cross correlation of objects h_1 and h_2 shifted in space by X.

References

[1] Gabor D., "A New Microscopic Principle," *Nature* **161**, 777–778 (1948).

[2] Goodman J. W., *Introduction to Fourier Optics* (Roberts and Company Publishers, 2005).

[3] Leith E. N. and J. Upatnieks, "Reconstructed wavefronts and communication theory," *JOSA* **52**, 1123–1128 (1962).

[4] Liu G. and P. D. Scott, "Phase retrieval and twin-image elimination for in-line Fresnel holograms," *JOSA A* **4**, 159–165 (1987).

[5] Onural L. and P. D. Scott, "Digital decoding of in-line holograms," *Optical Engineering* **26**, 261124–261124 (1987).

[6] Zhang Y. and X. Zhang, "Reconstruction of a complex object from two in-line holograms," *Optics Express* **11**, 572–578 (2003).

[7] Jain P. K., K. S. Lee, I. H. El-Sayed, and M. A. El-Sayed, "Calculated absorption and scattering properties of gold nanoparticles of different size, shape, and composition: applications in biological imaging and biomedicine," *The Journal of Physical Chemistry B* **110**, 7238–7248 (2006).

[8] Brioude A., X. C. Jiang, and M. P. Pileni, "Optical properties of gold nanorods: DDA simulations supported by experiments," *J. Phys. Chem. B* **109**, 13138–13142 (2005).

[9] Susie E. and A. El-Sayed Mostafa, "Why gold nanoparticles are more precious than pretty gold: Noble metal surface plasmon resonance and its enhancement of the radiative and nonradiative properties of nanocrystals of different shapes," *Chemical Society Reviews* **35**, 209–217 (2006).

[10] Link S., M. B. Mohamed, and M. A. El-Sayed, "Simulation of the optical absorption spectra of gold nanorods as a function of their aspect ratio and the effect of the medium dielectric constant," *The Journal of Physical Chemistry B* **103**, 3073–3077 (1999).

[11] Kelly K. L., E. Coronado, L. L. Zhao, and G. C. Schatz, "The optical properties of metal nanoparticles: the influence of size, shape, and dielectric environment," *J. Phys. Chem. B* **107**, 668–677 (2003).

[12] Su K. H., Q. H. Wei, X. Zhang, J. J. Mock, D. R. Smith, and S. Schultz, "Interparticle coupling effects on plasmon resonances of nanogold particles," *Nano Letters* **3**, 1087–1090 (2003).

[13] Jain P. K., W. Huang, and M. A. El-Sayed, "On the universal scaling behavior of the distance decay of plasmon coupling in metal nanoparticle pairs: a plasmon ruler equation," *Nano Letters* **7**, 2080–2088 (2007).

[14] Browning L. M., T. Huang, and X. N. Xu, "Far-field photostable optical nanoscopy (PHOTON) for real-time super-resolution single-molecular imaging of signaling pathways of single live cells," *Nanoscale* **4**, 2797 (2012).

[15] Valeur B., *Molecular Fluorescence: Principles and Applications* (VCH Verlagsgesellschaft Mbh, 2002).

[16] Diaspro A., *Nanoscopy and Multidimensional Optical Fluorescence Microscopy* (Chapman & Hall, 2009).

[17] Gur A., D. Fixler, V. Micó, J. Garcia, and Z. Zalevsky, "Linear optics based nanoscopy," *Optics Express* **18**, 22222–22231 (2010).

[18] Inouye Y. and S. Kawata, "Near-field scanning optical microscope with a metallic probe tip," *Optics Letters* **19**, 159–161 (1994).

[19] Andersson S. B., "Precise localization of fluorescent probes without numerical fitting," in *4th IEEE International Symposium on Biomedical Imaging: From Nano to Macro, 2007. ISBI 2007* (IEEE, 2007), pp. 252–255.

[20] Cheezum M. K., W. F. Walker, and W. H. Guilford, "Quantitative Comparison of Algorithms for Tracking Single Fluorescent Particles," *Biophysical Journal* **81**, 2378–2388 (2001).

[21] Carter B. C., G. T. Shubeita, and S. P. Gross, "Tracking single particles: a user-friendly quantitative evaluation," *Physical Biology* **2**, 60 (2005).

[22] Pertsinidis A., Y. Zhang, and S. Chu, "Subnanometre single-molecule localization, registration and distance measurements," *Nature* **466**, 647–651 (2010).

[23] Zalevsky Z., E. Gur, J. Garcia, V. Micó, and B. Javidi, "Superresolved and field-of-view extended digital holography with particle encoding," *Optics Letters* **37**, 2766–2768 (2012).

[24] Weaver C. S. and J. W. Goodman, "A technique for optically convolving two functions," *Applied Optics* **5**, 1248 (1966).

Part Three

Multi-Dimensional Imaging and Display

11

Three-Dimensional Integral Imaging and Display

Manuel Martínez-Corral[1], Adrián Dorado[1], Anabel LLavador[1],
Genaro Saavedra[1] and Bahram Javidi[2]

[1]*Department of Optics, University of Valencia, Spain*
[2]*Department of Electrical and Computer Engineering, University of Connecticut, USA*

11.1 Introduction

Conventional photographic cameras do not have the capacity to record all of the information carried by the rays of light passing through their objective lens, which indeed impacts on the image sensor pixels from many directions [1]. The irradiance collected by any pixel is proportional to the sum of radiances of all rays, regardless of their incidence angles. Thus, a typical picture obtained with a conventional camera contains, of course, a 2D image (or in other words a pixelated 2D irradiance map) of the original 3D object, in which any information about intensity and angle of impinging light rays is lost.

Much more interesting would be to have a system with the capacity of registering a map with the radiance and direction of all the rays proceeding from the 3D scene. Such a map has been named in different ways, such as integral photography, integral imaging, lightfield map, or even a *plenoptic* map.

Interest in the capture and display of 3D images is not a modern issue. In fact in 1838, Wheatstone [2] tackled the problem of displaying 3D images through the first stereoscope. Soon later, Rollman faced the same problem, but proposed the use of anaglyphs [3]. But the main problem of these techniques, and also of all the stereoscopic techniques developed since then, is that they do not actually produce a 3D reconstruction of a 3D scene. Instead, they produce a stereoscopic pair of images which, when projected onto the retinas of the left and right eyes, provides the brain with the sensation of perspective vision and depth discrimination. Apart from the problem that stereoscopy provides the same perspective to different observers, whatever their relative position to the screen, the main drawback comes from the conflict between

Multi-dimensional Imaging, First Edition. Edited by Bahram Javidi, Enrique Tajahuerce and Pedro Andrés.
© 2014 John Wiley & Sons, Ltd. Published 2014 by John Wiley & Sons, Ltd.

convergence and accommodation [4]. This conflict occurs when the accommodation of the eye lens is fixed to one distance, whereas the convergence of the eyes' axes is set to a different distance. This is a strongly unnatural physiological procedure that gives rise to visual discomfort after prolonged observations.

The first scientist to propose a method for displaying/observing 3D images without the need for any special glasses was Gabriel Lippmann, who proposed the integral photography (IP) in 1908 [5–7]. Lippmann's idea was that it is possible to record the information of a 3D scene with a 2D sensor, provided that many perspectives of the scene are stored. To do that, Lippmann proposed to remove the objective from the camera and, instead, to insert a microlens array (MLA) in front of the sensor. This way of proceeding permits to record an integral image (InI), that is, a collection of elemental images each with a different perspective of the scene. The InI captured following Lippmann's recipe can be used for many proposals. One is for the implementation of an IP digital monitor [8–10]. For this task it is necessary to project the set of elemental images onto a pixelated display, like a LCD or LED monitor, and insert a MLA, similar to the one used in the capture, just in front. In the display process, any pixel of the LCD panel emits a light cone, which after passing through the corresponding microlens produces a light cylinder in free space. It is the intersection of all the ray cylinders that produces an irradiance distribution in front of the monitor, which reproduces the irradiance of the original 3D scene. This distribution is perceived as 3D by the observer, whatever his/her position, with a continuous viewpoint and also with full parallax. Since the observer is watching a real light concentration, the observation is produced without conflict between convergence and accommodation.

IP, like holography, is usually catalogued as an auto-stereoscopic technique due to the fact that it does not need the use of special glasses. However, auto-stereoscopic is a misleading name since it implies IP is based in stereoscopy, which is clearly not the case.

An alternative architecture for the capture of the integral image was proposed by Davies and McCormick [11], and later by Adelson and Wang [12]. The technique consisted of inserting a microlens array at the back-focal plane of a conventional photographic camera, in front of the sensor. Proceeding in this way, it is possible to capture the radiance map (or plenoptic picture) of far 3D scenes. This alternative architecture has been named as plenoptic camera [1,13,14], lightfield camera [14,15] and also far-field IP camera [16,17].

The IP concept was initially intended for the capture and display of 3D pictures or movies. In this sense, many important advances have been achieved in the search for the improvement of display resolution [18–22], viewing angle [23–25], or the depth of field [26–29]. However in the past few years, the computational reconstruction of the irradiance distribution of incoherent 3D scenes has become a major application [30–36]. This reconstruction can be very useful for depth segmentation of 3D objects [37–40], or for the sensing and recognition of 3D objects in normal [41–44] or in low light levels [45,46]. Other interesting applications of the IP concept are, for example, the tracking of moving objects in a 3D environment [47,48], the polarimetric discrimination [49], the integration of 3D hyperspectral information [50], the 3D microscopy [51–53], or even wavefront sensing [54].

The aim of this chapter is to expose in a simple manner the theory behind the IP concept, the relation between the different forms of capturing the radiance map, and the algorithms and methods for 3D depth reconstruction in the display of 3D scenes. To this end we have divided the chapter into eight sections. In Section 11.2 we expose the basic theory of conventional 2D image capturing. In Section 11.3 we define the plenoptic function and corresponding transformation rules. In Section 11.4 we describe the different methods for capturing the plenoptic map. In Section 11.5 we study the transformations suffered by the plenoptic map when changing the reference plane, and also calculate the 2D picture associated to any plane. In Section 11.6 we calculate the reconstructed 3D scene for different methods of plenoptic map capture. In Section 11.7 we tackle the implementation of 3D IP displays. Finally, in Section 11.8 we summarize the main outcomes of this chapter.

11.2   Basic Theory

A diffusing or self-luminous object can be treated as a continuous distribution of point sources that emit light isotropically. Although the natural way of describing this emission is through the concept of a spherical wavefront, in what follows we will assume that the phenomena we deal with are well described by the ray-optics theory, in which rays are the carriers of light energy and propagate in a straight line. We also assume that the point sources are mutually incoherent so that light emitted by them cannot interfere.

The most common device for capturing light emitted by a 3D scene is the photographic camera. As shown in Fig. 11.1, to obtain a picture of a 3D scene the camera is set so that the sensor is conjugate with a plane of the object space, the *reference plane*. The light emitted by points on such a plane is registered sharply in the pixelated sensor. The sensor also records the light emitted by out-of-focus points but with some blurring, which depends on the geometrical parameters of the capture system. Since the behavior of the electronic sensor is, in good approximation, linear, any pixel register the sum of the radiances of all the rays impacting the pixel. Thus, the recorded picture contains the sharp image of the parts of the scene that are on the reference plane, plus blurred information from the rest of the scene. However, it does not contain any individualized information about the radiance of the rays that reach the pixels. The loss of this angular information prevents the recorded picture from recovering the 3D structure of the scene from a conventional picture.

The magnitude that allows an accurate description of angular and spatial information of propagated rays is the radiance, defined as the radiant flux, Φ, per unit of area, A, and unit of

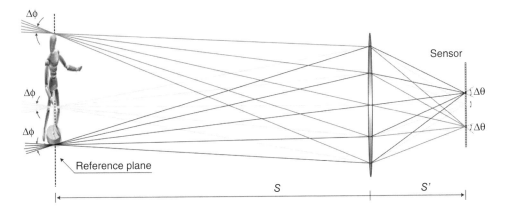

Figure 11.1   Optical scheme of a conventional photographic camera

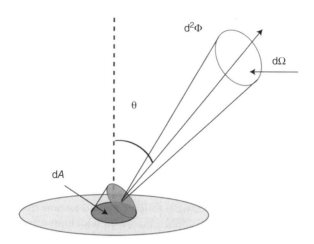

Figure 11.2 Scheme for the definition of the radiance

solid angle, emitted by a differential surface in a given direction. In mathematical terms the radiance

$$L = \frac{d^2\Phi}{d\Omega \, dA \, \cos\theta},$$ (11.1)

where θ is the angle between the normal to the surface and the specified direction. This definition permits to establish the concept of ray of light as the light cone delimited by an infinitesimal solid angle, $d\Omega$, see Fig. 11.2. The 3D spatial distribution of radiance emitted by a 3D scene is usually named the *lightfield* or equivalently, the *plenoptic field*.

A ray of light can be parameterized through the spatial coordinates $\mathbf{r} = (x, y, z)$ of one point of its trajectory, and the angular coordinates $\boldsymbol{\theta} = (\theta, \phi)$ that describe its inclination. Then, given a region in the space, the plenoptic field is described by a 5D function. If we consider the case in which the rays propagate in a region free of occlusions and diffusing media, the direction and radiance of rays do not change with propagation. This redundancy permits saving one dimension in the description of the plenoptic field, and then parameterizing it through a 4D function. Following the paper by Georgiev and Lumsdaine [13] we can represent the plenoptic function as $L(\mathbf{x}, \boldsymbol{\theta})$, where $\mathbf{x} = (x, y)$. However, and for the sake of the simplicity in the forthcoming graphic representations, we will consider the case in which the plenoptic function is 2D, $L(x, \theta)$, so that we only will consider rays propagating on a plane. Naturally, this simplification does not subtract any generality to the forthcoming study.

11.3 The Plenoptic Function

The simplest example of the plenoptic field is that generated by a monochromatic point source. To build the 4D plenoptic function, first one has to select the plenoptic reference plane. This plane is typically perpendicular to the chief direction of propagation of the light beam. Then, one has to evaluate the inclination of rays when impacting the plenoptic reference plane. Assuming that the point source radiates isotropically, the plenoptic field at a plane placed at a distance z_0 from the source, is represented with a straight line, of slope $\mu = 1/z_0$, since there

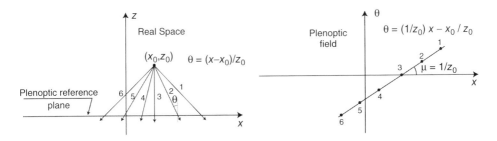

Figure 11.3 Plenoptic field generated by a monochromatic point source

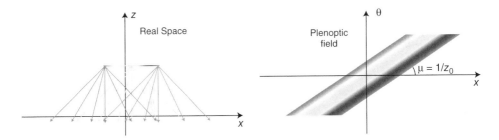

Figure 11.4 Plenoptic field generated by a monochromatic plane source

is direct proportionality (within the paraxial approximation) between the spatial coordinate of the impact point and the inclination angle (see Fig. 11.3).

Of similar complexity is the plenoptic field produced by a plane object that is set parallel to the plenoptic reference plane. Any point of the object is represented in the plenoptic diagram by a straight line. The bundle composed by the inclined lines produced by all the points of the plane object constitutes the plenoptic field, as shown in the Fig. 11.4.

It is interesting to find out how the plenoptic function changes when light propagates in free space, or when it passes through a converging lens. These transformations can be formalized by means of a transfer matrix [55]. The propagation in free space implies a change in the spatial coordinates, but not in the inclination angles, as shown in Fig. 11.5(a). The transfer matrix is of the form

$$T = \begin{pmatrix} 1 & t \\ 0 & 1 \end{pmatrix},$$ (11.2)

where t is the propagation distance. The propagated plenoptic field can be obtained from the original one through the coordinates' transformation

$$\begin{pmatrix} x' \\ \theta' \end{pmatrix} = \begin{pmatrix} 1 & t \\ 0 & 1 \end{pmatrix} \begin{pmatrix} x \\ \theta \end{pmatrix} = \begin{pmatrix} x + t\theta \\ \theta \end{pmatrix}.$$ (11.3)

This can be understood as a shearing of the plenoptic function in the direction of the spatial coordinates, as shown in Fig.11.6(a).

Another interesting issue is to find out how the plenoptic field is modified when passing through a converging lens. As shown in Fig. 11.5(b), this transition implies a change in the

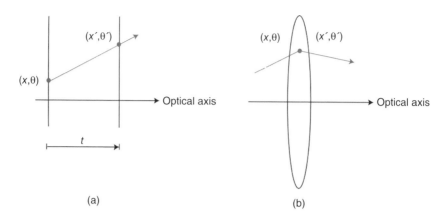

Figure 11.5 Coordinate transformation associated with (a) free space propagation and (b) refraction in a lens

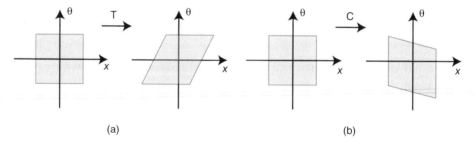

Figure 11.6 Shearing of the plenoptic function as result of (a) free-space propagation and (b) refraction in a lens

inclination angle of impinging rays, but not in the spatial coordinate. According to [55], the transfer matrix is

$$C = \begin{pmatrix} 1 & 0 \\ -1/f & 1 \end{pmatrix}, \tag{11.4}$$

where f is the focal length of the lens. The plenoptic field after refraction through the lens can be obtained from the impinging one by

$$\begin{pmatrix} x' \\ \theta' \end{pmatrix} = \begin{pmatrix} 1 & 0 \\ -1/f & 1 \end{pmatrix} \begin{pmatrix} x \\ \theta \end{pmatrix} = \begin{pmatrix} x \\ -x/f + \theta \end{pmatrix}. \tag{11.5}$$

Also in this case the plenoptic function suffers a shearing, but now in the angular direction, as shown in Fig. 11.6(b).

Let us remark that, from the plenoptic function evaluated in a plane, it is possible to calculate the "picture" by simply summing up the radiances at any point of that plane. In mathematical terms this can be made by calculation of the Abel transform [56] (or angular projection onto

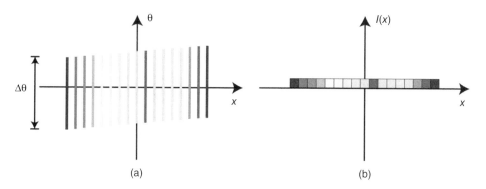

Figure 11.7   (a) Plenoptic filed captured by a conventional photographic camera; (b) registered picture, which is obtained as the angular projection of the plenoptic field

the spatial axis) of the plenoptic function; that is

$$I(\mathbf{x}) = \int L(\mathbf{x}, \boldsymbol{\theta}) d\boldsymbol{\theta}. \tag{11.6}$$

When a conventional photographic camera is used to record the light emitted by a 3D scene, any pixel captures all the rays passing through its conjugate in the object's reference plane (see Fig. 11.1). Expressed in terms of the plenoptic field, any pixel captures the plenoptic field contained in a vertical segment whose length is equal to the angle subtended by the aperture of the camera lens; see Fig. 11.7. Of course, the recorded picture is given by the Abel transform of the plenoptic function. It is apparent that, in the photographic shot, the angular information, and therefore the 3D information, is lost.

11.4   Methods for the Capture of the Plenoptic Field

11.4.1   Integral Photography

A very clever way of recording sampled information of the plenoptic field produced by 3D objects is the use of multiview camera systems. Such kind of systems can be implemented, as proposed by Lippmann [5], by placing a *microlens array* (MLA) in front of a light sensor. Other ways of implementing the IP capturing system, mainly useful when aiming the recording of the plenoptic field produced by large 3D scenes, is to use a set of digital cameras arranged in a rectangular grid. The picture captured by any camera is named here as an elemental image. As shown in Fig. 11.8, an array of equidistant cameras placed at the plenoptic reference plane can acquire sampled information of the plenoptic field. Any elemental image contains discrete information about the angles of rays passing through the center of the entrance pupil of the camera. Thus, every elemental image contains sampled information of a vertical line in the plenoptic function (see the black envelope in Fig. 11.8b). On the contrary, a horizontal line in the plenoptic diagram corresponds to a set of rays passing through the reference plane, equidistant and parallel to each other (see the white envelope in Fig. 11.8b). The pixels of a horizontal line can be grouped to form a sub-image of the 3D scene. These sub-images, which will be named *micro-images* here, are orthographic views of the 3D scene. Orthographic means

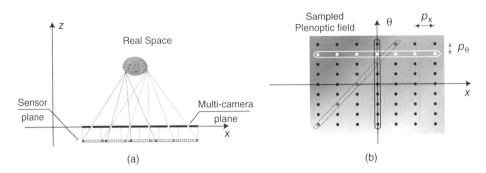

Figure 11.8   (a) Capture of the plenoptic field with a multi-camera system; (b) registered plenoptic field

Figure 11.9   Experimental set up for the acquisition of the elemental images

that the scale of the image does not depend on the distance from the object to the lens. Finally, an inclined line (see the dotted envelope) corresponds to the plenoptic field radiated by a point of the 3D scene.

Next, we describe a typical experiment for the capture of the plenoptic field with an experimental setup based on the IP concept (see Fig. 11.9). Although there are already some compact realizations of IP capturing setups [57,58], we used the so-called synthetic aperture method [59], in which all the elemental images are picked up with just one digital camera that is mechanically translated. The synchronized positioning, shooting, and recording of the elemental images was controlled by a LabVIEW® code. The digital camera (Canon 450D) was focused at the wooden panel, which was placed at a distance of 630 mm. The camera parameters were fixed to $f = 18$ mm and $f_\# = 22.0$, so that the depth of field was large enough obtain sharp pictures of the entire 3D scene. With the setup there is a set of $N_H = N_V = 11$ elemental images with pitch $P_H = P_V = 10.0$ mm. Since the pitch is smaller than the size of the CCD sensor (22.2 × 14.8 mm), we cropped every elemental image to remove the outer parts. In

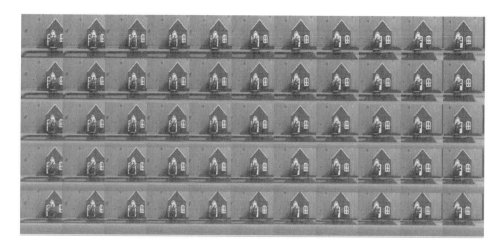

Figure 11.10   Subset of the elemental images obtained experimentally

addition, we resized the elemental images so that any image was composed by $n_H = n_V = 300$ pixels.

Next, in Fig. 11.10 we show a subset ($N_H = 11$ and $N_V = 5$) of the captured elemental images. Note that each elemental image stores different perspective information about the 3D scene. From the elemental image set (the InI) one can easily calculate the micro-images by simply extracting and composing the pixels with the same local position in every elemental image. This transposition of the plenoptic information allowed for calculation of $N_H = 300 \times N_V = 300$ micro-images composed each by $n_H = 11 \times n_V = 11$ pixels. In the Fig. 11.11 we show the complete micro-image collection.

11.4.2   The Plenoptic Camera

Based on the concepts reported by Lippmann, some research groups (Davies and McCormick [11], Adelson and Wang [12], and later Okano *et al.* [17]) proposed a technique for recording, after a single snapshot, the radiance emitted by a 3D scene. Due to Adelson and Wang, this setup is known as the *plenoptic camera*.

A scheme of this camera is shown in Fig. 11.12. In the plenoptic camera an array of microlenses is inserted in front of the sensor. This new architecture, in which the 3D scene is not directly in front of the MLA but projected close to it, is useful in the case of far objects, or for special applications like microphotography or ophthalmoscopy. The conjugation relations are of great importance in the plenoptic scheme. Specifically, the system is adjusted in such a way that the reference plane is conjugated with the MLA through the camera lens [1]. On the other hand, the pixelated sensor is conjugated with the camera lens (or, more specifically, with its exit pupil) through the microlenses. The images recorded by the sensor will be named *micro-images* here. In order to avoid overlapping between micro-images, the aperture angle of the camera lens must be equal to that of the microlenses.

Figure 11.11 Micro-images calculated from the integral image. In the inset we show a subset of 6×6 micro-images

As shown in Fig. 11.12, the plenoptic camera does not directly capture the plenoptic field emitted by the 3D scene, but the one imaged by the camera lens. However, since there is a simple scaling relation between them, it is not difficult to calculate the plenoptic field in the object reference plane. Next, in Fig. 11.13 we depict the plenoptic field captured by the plenoptic camera in Fig. 11.12. Note that only the central pixel of the central microlens captures a ray with inclination $\theta = 0$, the rest of the central pixels capture rays whose inclination angle is proportional to the spatial coordinate of the center of the corresponding microlens. This is the reason for the sheared aspect of the captured plenoptic field.

Similar to that we explained in Section 11.4.1, the pixels in any vertical stack correspond to the micro-images. Here we can also compute sub-images, which are obtained by grouping pixels that have the same relative position in their respective micro-image (see the dotted envelope in Fig. 11.13). As we will see later, it is reasonable to name elemental images corresponding to the sub-images of the micro-images, and vice-versa.

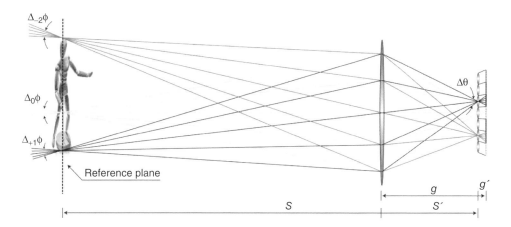

Figure 11.12   Scheme of the plenoptic camera

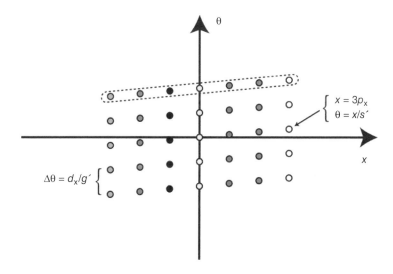

Figure 11.13   The sampled plenoptic field captured with the plenoptic camera of previous figure

Although there are already some commercial realizations of the plenoptic camera [60,61], we prepared our own plenoptic device in an open configuration in our laboratory. In Fig. 11.14 we show the experimental setup. A camera lens of $f = 100$ mm was used to conjugate the object reference plane with the MLA. The MLA was composed of a 94×59 lenslets of focal length $f_L = 0.93$ mm arranged in square grid of pitch $p_x = p_y = 0.222$ mm (APO-Q-P222-R0.93 model from AMUS). A digital camera with a macro objective 1:1 was used as the relay system that imaged the micro-images onto the sensor.

After the snapshot we obtained the plenoptic frame composed of 94×59 micro-images with 31×31 pixels each. The plenoptic frame is shown in Fig. 11.15. A pair of the computed sub-images is shown in Fig. 11.16.

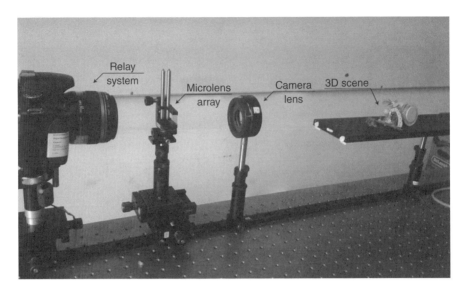

Figure 11.14   Experimental setup used for the capture of the plenoptic frame

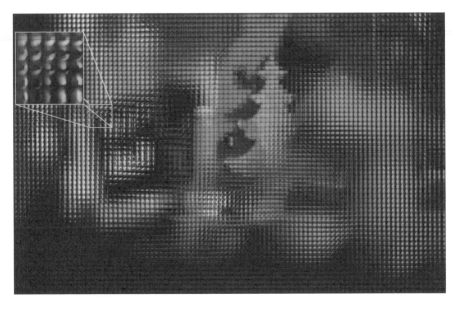

Figure 11.15   Captured plenoptic frame composed by 94×59 micro-images with 31×31 pixels each. In the inset we show a subset of 5×5 micro-images

Figure 11.16   Two of the subimages (also named *elemental images*) computed from the captured micro-images. From the plenoptic frame in Fig. 11.15 we could compute 31×31 elemental images composed of 94×59 pixels each. Note that any subimage observes the 3D scene from a different perspective

11.5   Walking in Plenoptic Space

As explained in Section 11.3, from the plenoptic function evaluated, or captured, in a given plenoptic reference plane, it is algebraically easy to calculate the plenoptic function in other planes, along with calculating the 2D pictures in such planes. To exemplify this, let us mark some specific planes in the plenoptic scheme depicted in Fig. 11.17(a) (Plate 18). To simplify the calculations, in this figure we have considered that the MLA is conjugated, through the camera lens, with infinity, so that $g = f$. Besides, since f is much larger than the focal length of the microlenses, the distance g' can be approximated by f_L.

Let us calculate first the plenoptic field at a plane (x_0, θ_0) placed in front of the MLA, at a distance t,

$$\begin{pmatrix} x_1 \\ \theta_1 \end{pmatrix} = T^{-1} \begin{pmatrix} x_0 \\ \theta_0 \end{pmatrix} = \begin{pmatrix} x_0 - t\theta_0 \\ \theta_0 \end{pmatrix}, \tag{11.7}$$

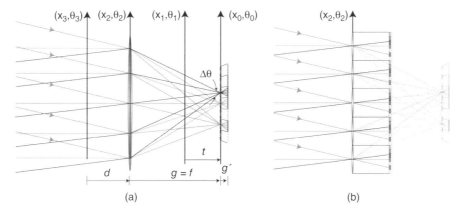

Figure 11.17 (Plate 18)   (a) Scheme for the calculation of the plenoptic function in planes parallel to the MLA; (b) the plenoptic field as evaluated in the plane of the camera lens is equivalent to the one captured with an IP setup. *See plate section for the color version*

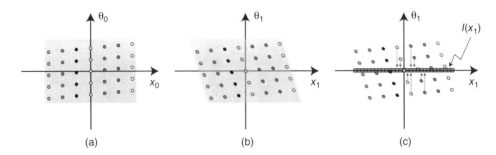

Figure 11.18 (a) Captured plenoptic field, (b) plenoptic field in the plane (x_1, θ_1), which is obtained after shearing the captured plenoptic function, and (c) irradiance distribution at the plane (x_1, θ_1), obtained after projecting the sheared plenoptic function

consequently,

$$L_1(x, \theta) = L_0(x - t\theta, \theta). \tag{11.8}$$

And therefore the picture in such plane can be calculated as

$$I_1(x) = \int L_0(x - t\theta, \theta)\, d\theta. \tag{11.9}$$

In Fig. 11.18 we illustrate this transformation. Note that Eqs (11.8) and (11.9) are valid for both positive and negative values of t, and therefore valid for calculating pictures in front or behind the MLA.

Now we calculate the plenoptic field in the plane of the camera lens (before the refraction),

$$\begin{pmatrix} x_2 \\ \theta_2 \end{pmatrix} = (T \cdot C)^{-1} \begin{pmatrix} x_0 \\ \theta_0 \end{pmatrix} = \begin{pmatrix} 1 & -f \\ 1/f & 0 \end{pmatrix} \begin{pmatrix} x_0 \\ \theta_0 \end{pmatrix} = \begin{pmatrix} x_0 - f\theta_0 \\ x_0/f \end{pmatrix}. \tag{11.10}$$

In Fig. 11.19 we illustrate this process.

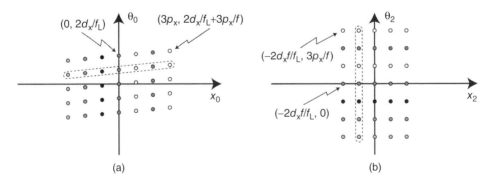

Figure 11.19 (a) Captured plenoptic field; (b) plenoptic field in the plane placed just before the refraction in the camera lens

As we can see from Eq. (11.10) and Fig. 11.19 (and have illustrated in Fig. 11.17(b)), the plenoptic field captured by the MLA placed at the back focal plane of the camera lens is nothing but a sheared and rotated version of the plenoptic field that could be captured by an adequate IP system placed at the camera-lens plane. A similar relation holds in the opposite direction; from the plenoptic field captured with an IP system, it is possible to calculate the plenoptic field captured by the proper plenoptic camera. The transposition relation of both plenoptic fields [15] can be useful for the calculation of elemental images from the micro-images captured with a plenoptic camera or vice versa. This property permits us, for example, to state that the micro-images calculated in Fig. 11.11 are similar to the micro-images captured directly with a proper plenoptic camera.

Finally, the plenoptic field at the plane (x_3, θ_3) can be calculated either from the plenoptic image or from the corresponding integral image,

$$\begin{pmatrix} x_3 \\ \theta_3 \end{pmatrix} = \begin{pmatrix} 1 & -d \\ 0 & 1 \end{pmatrix} \begin{pmatrix} x_2 \\ \theta_2 \end{pmatrix} = \begin{pmatrix} 1 - d/f & -f \\ 1/f & 0 \end{pmatrix} \begin{pmatrix} x_0 \\ \theta_0 \end{pmatrix}. \tag{11.11}$$

Therefore the picture in such plane can be calculated as

$$I_3(x) = \int L_2(x - \theta d, \theta) \, d\theta = \int L_0(x(1 - d/f) - f\theta, x/f) \, d\theta. \tag{11.12}$$

11.6   Reconstruction of Intensity Distribution in Different Depth Planes

As explained in Section 11.5, from the sampled version of the plenoptic field captured either with an IP setup or with a plenoptic camera, it is possible to calculate the irradiance distribution in different transverse sections of the original 3D scene. Although the integral image and the plenoptic image carry the same information, they are useful for different proposals. In particular, it is more convenient to use the integral image for implementing reconstruction algorithms, since it is composed of a low number of elemental images with a high number of pixels each.

The numerical reconstruction can be made by use of different algorithms. However, all are based on the same principle of shearing the plenoptic function and applying the Abel transform. This procedure can be visualized more intuitively, as shown in Fig. 11.20 (Plate 19), as projecting the pixels of any elemental image through a pinhole placed at the center of the corresponding microlens onto the corresponding reconstruction plane.

From the scheme, we can see that there is a univocal relation between the position of the reconstruction plane and the degree of overlapping between projected elemental images. Specifically, the degree of overlapping, M, is

$$M = \frac{z_R}{g} = \frac{N}{N - n}, \tag{11.13}$$

where N is the number of pixels per microlens and n is the number of pixels that overlap with the neighboring projected elemental image. It is apparent that the higher the degree of overlapping, the smaller the number of pixels of reconstructed images. To avoid unbalanced values of pixels of reconstructed images, the reconstructed pixels must be normalized, taking

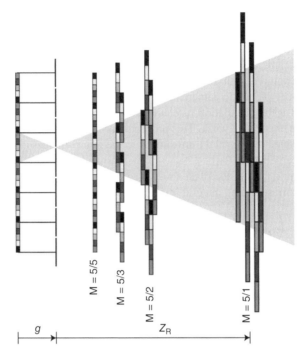

Figure 11.20 (Plate 19)   Scheme of the conventional reconstruction algorithm. In this figure, the number of pixels per microlens is $N = 5$. *See plate section for the color version*

into account the number of projected pixels that contribute to any reconstructed pixel. In planes where the projected pixels do not match (i.e., in case of non-integer value of n) it is also possible to calculate the reconstruction. But the algorithm is slower, since it has to evaluate the percentage of contribution of any projected pixel to the pixels of the reconstructed images.

To show the power of reconstruction procedure, we have calculated, from the elemental images shown in Fig. 11.10, the irradiance distribution (i.e., the reconstructed picture) in some depth planes. The reconstructed pictures are shown in Fig. 11.21. Note that the resolution of the images is similar to the resolution of the elemental images [62]. The segmentation capacity, that is, the capacity of blurring images of out-of-focus planes, is determined by the number of elemental images and also by their parallax. In this experiment, the segmentation capacity is so high that the cook seems to disappear in the last reconstructed image.

As stated previously, the same reconstruction algorithm can be applied to the images obtained in the plenoptic experiment of Fig. 11.14. In this case the input for the algorithm is not the micro-images registered by the camera but the elemental images calculated after applying the transformation of Eq. (11.10). The reconstructed images are shown in Fig. 11.22. As in the previous case, the number of pixels of the elemental images (i.e., the number of micro-images) determines the resolution of reconstructed images. The number of elemental images (i.e., the number of pixels per micro-image) determines the segmentation capacity. These reconstructed images are poorer than the ones obtained here from the IP experiment. This is because the number of pixels in each elemental image, along with its parallax, is much lower.

(a)   (b)

(c)   (d)

Figure 11.21 Reconstruction of the 3D scene in: (a) the plane of the hand; (b) the plane of the eyes; (c) the plane of the house; (d) the wooden panel

(a)   (b)   (c)

Figure 11.22 Reconstruction of the 3D scene in (a) the plane of the flag, (b) the plane of the fir, and (c) the plane of the chart

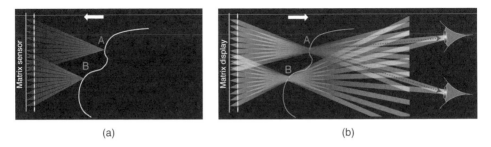

(a)   (b)

Figure 11.23 Scheme of capture and display process in integral photography. (a) In the capture stage any microlens stores different perspective information of the 3D scene. (b) In the display, any of the pixels in the panel can produce a light beam after passing through the corresponding microlens. The intersection of the light beams produces a light distribution that reconstructs the 3D scene

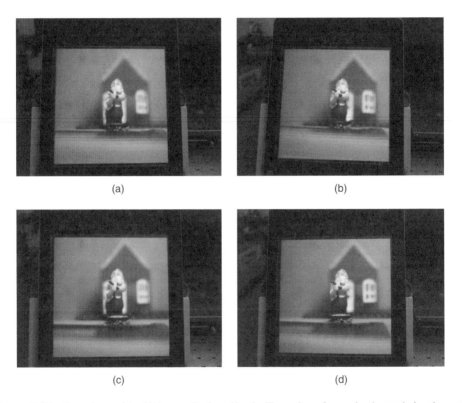

Figure 11.24 Four views of the 3D image displayed by the IP monitor after projecting on it the plenoptic frame as shown Fig. 11.11

11.7   Implementation of the Integral Imaging Display Device

When the integral (or the plenoptic) photography is acquired with the aim of projecting it onto an IP monitor, one should take into account the following facts. (1) The IP monitor is composed by a pixelated display (like a LCD or a LED display) and a MLA adjusted so that the display panel is at the front focal plane of the microlenses. (2) Since the resolution unit of an InI monitor is the pitch of the MLA [63], one should select a MLA with a high number of small microlenses. (3) Since the perspective resolution is determined by the number of pixels behind any microlens, one should arrange the display so that the number of pixels per microlens is about $12-16$. (4) The images projected behind the MLA should be processed in a way that the monitor would project an orthoscopic 3D image. (5) The displayed 3D image should be centered at the display plane, so that some parts of the 3D image should be floating in front of the panel, and other parts should be behind it.

Next, in Fig. 11.23 we show a scheme of the display process in an InI monitor.

The easiest way of satisfying all the constraints listed previously is by projecting a plenoptic frame onto the display panel. As explained in previous sections, a plenoptic frame can be recorded directly with a plenoptic camera. Another solution is to record an IP with an array of digital cameras and later calculate the transposed plenoptic frame. Both, the IP and the plenoptic capture should take into account the parameters of the display. For example, we have prepared an IP monitor by using an iPad with retina display, consisting of 2048×1536 RGB pixels with width $\Delta_x = \Delta_y = 89.0\,\mu m$, and a MLA composed by lenses with $f_L = 3.0\,mm$ and pitch $p_x = p_y = 1.0\,mm$. Thus, the plenoptic frame should be composed by up to 186×140 micro-images with 11×11 pixels each.

The calculated plenoptic frame shown in Fig. 11.11 is precisely composed by micro-images with 11×11 pixels each. Thus, it is ready to be projected onto the display device. In Fig. 11.24 we show some views of the IP display. The horizontal and vertical parallax of the 3D image

(a)   (b)

Figure 11.25   Two views of the 3D image displayed by the IP monitor after projecting on it the plenoptic frame shown Fig. 11.15

displayed by the monitor can easily be seen. Although it is not possible to show it here, some parts of the displayed scene were floating out from the monitor and others into it. This was clearly seen by binocular observers.

In our second display experiment we projected onto the panel the plenoptic frame captured directly with the plenoptic camera, see Fig. 11.15. Note that although in Fig. 11.25 we show only the horizontal parallax, the displayed images also have vertical parallax. As in Section 11.6, the quality of this image is much poorer than that of the previous one. This is because the total number of pixels involved is much smaller.

Naturally there are some computational methods, not used here, that would permit to recalculate the plenoptic frame so that the number of micro-images and the number of pixels per microimage could be adapted to the parameters of the IP monitor [64,65].

11.8   Conclusion

The aim of this chapter is to expose integral photography; a technique that, although reported more than a century ago, has been recently revealed as the most effective method for the aim of projecting 3D pictures or movies to audiences or more than one person. We have explained the physics behind the IP concept and the relation to other realizations of the same concept. We have shown that there are mainly two different ways of implementing IP. In both cases, it is possible to apply computational reconstruction algorithms and also to implement IP monitors. Although the level of development of IP is currently high, a significant improvement in resolution, segmentation capacity, viewing angle, and development of new applications is expected in the next few years.

Acknowledgments

This work was funded by the Plan Nacional I+D+I (grant DPI2012-32994) Spain, and by the Generalitat Valenciana (grant PROMETEO2009-077).

References

[1] Ng R., *Digital light field photography*, Ph.D. Thesis (Stanford University, 2006).

[2] Wheatstone C., "Contributions to the physiology of vision", *Philosophical Transactions of the Royal Society of London* **4**, 76–77 (1837).

[3] Rollmann W., "Notiz zur Stereoskopie", *Ann. Phys.* **165**, 350–351 (1853).

[4] Kooi F. L. and A. Toet, "Visual comfort of binocular and 3D displays," *Displays* **25**, 99–108 (2004).

[5] Lippmann G., "Epreuves reversibles donnant la sensation du relief", *J. of Phys.* **7**, 821–825 (1908).

[6] Ives H. E., "Optical properties of a Lippman lenticulated sheet," *J. Opt. Soc. Am.* **21**, 171 (1931).

[7] Burckhardt C. B., "Optimum parameters and resolution limitation of integral photography," *J. Opt. Soc. Am. A* **58**, 71–74 (1968).

[8] Martínez-Corral M., H. Navarro, R. Martínez-Cuenca, G. Saavedra and B. Javidi, "Full parallax 3-D TV with programmable display parameters," *Opt. Phot. News* **22** (12), 50 (2011).

[9] Arai J., F. Okano, M. Kawakita, M. Okui, Y. Haino, M. Yoshimura, *et al.*, "Integral three-dimensional television using a 33-megapixel imaging system," *J. Display Technol.* **6**, 422–430 (2010).

[10] Miura M., J. Arai, T. Mishina, M. Okui, and F. Okano, "Integral imaging system with enlarged horizontal viewing angle," *Proc. SPIE* **8384**, 83840o (2012).

[11] Davies N., M. McCormick, and L. Yang, "Three-dimensional imaging systems: a new development," *Appl. Opt.* **27**, 4520–4528 (1988).

[12] Adelson E. H. and J. Y. A. Wang, "Single lens stereo with plenoptic camera," *IEEE Trans. Pattern Anal. Mach. Intell.* **14**, 99–106 (1992).

[13] Georgiev T. and A Lumsdaine, "The focused plenoptic camera and rendering," *J. Elect. Imaging* **19**, 2 (2010).

[14] Ng R., M. Levoy, M. Brédif, G. Duval, M. Horowitz, and P. Hanrahan, "Light field photography with a hand-held plenoptic camera", *Tech. Rep. CSTR 2*, (2005).

[15] Levoy M., R. Ng, A. Adams, M. Footer, and M. Horowitz, "Light field microscopy," *ACM SIG-GRAPH* 924–934 (2006).

[16] Navarro H., J. C. Barreiro, G. Saavedra, M. Martínez-Corral, and B. Javidi, "High-resolution far-field integral-imaging camera by double snapshot," *Opt. Express* **20**, 890–895 (2012).

[17] Okano F., J. Arai, H. Hoshino, and I. Yuyama, "Three-dimensional video system based on integral photography," *Opt. Eng.* **38**, 1072–1077 (1999).

[18] Park J.-H., K. Hong, and B. Lee, "Recent progress in three-dimensional information processing based on integral imaging," *Appl. Opt.* **48**, H77–H94 (2009).

[19] Arai J., M. Kawakita, T. Yamashita, H. Sasaki, M. Miura, H. Hiura, *et al.* "Integral three-dimensional television with video system using pixel-offset method," *Opt. Express* **21**, 3474–3485 (2013).

[20] Javidi B. and J.-S. Jang, "Improved resolution 3D TV, video, and imaging using moving microoptics array lens techniques and systems (MALTS)," *Proc. SPIE* **4902**, 1–12 (2002).

[21] Lim Y.-T., J.-H. Park, K.-C. Kwon, and N. Kim, "Resolution-enhanced integral imaging microscopy that uses lens array shifting," *Opt. Express* **17**, 19253–19263 (2009).

[22] Navarro H., R. Martínez-Cuenca, A. Molina-Martín, M. Martínez-Corral, G. Saavedra, and B. Javidi, "Method to remedy image degradations due to facet braiding in 3D integral imaging monitors," *J. Disp. Technol.* **6**, 404–411 (2010).

[23] Lee B., S. Jung, and J.-H. Park, "Viewing-angle-enhanced integral imaging by lens switching," *Opt. Lett.* **27**, 818–820 (2002).

[24] Choi H., S.-W. Min, S. Jung, J.-H. Park, and B. Lee, "Multiple-viewing-zone integral imaging using a dynamic barrier array for three-dimensional displays," *Opt. Express* **11**, 927–932 (2003).

[25] Martínez-Cuenca R., H. Navarro, G. Saavedra, B. Javidi, M. Martínez-Corral, "Enhanced viewing-angle integral imaging by multiple-axis telecentric relay system," *Opt. Express* **15**, 16255–16260 (2007).

[26] Zhang L., Y. Yang, X. Zhao, Z. Fang, and X. Yuan, "Enhancement of depth-of-field in a direct projection-type integral imaging system by a negative lens array," *Opt. Express* **20**, 26021–26026 (2012).

[27] Bagheri S., Z. Kavehvash, K. Mehrany, and B. Javidi, "A fast optimization method for extension of depth-of-field in three-dimensional task-specific imaging systems," *J. Display Technol.* **6**, 412–421 (2010).

[28] Tolosa A., R. Martínez-Cuenca, A. Pons, G. Saavedra, M. Martínez-Corral, and B. Javidi, "Optical implementation of micro-zoom arrays for parallel focusing in integral imaging," *J. Opt. Soc. Am. A* **27**, 495–500 (2010).

[29] Xiao X., B. Javidi, M. Martínez-Corral, and A. Stern, "Advances in three-dimensional integral imaging: sensing, display, and applications," *Appl. Opt.* **52**, 546–560 (2013).

[30] McMillan L. and G. Bishop, "Plenoptic modeling: an image-based rendering system," *Proc. ACM SIGGRAPH Conf. on Comp. Graphics* 39–46 (1995).

[31] Chai J.-X., X. Tong, S.-C. Chan, and H.-Y. Shum, "Plenoptic sampling," *Proc. ACM SIGGRAPH Conf. on Comp. Graphics* 307–318 (2000).

[32] Kishk S. and B. Javidi, "Improved resolution 3D object sensing and recognition using time multiplexed computational integral imaging," *Opt. Express* **11**, 3528–3541 (2003).

[33] Hong S. H., J. S. Jang, and B. Javidi, "Three-dimensional volumetric object reconstruction using computational integral imaging," *Opt. Express* **1**, 483–491 (2004).

[34] Levoy M., "Light fields and computational imaging", *IEEE Computer* **39**, 46–55 (2006).

[35] Cho M. and B. Javidi, "Computational reconstruction of three-dimensional integral imaging by rearrangement of elemental image pixels," *J. Disp. Technol.* **5**, 61–65 (2009).

[36] Navarro H., E. Sánchez-Ortiga, G. Saavedra, A. Llavador, A. Dorado, M. Martínez-Corral, and B. Javidi, "Non-homogeneity of lateral resolution in integral imaging," *J. Display Technol.* **9**, 37–43 (2013).

[37] Park J.-H., S. Jung, H. Choi, Y. Kim, and B. Lee, "Depth extraction by use of a rectangular lens array and one-dimensional elemental image modification," *Appl. Opt.* **43**, 4882–4895 (2004).

[38] DaneshPanah M. and B. Javidi, "Profilometry and optical slicing by passive three-dimensional imaging," *Opt. Lett.* **34**, 1105–1107 (2009).

[39] Saavedra G., R. Martínez-Cuenca, M. Martínez-Corral, H. Navarro, M. Daneshpanah and B. Javidi, "Digital slicing of 3D scenes by Fourier filtering of integral images," *Opt. Express* **16**, 17154–17160 (2008).

[40] Park J. H. and K. M. Jeong, "Frequency domain depth filtering of integral imaging," *Opt. Express* **19**, 18729–18741 (2011).

[41] Park J.-H., J. Kim, and B. Lee, "Three-dimensional optical correlator using a sub-image array," *Opt. Express* **13**, 5116–5126 (2005).

[42] Matoba O., E. Tajahuerce, and B. Javidi, "Real-time three-dimensional object recognition with multiple perspectives imaging," *Appl. Opt.* **40**, 3318–3325 (2001).

[43] Cho M. and B. Javidi, "Three-dimensional visualization of objects in turbid water using integral imaging," *J. Disp. Technol.* **6**, 544–547 (2010a).

[44] Hong S. H. and B. Javidi, "Distortion-tolerant 3D recognition of occluded objects using computational integral imaging," *Opt. Express* **14**, 12085–12095 (2006).

[45] Cho M. and B. Javidi, "Three-dimensional visualization of objects in turbid water using integral imaging," *J. Disp. Technol.* **6**, 544–547 (2010b).

[46] DaneshPanah M., B. Javidi, and E. A. Watson, "Three dimensional object recognition with photon counting imagery in the presence of noise," *Opt. Express* **18**, 26450–26460 (2010).

[47] Zhao Y., X. Xiao, M. Cho, and B. Javidi, "Tracking of multiple objects in unknown background using Bayesian estimation in 3D space," *J. Opt. Soc. Am. A* **28**, 1935–1940 (2011).

[48] Lynch K., T. Fahringer, and B. Thurow, "Three-dimensional particle image velocimetry using a plenoptic camera," *AIAA*–**1056**, 1–14 (2012).

[49] Xiao X., B. Javidi, G. Saavedra, M. Eismann, and M. Martinez-Corral, "Three-dimensional polarimetric computational integral imaging," *Opt. Express* **20**, 15481–15488 (2012).

[50] Latorre-Carmona P., E. Sánchez-Ortiga, X. Xiao, F. Pla, M. Martínez-Corral, H. Navarro, G. Saavedra, and B. Javidi," Multispectral integral imaging acquisition and processing using a monochrome camera and a liquid crystal tunable filter," *Opt. Express* **20**, 25960–25969 (2012).

[51] Jang J.-S. and B. Javidi, "Three-dimensional integral imaging of micro-objects," *Opt. Lett.* **29**, 1230–1232 (2003).

[52] Levoy M., Z. Zhang, and I. McDowall, "Recording and controlling the 4D light field in a microscope using microlens arrays," *J. Micros.* **235**, 144–162 (2009).

[53] Lim Y.-T., J.-H. Park, K.-C. Kwon, and N. Kim, "Resolution-enhanced integral imaging microscopy that uses lens array shifting," *Opt. Express* **17**, 19253–19263 (2009).

[54] Rodríguez-Ramos L. F., I. Montilla, J. J. Fernández-Valdivia, J. L. Trujillo-Sevilla, and J. M. Rodríguez-Ramos, "Concepts, laboratory and telescope test results of the plenoptic camera as a wavefront sensor," *Proc. SPIE* **8447**, 844745 (2012).

[55] Saleh B. E. A. and M. C. Teich, *Fundamentals of Photonics*. Chichester: John Wiley & Sons, Ltd (1991).

[56] Gorenflo R. and S. Vessella, *Abel Integral Equations: Analysis and Applications*, Lect. Notes Math., Vol. **1461**, Berlin, Heidelberg, New York: Springer (1991).

[57] Tanida J., K. Yamada, S. Miyatake, K. Ishida, T. Morinoto, N. Kondou, *et al.*, "Thin observation module my bound optics (TOMBO): concept and experimental verification," *Appl. Opt.* **40**, 1806–1813 (2001).

[58] ProFUSION25. 5×5 Digital Camera Array. Website, available at: www.ptgrey.com/products/profusion25/ProFUSION_25_datasheet.pdf (accessed December 6, 2013).

[59] Jang J. S. and B. Javidi, "Three-dimensional synthetic aperture integral imaging," *Opt. Lett.* **27**, 1144–1146 (2002).

[60] 3D lightfield camera. Website, available at: www.raytrix.de/ (accessed December 5, 2013).

[61] Lightfield based commercial digital still camera. Website, available at: www.lytro.com (accessed December 5, 2013).

[62] Kavehvash Z., M. Martinez-Corral, K. Mehrany, S. Bagheri, G. Saavedra, and H. Navarro, "Three-dimensional resolvability in an integral imaging system," *J. Opt. Soc. Am. A* **29**, 525–530 (2012).

[63] Martínez-Cuenca R., G. Saavedra, M. Martínez-Corral, and B. Javidi, "Progresses in 3-D multiperspective display by integral imaging," *Proc. IEEE* **97**, 1067–1077 (2009).

[64] Navarro H., R. Martínez-Cuenca, G. Saavedra, M. Martínez-Corral, and B. Javidi, "3D integral imaging display by smart pseudoscopic-to-orthoscopic conversion," *Opt. Express* **18**, 25573–25583 (2010).

[65] Jung J.-H., J. Kim, and B. Lee, "Solution of pseudoscopic problem in integral imaging for real-time processing," *Opt. Lett.* **38**, 76–78 (2013).

12

Image Formats of Various 3-D Displays

Jung-Young Son[1], Chun-Hea Lee[2], Wook-Ho Son[3], Min-Chul Park[4] and Bahram Javidi[5]

[1]*Biomedical Medical Engineering Department, Konyang University, Korea*
[2]*Industrial Design Department, Joongbu University, Korea*
[3]*Content Platform Research Department, Electronics and Communication Technology Research Institute, Korea*
[4]*Sensor System Research Center, Korea Institute of Science and Technology, Korea*
[5]*Department of Electrical and Computer Engineering, University of Connecticut, USA*

12.1 Chapter Overview

The images displayed on image panels and screens to generate various forms of three dimensional (3-D) images in 3-D displays are formatted many different ways based on spatial, temporal, and spatiotemporal multiplexing, in order to deal with the required amount of image data to be displayed and to fit onto an available display. The image formats of different 3-D imaging methods such as multiview, volumetric, and holographic are unique from each other: The image format for multiview includes a set of different view images for generating the virtual arrangement of voxels in imaging space; for volumetric, a set of depth-wise images generate a spatial arrangement of voxels in space; and for holographic methods a set of images laden with fringe patterns generate a spatial image with a continuous depth. Currently, the main display means of these images for the three methods is the flat panel display for plane images, and display chips like DMD and LCoS. It is expected that this trend will continue in the future because the means are rapidly developing to have more pixel density and resolution.

In this chapter, image formats of various 3-D imaging methods, such as multiview, volumetric, and holographic methods, which will fit into flat panel displays are presented based on their multiplexing schemes. Regarding the image format, methods of loading multiview

Multi-dimensional Imaging, First Edition. Edited by Bahram Javidi, Enrique Tajahuerce and Pedro Andrés.
© 2014 John Wiley & Sons, Ltd. Published 2014 by John Wiley & Sons, Ltd.

images on the unit image cell and creating the image cells of different shapes, and 3-D images obtained by each image format, are introduced.

12.2 Introduction

The same object or scene can be captured as images with several different forms. These images of different forms can reproduce the original object/scene as long as they contain object information in a certain fashion and a proper display mechanism appropriate for each of the image forms is available, as demonstrated by CGH (*Computer Generated Hologram*). In the hologram, the phase information of object points is the most important data because it preserves the depth information of each object point. The phase information is preserved as interference fringes in the hologram but the interference fringes are not the only way of preserving this information. An open hole in each of the boxes in a checker board pattern can preserve the phase information if its relative position in the box can be changed according to the phase of a point represented by the box, as demonstrated by the Lohmann hologram [1]. Three-dimensional imaging methods have been developed by finding new image formats for preserving a scene and/or object with its depth information more accurately, and inventing an appropriate mechanism of reproducing the object and/or scene and its depth information as it is in the image from each of the image forms.

Three-dimensional imaging methods can be grouped into three based on their image forms and reproducing mechanisms to achieve a depth sense with the images. They are: *multiview imaging* including stereoscopic images, *volumetric* and *holographic* imaging [2]. The governing factors defining the image format of the first methods are (1) the total number of multiview images [3], which will be loaded in the display panel/projection screen where the multiview images represent a set of images viewed at different viewing directions of (an) object(s) or a scene; (2) the shape of an image cell, which will work as a unit of loading the multiview images; and (3) multiplexing schemes of the multiview images. The multiview image defines the scene/object space that will be presented through a display panel/projection screen, and neighboring images within the multiview images should be fused as an image with a certain depth. The basic image cell for loading multiview images on a display panel is a *pixel cell* [3,4]. The display panel consists of an array of pixel cells. The pixel cell can be made to have many different shapes in order to enhance image quality [5]. The minimum number of different view images in multiview images is two for stereoscopic images [6]. The image format is not even related to the number of different view images but is related to their presentation in the panel/projection screen. The multiview images can be presented simultaneously, time-wise, or part of the images simultaneously and the others time-wise, in order to make them be perceived with depth by viewers within a given time period. These image presenting practices are called *multiplexing schemes*. There are three different multiplexing schemes of spatial, time, and spatiotemporal [7]. This order corresponds to the order in the previous statement. The image cell is a spatial multiplexing method of the multiview images. There is another image format that doesn't require any multiplexing scheme, though it is far coarser than that based on the pixel cell [8]. This format is defined by the pixel patterns in the display panel corresponding to the virtual voxels formed in the viewing zone-forming geometry of contact-type multiview 3-D imaging [3].

The volumetric imaging has two different image formats: One is composed of a set of images like the multiview. However, the images are not from different viewing directions but from different depth-wise positions of a scene or (an) object(s) [9]. These images can be displayed using the same multiplexing schemes as the multiview. Another has a completely different image format from the first because it is composed of a set of voxels that will form a spatial image in a given imaging space as the image on a glass block [10]. These voxels can be created one by one in a time sequentially [11], or spatially, and spatiotemporally [12]. Holographic imaging has a completely different image format to the multiview and the volumetric because most of the holographic images are composed of interference fringes. The interference fringes are presented as an acoustic wave train [13,14], on a display panel, or display chips [15–17]. The acoustic wave train is spatiotemporally multiplexed to make a hologram frame the panel/chips by either spatial or time multiplexing. There are other image formats for the holographic imaging. One of them is composed of open holes of different heights, as in the Lohmann hologram. The Lee hologram has slightly different image format to the Lohmann hologram [18]. This hologram divides the box in the Lohmann hologram into four equal segments in horizontal direction, and fills the segments with real and imaginary parts of amplitude and phase. The zebra hologram is a stereo-hologram that has the same image format as the multiview [19]. The multiview images are spatially multiplexed in this hologram. FLA (*Focused Light Array*) imaging [20] is another stereo-hologram type of imaging that has the same image format as the multiview but the images are time-multiplexed.

12.3 Multiplexing Schemes

Multiplexing is arranging scheme of image data for display and transmission. It defines the image format for different 3-D imaging methods. The multiplexing schemes in 3-D imaging present required image data in a display device within a given time slot in order for viewers to perceive depth senses. Since 3-D imaging requires much more image data than the plane image, a proper multiplexing scheme is naturally needed to display all this data within the pre-determined time slot. However, the selection of a proper multiplexing scheme solely depends on an available display means. So far, many different display means have been introduced [3,7]. But the flat panel display is still considered best display means for 3-D imaging. The reason is simple: no moving components and compatibility with all image formats. The flat panel display will be the dominant display means for 3-D imaging in future, as long as the compatibility exists. However, the currently available flat panel displays are not fit for displaying multiview 3-D and holographic images. The most important parameters of flat panel displays for 3-D imaging are number of pixels, pixel size, and operating speed. The number of pixels, that is, the resolution, indicates how much detail in the scene/object can be displayed and the data amount that can be displayed. The pixel size reveals the minimum resolvable object/scene details. The operating speed will be the same as the resolution. There is no strict requirement for resolution in 3-D imaging, but the recommended resolution is the total resolution of multiview images to be loaded on the display. The resolution requirement of holographic imaging depends on the viewing angle and hologram size. For the case of a hologram of $10\,cm^2$ with a 30° viewing angle, $(166\,667)^2$ pixels are needed. The operating speed of the display can effectively increase the display's resolution. When the speed becomes twice the typical display panel, the effective resolution will be twice its resolution by time multiplexing. The pixel size is especially important for holographic display because viewing angle is determined by pixel size of the flat panel.

Hence, no flat panel displays currently available can be used for 3-D imaging, except for stereoscopic imaging. This lack of a proper display panel leads to interest in display chips. Display chips have small active surface areas but their resolutions are higher and pixel sizes are smaller than those of flat panel displays. Furthermore, one of the chips, the DMD (*Digital Micromirror Device*) has a very high operation speed. Its speed allows the display of several 100 000 frames/sec [21]. These properties of display chips allow use of all three multiplexing schemes in 3-D imaging. Projection type 3-D imaging is the most typical application of the display chips.

As mentioned before, there are three different multiplexing schemes: time, spatial, and spatiotemporal. In time multiplexing, the image data is arranged sequentially by time and is arranged in parallel for spatial multiplexing. The spatiotemporal scheme uses both time and spatial multiplexing: The time multiplexing scheme has been mostly applied to the 3-D projection type, and volumetric imaging with use of a high speed projector. Each view image, that is, a component image consisting of multiview images, is projected at the same time as a time sequence to a screen with an optical power for the projection-type and of different layer images, to a rotating screen with a certain surface curvature, or a translating flat screen for volumetric imaging systems. In this scheme, either each view image of multiview images or each layered image from all layered images are sampled for a short time period and then projected in a specific order in a time sequence to the screen. This projection will be repeated with different time slot images of the multiview or the layered images for a number of times/sec so as not to cause any image flickering to viewers. This scheme is the principle behind displaying stereoscopic images in the commercial eyeglass type of stereoscopic displays based on a high speed LCD [22]. It is also applied to the volumetric imaging generating an object's contour image with use of a set of voxels that are formed by a scanning laser beam as a programed time sequence in synchronization with the screen movement or on the imaging space [23–26].

Spatial multiplexing is mostly applied to 3-D imaging when high speed display devices are not available. This multiplexing scheme displays all required image data simultaneously on (1) multiple display devices as in the projection-type systems with normal speed projectors, and volumetric and electro-holographic imaging based on many display panels/chips [16,27], and (2) a display panel by reducing the resolution of each view image in contact-type multiview 3-D imaging. For the case in (1), a complete image frame is divided into several parts and each part is displayed in a separate display panel. The image from each display panel should be combined spatially as a large size image with the images from other display panels. For the case in (2), either a specific image column or a pixel from each view image or different view images in the multiview images to be displayed are sampled periodically, and then they are rearranged as a spatial image sequence to be a full frame image.

Spatiotemporal uses both time and spatial multiplexing simultaneously to deal with more different view images than the high speed projector can deal with in multiview imaging [28], and a large amount of image data such as $(166\ 667)^2$ pixels as in electro-holography. Typical examples of spatiotemporal multiplexing are electro-holographic systems based on a single AOM (*Acousto-Optic Modulator*) with many parallel input channels [29] and multiple AOMs aligned in parallel [30]. The signals in these AOMs are time multiplexed to generate a frame of a hologram and to increase hologram size. Another example of multiplexing is the multiview image system based on combining two time multiplexed multiview image channels spatially, such that the viewing region of each channel is joined to another without any overlap between them, in order to make a single viewing zone [31]. The multiplexed image is loaded to the display devices of the 3-D imaging systems in such a way that each component image is easily separated from the others in the multiplexed image sequences for the multiview imaging, because its depth cue is the other parallaxes. For volumetric and holographic imaging, the multiplexed image should be loaded to generate a desired spatial image with a volume. The typical practice of loading the image sequence to the display devices is: (1) allocating each separated image to its corresponding projector in the spatial multiplexing scheme, or for a fixed time period to a high speed projector; (2) changing the image data as a line image sequence in time, such as transforming the sequence as an analog type line image signal as in an electro-holographic system to fit on the available display device; and (3) arranging multiview images on the display panel as a pixel cell array. In the electro-holography system based on AOM, each line of CGH should be transformed to an analog type line image signal with a chirp signal pattern to excite the AOM. To display multiview images simultaneously in a display panel, these images should be arranged as the pixel cell unit is. The pixel cell is a basic unit of loading multiview images on a display panel. The contact-type multiview images can be arranged either in an image base as in IP (*Integral Photography*) [32–34], or in a pixel or pixel line base as in the MV (*multiview*). The IP and the pixel base MV format are for displaying full parallax 3-D images. The difference between IP and MV in their equivalent optical geometries is that IP has a parallel projection configuration and MV a radial projection configuration [35]. The difference between MV and IP type projection configurations and typical projection type multiview 3-D imaging with radial and parallel projection configurations is that the images are focused at the screen plane for the projection type but there is no focused plane for the IP and MV. Each projector image is continuously expanded with distance from the display panel, though there is a plane where all projector images are completely superposed together in MV. This plane is called the *viewing zone cross section* and it defines the viewing distance in MV. In this plane, all multiplexed images are individually separated from others. The space surrounding this plane is defined as the viewing zone because viewers can perceive a depth sense from the multiview images on the display panel here.

12.4   Image Formats for 3-D Imaging

The factors affecting image formats for 3-D imaging are not only the multiplexing schemes but also the content of the component image, and the specific parallax direction to which each 3-D imaging is intended for. There are two parallax directions: full and horizontal. The *full* means parallaxes are given to both horizontal and vertical directions. The *horizontal*, the parallaxes exist only in the horizontal direction. The contents of each of the component images in multiview, volumetric, and holographic imaging are completely different, as mentioned before. The component image for multiview imaging is obtained mostly with a camera. For volumetric imaging, the image can also be obtained with a camera, but extra effort is necessary to extract depth-wise scenes from the image. These depth-wise scenes can also be used to create voxel points for contour images. The component image for holographic imaging is completely different from that of the multiview and the volumetric but still can be obtained with a camera, as in digital holography [36]. However, in most cases, the component image is calculated by a computer, that is, CGH. As indicated, no camera can record the data amount. Furthermore, the laser beam, which is the main light source for recording holograms, doesn't have enough power to illuminate a natural scene and the requirements of recording the hologram are too difficult to be fulfilled in outdoor conditions. This is the reason why CGH is the main method of obtaining a hologram for holographic imaging. The stereo-hologram makes it possible to display natural scenes holographically [37] but its reconstructed image consists of one- or two-dimensional arrangements of multiview images as in the IP type of multiview 3-D imaging. There is no difference between images projected to viewers' eyes if the multiview images have the same resolution between them. But in practice, the resolution of IP is much smaller than that of the stereo-hologram. In this section, various image formats for 3-D imaging will be described.

12.4.1 Image Formats for Multiview 3-D Imaging

As mentioned before, multiview 3-D imaging requires a set of images that are called multiview images and originate from a multiview camera array. This camera array can have a 1- or 2-D form, which is aligned in parallel or radial with an equal distance between cameras in both horizontal and vertical directions. The component cameras in this array have the same optical characteristics. The multiview images can be numbered $1-k$, depending on their corresponding camera orders from left to right in the camera array. For the 2-D array, the images will be numbered from $(\ell - 1)k + 1$ to ℓk, where $\ell = 1$ to L (L = the number of vertical lines) for further description. There are several image formats for this type of imaging. In this sub-section, the image formats for projection-type, for MV and IP, for virtual voxel-based and intensity sharing, are different to each other. Furthermore, the shapes of pixel cells also bring different image formats in the type of contact.

12.4.1.1 Projection and High Speed Display Type

Projection type multiview imaging can be divided into three groups based on the multiplexing schemes they adopt. This type of display can use all three multiplexing schemes. The first group is based on the *time-multiplexing scheme* with use of a display device that is capable of projecting images at high speed.

In the high speed projection and display-type, to display k different view images time sequentially, display devices should have the speed of no less than $60\,k$ fields/sec for the interlace (high-low) scanning type as in the high speed CRT and $60\,k$ frames/sec in the high speed display chip-like DMD in order to display a flicker-less image at the usual TV brightness conditions. Hence, the sampling period for each field should be less than $1/60\,k$/sec (for each second, k views × 30 frames/sec × 2 fields/frame should be displayed). For example, to display five view images, a 300 fields (frames)/sec sampling rate is required. Therefore each field should be sampled within 3.33 ms. This type uses three different image formats deduced from the dividing methods of each component image; such as interlacing, high-low, and full frame. The first one is the *interlacing* format. This format divides each frame of each component image into two parts by scanning all odd lines in the frame first and then all even lines in the frame. Hence, all odd lines in the frame in each of the component images in multiview are displayed with the order of $1-k$ and then even lines in the frame in the same order as the odd lines. Following this frame, the second, third, and so on, frames of each component image follow in the same sequence as the first frame. This format was popular for the high speed CRT (*Cathode Ray Tube*) based 3-D imaging, such as eyeglass-type stereoscopic imaging methods based on color filters (anaglyph), polarization and high speed shutter glasses, and multiview

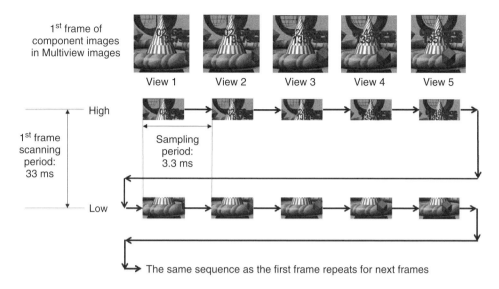

Figure 12.1   High-low type time multiplexing

imaging based on aperture-sharing with high speed shutters [38]. However, CRT has almost completely been replaced by flat panel displays.

The second format is the *high-low* type. In the high-low format, each frame of each component image is divided into two equal parts; that is, top half and bottom half of the frame. All the rest of the sequence is the same as in the interlacing format. The operation of the high-low format is shown in Fig. 12.1 for the case of $k = 5$. The sampling time of each field should be not more than 3.33 ms in this case for a 60 k field rate. The first frame scanning period for all five different component images cannot exceed 33 ms. This high-low image format can be equally applied to the all imaging methods for the interlace format.

The high-low format was also applied to the CRT but this format fits better to the progressive scanning type of display, such as the LCD, because image lines are scanned from top to bottom. This image format is used in the shutter-glass type of high speed LCD based stereoscopic imaging system [22]. The LCD in this system works at a 120 frames/sec rate but the operating speed was increased to 240 Hz/sec with use of the high-low format. The full image format can be realized when the high speed display chip like DMD is used instead of the high speed CRT. Since DMD can operate at more than 100 000 frames/sec, this frame speed allows 1667 different view images to be displayed with the full frame rate of 60 frames/sec. The full frame image format can be used instead of interlacing or high-low formats. The frame sequence is the same as the interlacing. Necessary to use this format is to replace the even/odd fields or high-low fields with the full frames.

The second group is based on the *spatial multiplexing scheme*. In this group, each component image is projected by its own projector. Hence, if there are ℓk different view images, ℓk projectors should be aligned either in parallel or radially along both horizontal and vertical directions for full parallax image generation. The $\ell = 1$ case, that is, a horizontal projector array for a *horizontal parallax only* (HPO) image is typical for this scheme. All ℓk projectors project the focused component images on the projection screen and their optical axes are directed to the

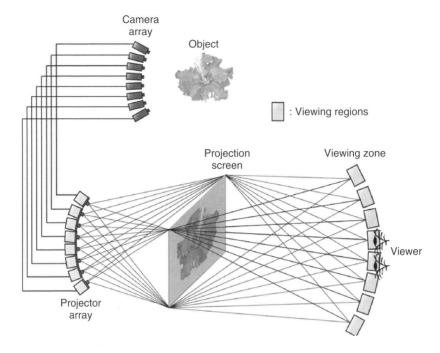

Figure 12.2   Radial type spatial multiplexing

center of the projection screen for the radial arrangement as shown in Fig. 12.2. The component images on the screen will be overlapped together with a certain distance between them for the parallel and with the same image center for the radial, Hence, all images will be mixed as shown in Fig. 12.3. Therefore, the image format of this group is difficult to define. However, each component image will be separately viewed at the viewing zone of the display system because the optical power held by the projection screen images is separate from the output pupil of each projector's objective at the viewing zone as shown in Fig. 12.2.

The third group is based on *spatiotemporal multiplexing*. When available, a high speed projector can display images at a rate of 60 kfields/sec, this speed allows display of only k different view images at most, where k can be any integer. To display more than k view images with the same projector, more than two projectors of the same kind should be combined spatially. One example of the scheme is spatially combining two time-multiplexed multiview image channels by joining their viewing zones without overlap, as shown in Fig. 12.4. In this scheme, two time multiplexed channels are merged at the input pupil of an objective by a triangle prism. With this combination, at most, 16 different view images can be displayed with the eight-view image capable display device [31]. Figure 12.4 also shows the image format of a channel for this display system. In this case $k = 8$. The sequence is not different from Fig. 12.1, except for the image numbers. The same sequence in Fig. 12.4 will be repeated for 9–16 view images for channel 2. These two channels work simultaneously.

Figure 12.3   The image on the projection screen. All projector images are mixed

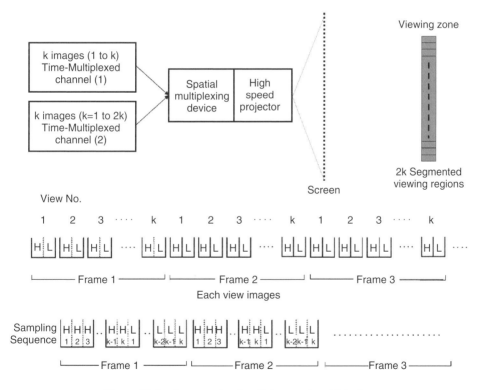

Figure 12.4   An example of spatiotemporal multiplexing

12.4.1.2 MV and IP Types

The typical contact-type multiview 3-D imaging consists of a flat panel display and *viewing zone forming optics* (VZFO). The active surface of the flat panel display consists of image cells named pixel cells [39]. This cell is the unit of loading multiview images on the panel. The pixel cells can be arranged 1- and 2-D for HPO image and full parallax images, respectively. The VZFO is formed by an array of elemental optics. Each elemental optic has the property of a lens. The array dimension of the elemental optics in VZFO is the same as that of pixel cells in the panel. However, the dimensions of a pixel cell and elemental optics can be equal or can be different. This difference between sizes of the pixel cell and elemental optic divides the imaging into IP and MV. IP is the equal case and MV is the "different" case. This difference makes the optical appearances of IP and MV similar to parallel and radial projection configurations, respectively. IP and MV also have other differences. They have a different way of loading images on pixel cells, and IP has originally been developed for full parallax imaging, but MV is used for HPO. However, there is no problem making MV have a full parallax image. Each pixel cell in IP is filled with an entire view image; hence, the cell has another name of *elemental* image. For the MV, the pixel cell is filled with a pixel from each view image. In Fig. 12.5, the image formats of IP and MV are compared. When there are 15 (5×3) multiview images with 3×3 pixels for each view image, all these are loaded on the panel in the same

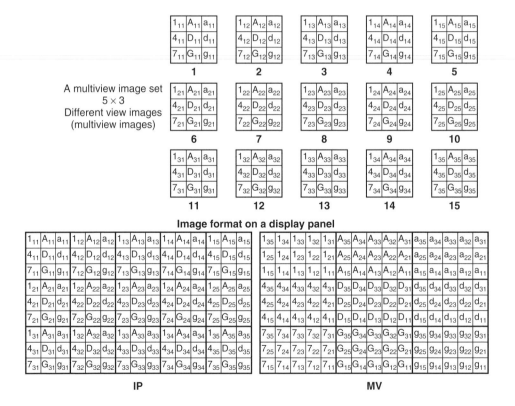

Figure 12.5 Comparisons of image arrangements in MV and IP

image order as in the multiview image set. To load all these images, the image panel resolution should be not less than 15×9, which corresponds to the combined resolution of the multiview images. Since each of the multiview images works as an elemental image, there are 5×3 elemental images on the panel.

So, VZFO should consist of a 5×3 elemental optic array. For the MV case, the image format is very different to IP. The 15 multiview images are grouped into a 3×3 pixel cell array. The array numbers are the same as the pixel array in each view image. Each pixel cell consists of a 5×3 pixel array. These numbers are the same as the image array of the multiview image set. The panel consists of a 3×3 pixel cell array. So, VZFO should consist of a 3×3 elemental optic array. The pixel cell consists of the same number pixels from the multiview images when pixels in each view image are numbered as they are in Fig. 12.5. The pixels in each view image should be numbered the same way as those in other different view images. The same number of pixels are arranged first in their image order in the multiview image set, then this arrangement is rotated $180°$ to count the image inversion by the elemental optics. The pixel cells are arranged in the display panel in their pixel number order, as in each view image. These two image formats indicate that the resolution of each view image should be reduced by 5×3 times to fit onto the display panel when the original resolution of each view image is considered to have the resolution of the panel; that is, the resolution of each view image is reduced five times in the horizontal direction and three times in the vertical direction. It is typical that the horizontal resolution is reduced more than the vertical because more different view images are necessary in the horizontal direction than the vertical in order for viewers to perceive a smooth parallax change in the horizontal direction. The 5×3 corresponds to the image array on the multiview image set. This resolution reduction scheme is shown in Fig. 12.6. A 5×3 pixel array is reduced to a pixel. Hence there is no doubt that the image details are lost by this resolution reduction for each view image. This means that the image quality of each view image will deteriorate tremendously. This is the main reason why the multiview imaging system could not stay on the market. To improve the resolution of each view image, each of the R (red), G (green), B (blue) sub-pixels are also used as independent

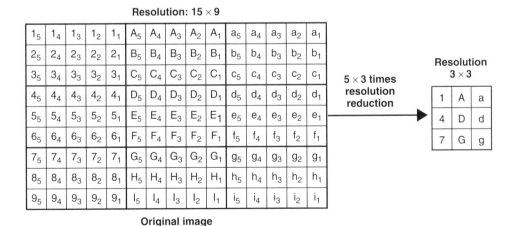

Figure 12.6 Image resolution reduction

pixels. In this way, the resolution of the display panel can be effectively increased three times, but this doesn't help much because the desired number of different view images is far more than 5 × 3. Figure 12.5 pertains to the full parallax imaging case. IP can also work as HPO imaging [40].

For the case of HPO imaging, the multiview image set has a 1-D image array. To match with this image array, the pixel cells/elemental images are also arranged one-dimensionally in the display panel. In Fig. 12.5, the 1-D image array corresponds to 5 × 1. Hence, the image formats for HPO are the same as the first three lines of the images on the display panel for both cases in Fig. 12.5. There are five elemental images in IP and three pixel cells in MV. Each pixel cell is composed of a vertical image line from each view image in MV. Since the height of each view image equals three pixel heights, it cannot fill the display panel. To fill the display panel, the height of each view image should be equal to nine pixel heights. This means that the vertical resolution of each view image should keep the original resolution as shown in Fig. 12.6. So the pixel resolution of each view image should be 3 × 9. Only the horizontal resolution of each view image is reduced five times. The image formats for HPO case are shown in Fig. 12.7. When a high speed LCD is used for both image panel and VZFO to display a stereoscopic image, it is possible to give a full panel resolution to each view image. In this method, VZFO is an active element for which characteristics can be controlled electronically. Each view image is divided into two parts of odd and even column images. These four images are combined on the display panel such that odd image from view 1 and the even image from view 2 are arranged

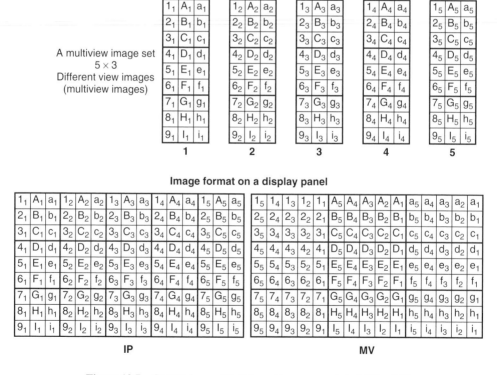

Figure 12.7 Comparisons of 1-D image arrangements in MV and IP

as in the MV image format, and then the even image from view 1 and the odd image from view 2 are combined as the MV image format: but in this case, the view 1 and view 2 order should be reversed. So the odd image of view 2 and the even image of view 1 are loaded to odd and even column lines of the display panel, respectively. Furthermore the positions of all elemental optics in the VZFO are electronically shifted a half a period to the right or left, synchronized with the second image format. By this shifting, the viewing regions of the view image 1 and 2 don't change, although the image order is changed. Hence, each view image is provided with the full display resolution [41].

There are several more image formats for HPO MV [42–44]: The multiview images can also be arranged in either a zigzag [42] or slanted line style [43]. These schemes use each of the RGB sub-pixels as a pixel. In the zigzagging line style, the height and width of each sub-pixel is designed to be the equal to and 1/6 of a pixel width, respectively. Hence the gap between sub-pixels is also 1/6 of a pixel width. In the display panel, the sub-pixels are aligned vertically. When there are k different view images, the sub-pixels in each pixel cell are aligned in such a way that if the view ℓ pixel is aligned at the sub-pixel, view ℓ-1 pixel can be either the sub-pixel directly above, or one sub-pixel down and one sub-pixel distance to the right from the view ℓ. When $\ell = 1$, the direct above and the down-right sub-pixels have the view k. And then the same procedure repeats. This is a rule of arranging pixels in a pixel cell in this scheme. By this rule, if pixels from odd numbered view images within k view images are aligned in a linear fashion, pixels from even numbered view images immediately follow the odd numbered view images along the same line. The pixels from even numbered view images are also followed immediately by those from the odd numbered view images. These sequences will be repeated until the sequence reaches the right side edge of the display panel. So pixels in a horizontal line of the pixel cell in Fig. 12.7 are arranged in two horizontal lines in such a way that one line is for pixels from odd (even) numbered view images, then the other line the pixels from even (odd) numbers. Hence, the pixels from different view images are arranged in a zigzag way. This scheme uses a lenticular plate composed of slanted lenslets with a negative slope of 1/6, as the VZFO. This slope corresponds to the slope angle of −9.46°. Figure 12.8 shows the image format for the zigzag scheme in the case of $k = 7$. The number on each sub-pixel represents a different view image number. The pixel cell in this scheme is defined as the area under each lenslet. Figure 12.8 shows that the sub-pixel in directly above the sub-pixel marked 2 in the second line is marked 3, and the sub-pixel at the one sub-pixel down and sub-pixel distance to the right from the sub-pixel marked 2 in the second line, is marked 3 too. It also shows that the sub-pixels found by this rule are marked 1 when 7 is reached. Furthermore, the pixels from odd numbered view images, 7, 5, 3, and 1 appear at odd number lines and those from even number, 6, 4, and 2 at even number lines for the first pixel cell, but the pattern is reversed for the second pixel cell. This indicates that the pixels from odd and even numbered view images repeatedly appear along each pixel line of the display panel.

If the first line repeats with the order of odd and even, then the second line goes even and odd, and the third line odd and even, and so on. This scheme allows increase in the horizontal resolution of each view image twice by scarifying the vertical resolution twice. At the viewing zone of this scheme, the viewing region of each view image appears in the order of 1–7 from left to right. The lenslet slope angle, of −9.46° allows the appearance of viewing regions of even numbered view images at the gap between the odd numbered view images' viewing regions, and pixels from each view image get RGB colors vertically for every five rows, as specified by broken lines for the pixel of view number 5. However, this image format forces

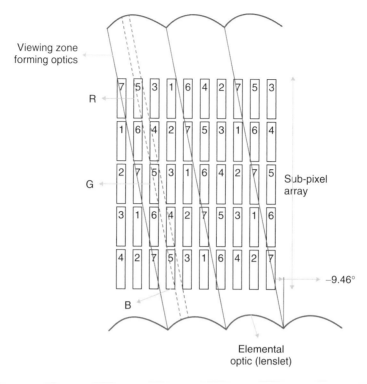

Figure 12.8   A zigzag type image arrangement

the active area of the display panel to be reduced to half the typical size due to the gaps between sub-pixels.

In the slanted line style arrangement, the pixels are arranged in the same way as in Fig. 12.7, except shifting N pixels for Nth row to the left. These shifted pixels will be eliminated from the format. In this way, each view image is shifted one sub-pixel to the left from its previous line. This arrangement results in a virtually slanted type image format. To fill the empty surface created in the right side by this left side shifting, different view images with parallelogram shapes should be prepared. The slanting angle is calculated as in following: since a sub-pixel is shifted to the left for every next row, $\tan^{-1}(1/3) = 18.4^0$ because each sub-pixel has 1/3 of a pixel width. It has a positive slanting angle.

The image formats mentioned in this section share mostly the horizontal resolution of the display panel. But it is also possible to display multiview images by sharing the vertical resolution, without affecting horizontal resolution. The fashion of sampling each view image is the same as in Fig. 12.7, however, the sampling direction is different. But in this case, a VZFO, which is capable of separating each horizontal line image from the vertically multiplexed multiview images and then directing them to their corresponding horizontally divided viewing regions, is needed. The VZFO composed of horizontal line grating patterns can perform the function. Each line grating has its specific grating direction and period to perform the function described previously [44]. There is one more image format working with the grating pattern. This image format is itself the first line of the multiview image set shown in Fig. 12.5.

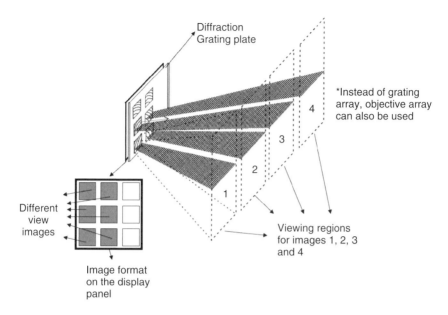

Figure 12.9   A multiview image arrangement in 3-D imaging based on a diffraction grating plate

If there are six different view images in the horizontal direction, the images can be aligned on a display panel such that three in the first line and three in the second line, or two in first and next two in the second, and the remainder in the third line. On each of these images, a 2-D diffraction grating directs the image it is under to a designated position in the viewing zone where all images are optically aligned to the order of $1-6$, from left to right, as shown in Fig. 12.9 [45]. Figure 12.9 shows four bottom images. The viewing regions of top two images appear at the right side of viewing region 4.

12.4.1.3   MV with Arbitrary Pixel Cell Shapes

For full parallax 3-D image generations, the pixel cells should have a 2-D shape. The shape doesn't have to be a rectangle/square as shown in Fig. 12.5, but can be any shape if it can be joined together with many of its kind, as shown in Fig. 12.10, and adapted to the periodic structure elemental optics in currently available VZFOs. Pixel cells with any rhombic or parallelogram shape can be effectively fitted to the structure and manipulated to have different numbers of pixels within the cells [46]. This manipulation is significant in minimizing the moiré fringes inherent in the contact-type 3-D imaging systems [47]. In these systems, moirés are naturally produced by overlapping the viewing zone forming optics and the display panel together, because the optics and the panel have regular patterns of comparable periods. The moirés can be minimized by changing the aspect ratio of the rhombus. The pixel cells with arbitrary can be designed on plotting paper because the pixel pattern in the display panel has the grid pattern of such paper. Figure 12.11(a) shows the design procedure. For the design, two parallel line groups with line slope angles of α and $-\alpha$ are crossed over each other. By this crossing, many rhombuses with the same shape are generated. The slope angle and the

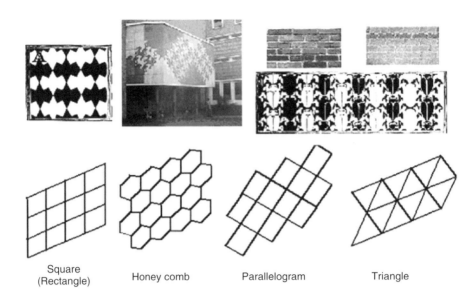

Square
(Rectangle)   Honey comb   Parallelogram   Triangle

Figure 12.10   Possible pixel cell shapes

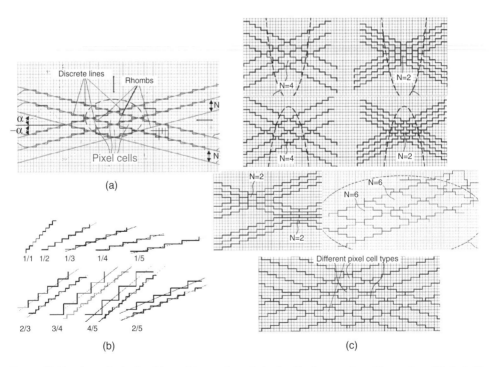

Figure 12.11   A method of making arbitrary shaped pixel cells. *Source*: Jung-Young Son, Vladmir V. Saveljev, Kae-Dal Kwack, Sung-Kyu Kim, and Min-Chul Park 2006. Reproduced with permission from the Optical Society

distance between lines in each line group are defined by considering the number of multiview images to be loaded on the display panel. Since the lines are straight, they cannot fit to the shape of the pixels. So the lines are approximated by discrete lines following the edge lines of the pixels. The pixel cells are defined by the discrete lines.

Figure 12.11(b) shows the discrete lines corresponding with many different slopes. When $\alpha = \pm\tan^{-1}0.75$, the discrete line can be drawn either by shifting four grids to the right and up (down) three grids, or combinations of one grid to the right and up (down) one or two times, and two grids to the right and up (down) one grid. The latter will better approximate the straight line. Figure 12.11(c) shows several pixel cell shapes possible with $\alpha = \pm\tan^{-1}0.5$ and $\alpha = \pm\tan^{-1}(1/3)$ for line distances of 2, 4, and 6 for the first α. Several different shape pixel cells within a rhombus can be designed by changing crossing points, even when the line distance is the same. Hence, there will be many different pixel cell shapes possible for different α values and line distances, and as a consequence, many different image formats can be designed with these rhombus type pixel cells.

Figure 12.12 (Plate 20) also shows the image format of the rhombus cell corresponding to a pixel cell with 4×4 pixels. This means that there are 16 different view images and the slope angle $\alpha = 45^0$. To design the image format, each view image is transformed into odd and even

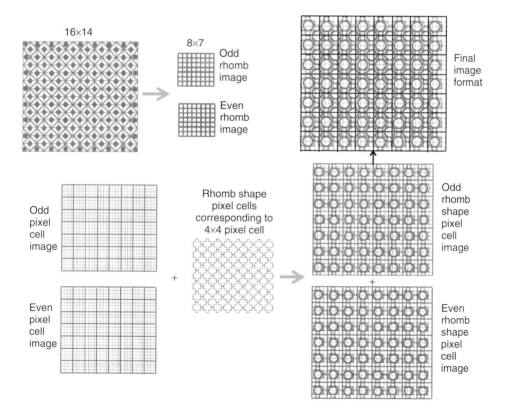

Figure 12.12 (Plate 20)   The image format of the rhombus cell corresponding to a pixel cell with 4×4 pixels. *See plate section for the color version*

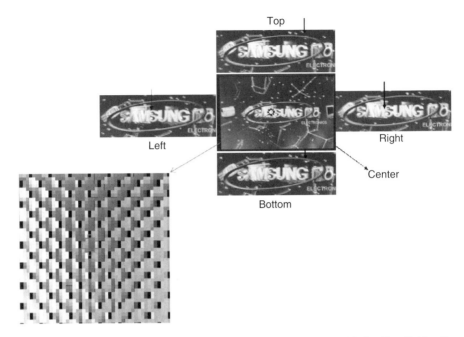

Figure 12.13 Three-dimensional images made by the rhombic pixel cells in Fig. 12.12 . *Source*: Jung-Young Son, Vladmir V. Saveljev, Kae-Dal Kwack, Sung-Kyu Kim, and Min-Chul Park 2006. Reproduced with permission from the Optical Society

rhombus images to reduce the resolution of each view image. Figure 12.12 (Plate 20) also shows how to make odd and even images with having 8×7 resolution from the original image of 16×14 resolution. The odd image is generated by taking the average intensity of four pixels surrounding the red dot, and the even four pixels surrounding the green dot. From the odd and even images of the 4×4 different view images, square pixel cell images corresponding to odd and even images are formed. Each of these is combined by rhombic shape pixel cell arrays corresponding to 4×4 square pixel cell arrays.

From each of these combined rhombic shape pixel cell images, the cells marked by red and green circles are combined as shown in the final image format. The 3-D image generated by the image format based on a rhombic shape pixel cell is shown in Fig. 12.13 with the piece of the image format to show the cell shape. The cell has the height of six pixels and a width of five pixels. It consists of 18 pixels.

12.4.1.4   MV Based on Virtual Voxels

In multiview 3-D imaging, the *voxel* is defined as a virtual spatial picture element supposedly composed of 3-D images [48,49]. In the viewing zone, forming geometry of MV, voxels plate are defined as the crossing points of rays from different pixel cells. Since these voxels are very uniformly distributed, it is possible to identify the pixels within specific pixel cells that are responsible for forming virtual voxels. Figure 12.14 shows the plane view of virtual pixels in the viewing zone forming geometry of the MV, The geometry is based on point light source

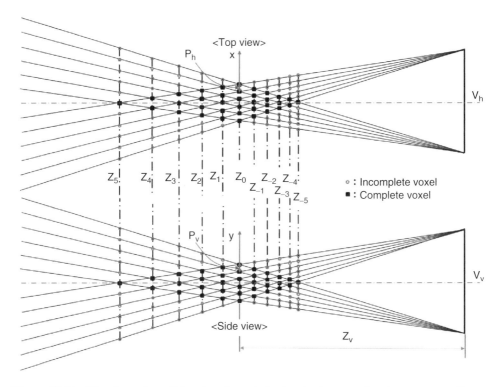

Figure 12.14 The plane view of the virtual voxels in the viewing zone forming geometry. *Source*: Adapted from Jung-Young Son, Vladmir V. Saveljev, Sung-Kyu Kim and Kyung-Tae Kim 2007

(PLS) array and each PLS illuminates a pixel in its front. This geometry leads the voxels to be distributed at planes formed by the crossing points and makes it easier to assign a coordinate value for each voxel. The dark points in Fig. 12.14 represent the virtual voxels.

The Z_0 plane is the PLS array plane and each PLS also behaves like a voxel. The circles represent incomplete voxels. The voxel planes are marked by Z_{-5}–Z_5. The image pattern for each voxel is shown in Fig. 12.15. The symbol I, the subscript and two superscript numbers represent the image pattern, the voxel plane, and relative positions in the x and y coordinates. In the pattern, each grid represents a pixel cell. The image patterns for voxels in the Z_0 and Z_{-1} planes have a square with the size of a pixel cell. However, the planes are further away from the Z_0 plane, so the square is divided into a smaller square array. The total size of squares in the array is the same as a pixel cell size. The image patterns for the incomplete voxels are also shown in Fig. 12.15. The complete image pattern is shifted to the edges and a part of the pattern is missed. Hence the patterns become incomplete. The image format from the image pattern for a triangle pyramid is shown in Fig. 12.16 (Plate 21). The left, center, and right side views of the 3-D image from the image format are also shown in Fig. 12.16. In this figure, K represents the voxel plane number. The differences are clear. With the rhombic pixel cell the image pattern for virtual pixels can also be determined. Figure 12.17 shows the image patterns for different voxels in the rhombic pixel cell, image patterns for 3-D images of five different Platonic solids displayed on a LCD monitor.

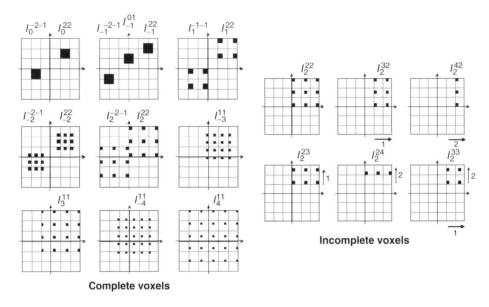

Figure 12.15   Image pattern for each voxel. *Source*: Jung-Young Son, Vladmir V. Saveljev, Bahram Javidi, Dae-Sik Kim, and Min-Chul Park 2006. Reproduced with permission from the Optical Society

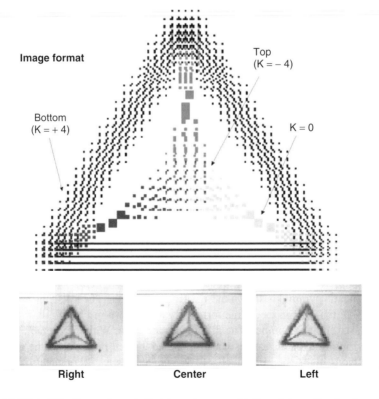

Figure 12.16 (Plate 21)   Image format of a triangular pyramid. *See plate section for the color version*

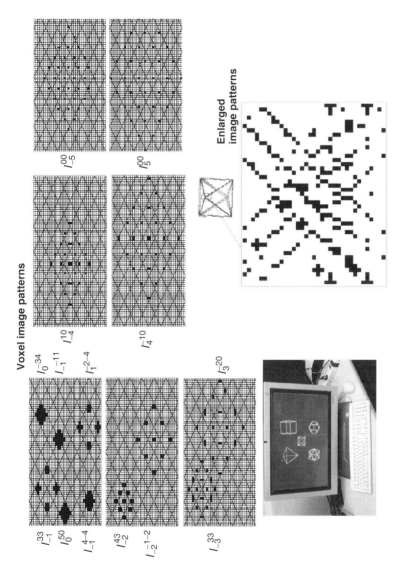

Figure 12.17 Image generated and the image patterns for rhombic pixel cells

12.4.1.5 Stereoscopic Image Based on Sharing Pixel Intensity

In the polarization eyeglass-type of stereoscopic imaging based on flat panel displays, a micro-strip polarization plate is used as the VZFO. However, this method can also be realized by using two high speed LCDs; one as the display panel, and another as an active polarization plate. This results in each view image with full panel resolution as in the viewing region shifting method [41]. Instead of using two high speed LCDs, two ordinary LCDs can also be used to make each view image have full panel resolution by making two corresponding position pixels from view images 1 and 2 share the intensity of the corresponding pixel on the display panel [50]. In this type of imaging, two corresponding pixel intensities are combined to represent the intensity of the corresponding pixel on the panel. The polarization direction of the light from the pixel in the panel is rotated by the angle defined by the original intensity ratio of two corresponding position pixels from view images 1 and 2 by the active polarization plate. Each pixel is designed to rotate the polarization angle of the light coming from its corresponding pixel on the panel by the amount determined by the ratio. Hence, view images 1 and 2 will be separated by the polarization eye glasses as the horizontal and vertical polarization components. The principle of this method, including the display structure, is depicted in Fig. 12.18. In this imaging, the intensity of each pixel, P_I^{ij} in the LC plate for

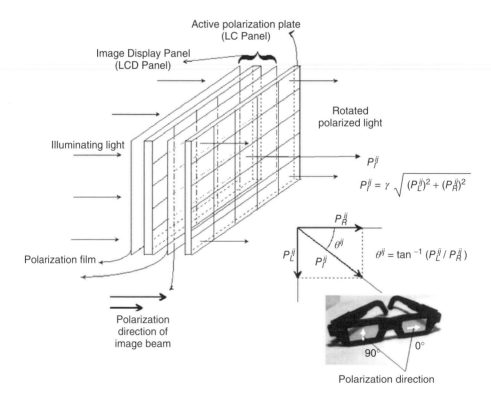

Figure 12.18   Image format in an intensity sharing type of stereoscopic display

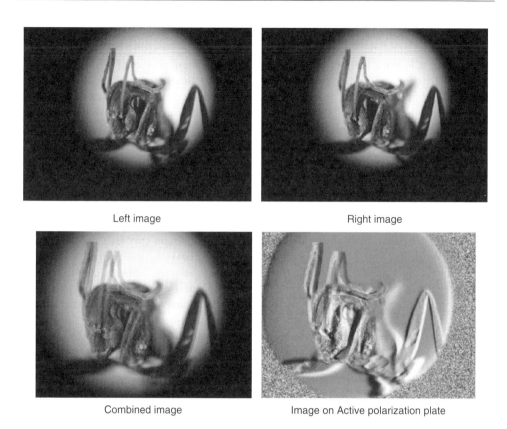

Left image   Right image

Combined image   Image on Active polarization plate

Figure 12.19   Images on two LCD panels in the intensity sharing type of stereoscopic display

displaying image is determined by $P_I^{ij} = \gamma \sqrt{\left(P_L^{ij}\right)^2 + \left(P_R^{ij}\right)^2}$ and the rotating angle of the incoming light's polarization direction, θ^{ij} by the corresponding pixel in the LC plate for active polarization plate $\theta^{ij} = \tan^{-1}\left(P_L^{ij}/P_R^{ij}\right)$. Where P_L^{ij} and P_R^{ij} are intensities of ij^{th} pixels in left and right eye images, respectively, and γ is a constant. Hence, the polarization direction and intensity of the light from ij^{th} pixel in the active LC plate will be rotated to θ^{ij} and equal to P_I^{ij}, respectively. The polarization direction will be divided into horizontal and vertical components by the polarization direction of polarizers in the eyeglasses. The intensities of the horizontal and vertical components are proportional to γP_L^{ij} and γP_R^{ij}, respectively. Figure 12.19 shows both left and right view images, the images on display panel, and the active polarization plate.

12.4.2   Image Formats for Volumetric Imaging

Volumetric imaging methods utilize rotating or translating screens to generate images with a spatial volume. The rotation and translating screens explicitly reveal that volumetric imaging is based on the time multiplexing scheme. On the screens, either each image in a set of images taken from different depth locations (i.e., layer images) is projected on to the screen when the screen position is at the image's corresponding depth location, or a set of voxels is projected or drawn continuously on the screen as raster, polygonal lines, and individual voxels, by a scanning laser beam when the screen is at its depth position. It is obvious that there will not be a difference between the layer and multiview images in usage of them. However, for the image format point of view, there will be difference between volumetric and multiview imaging because no resolution reduction is required for the volumetric. A set of layer images are simultaneously displayed on their corresponding displays aligned in a depth direction. For the case of voxel scanning, the 3-D image space created by the screen rotation or translation is filled with spatial line arrays by the raster scanning method, polygonal lines by the vector graphic method, or single voxels by the random access method [51]. Figure 12.20 shows the volumetric image system forming spatial images on the space created by a rotating screen. The contour lines are drawn by the raster scanning method to represent gray levels. The necessary projection or display speed of the layered images, each spatial line, each polygonal line, and a voxel for avoiding flicker, depends on the brightness of the images. For the case of a movie, images are projected at 48 frames/sec with screen brightness of about $60\,cd/m^2$ [52]. After introduction of a transparent display such as LCD, the volumetric image has been realized with use of many LCD layers with a certain distance between them [53]. Instead of laser scanning, a rotating LED array plate can be used [54]. Each LED in the rotating plate will make the trace of a circle. The number of voxels within the circle will be determined by the LED on/off times per/rotation.

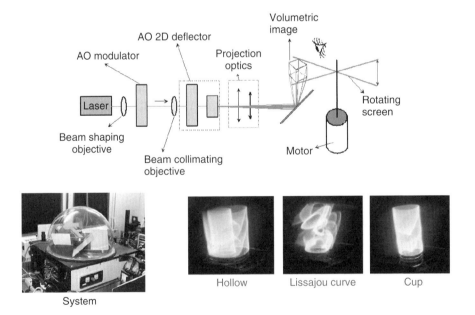

Figure 12.20   Laser scanning volumetric imaging system

12.4.3   Image Formats for Holographic Imaging

Here, holographic imaging means displaying a hologram electronically, that is, electro-holographic imaging. A hologram can also be displayed on a flat panel display or a projection chip and printed on a transparency like the image from a camera, but its image format is completely different from a camera image. The hologram is a 3-D photo of an object or a scene. A hologram records the phase variations in the illuminating light source induced by the surface shape of an object in the form of interference fringes. The interference fringe patterns on the hologram change with recording methods and the physical characteristics of the object, such as shape, transparency, surface texture, temperature, and so on. There are uncountable image patterns for holograms. In this sub-section, the image patterns induced in the process of making CGHs for fast calculations and to fit into a specific display device, as well as the fundamentals of generating CGHs, are described.

12.4.3.1   Recording Holograms

The hologram records the shape of the object/scene as the phase differences in the wave-front of the illuminating light beam, induced by the shape. Hence the images recorded on the hologram are fringe patterns. When a light beam, which has well-defined wavefronts, hits an object or a scene, the wavefronts of the reflected beam from the object/scene are modulated by their surface shapes and dimensions. This reflected beam is recorded together with a reference beam that has the same distance between wavefronts; that is, wavelength, and originates from the same light source as the illuminating beam, but the wavefronts have no modulation. These two beams interfere with each other and, as a consequence, an interference fringe pattern is recorded on the recording plane. Hence the interference fringe pattern comprises the object/scene's shape information. The periods of the fringes in the pattern are defined by the beam's wavelength and the crossing angle of the two beams. Currently, the light beam with well-defined wavefronts comes only from a laser. This phenomenon of interference can be expressed mathematically. In a Cartesian coordinate, the coordinate of each of the points forming the shape of the object/scene, the surface points on the recording plane, and the reference beam direction to the recording plane, can be determined. Hence the distance between an object point and a point on the recording plane can be calculated. This distance can be transformed to a phase by dividing it by the beam's wavelength. In this way, the shape of the object/scene can be transformed to a phase on each point on the recording plane. This phase information is recorded by interference with the reference beam. Since in each point of the recording plane beams from other object points are also superposed, each beam also interferes with the reference beam and other object point beams. Hence a lot of phase information is superposed at a point on the recording plane. This is the reason why a part of the recording plane can still preserve all of the object shape information. The recording plane in the CGH is the display panel but in optical holography, holographic photo plates/films, such as silver halide, chalcogenide, photopolymers, and others [55], are the recording planes. The CCD or CMOS chips are used as the recording plane in digital holography. Since the CGH calculates phase information mathematically with use of a computer and an object can be represented as points, any object or scene can be made a CGH if the coordinate values for each point of the points forming the shape is defined.

12.4.3.2 Mathematical Description of Interference

A hologram is a product of the light interference. To generate this interference phenomenon, at least two beams from the same light source are necessary and these beams have a well-defined wavefront at least for the distance between the object/scene and the recording plane.

In a hologram recording, the two beams are named the *object wave*, E_O, which illuminates the object/scene and then is reflected in the direction of the recording plane, and the *reference wave*, E_R, which is directly incident on the recording plane with an angle, φ_r as shown in Fig. 12.21. The object and reference waves are represented as,

$$E_O(x, y, z, t) = \sum_{n=1}^{\infty} \sum_{m=1}^{\infty} \sum_{j=1}^{\infty} O\left(x_j, y_m, z_n\right) e^{i\{\omega t + \varphi_O(x_j, y_m, z_n)\}}$$

$O\left(x_j, y_m, z_n\right)$: Amplitude of each object point

$\varphi_O\left(x_j, y_m, z_n\right)$: Initial Phase of each object point,   (12.1)

In Eq. 12.1, it is assumed that the object is composed of infinite number of points in all three directions for convenience.

$$E_R(x, y, z, t) = R(x, y, z)e^{i\{\omega t + \varphi_r(x, y, z)\}}$$

$R(x, y, z)$: Amplitude of Reference Wave

$\varphi_r(x, y, z)$: Initial Phase of Reference Wave

t : time.   (12.2)

As shown in Fig. 12.21, $\varphi_r(x, y, z)$ has a constant value defined by the beam incidence angle of the reference wave to the recording plane. The interference effect is mathematically

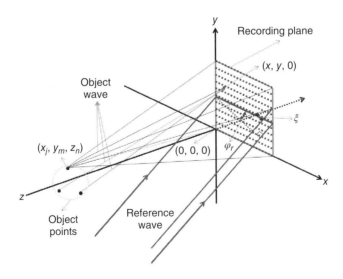

Figure 12.21   Recording geometry of a hologram

represented as [56],

$$I(x, y, z) = \left(E_O + E_R\right)\left(E_O + E_R\right)^*$$
$$= E_O E_O{}^* + E_R E_R{}^* + E_R E_O{}^* + E_O E_R{}^*, \tag{12.3}$$

where, * represents a conjugate term. The conjugate term has negative phase, that is, the phase on the exponential term has a negative sign. The interference produces four separate terms. The first two terms are interference between the same waves, hence they work as background and speckle noises, and the last two terms actually comprise the object's phase information. They are the most important terms in holography. By substituting Eqs. 12.1 and 12.2 into Eq. 12.3, the terms introduced by interference between their own waves are represented as,

$$E_R E_R{}^* = |R(x, y, z)|^2,$$

$$E_O E_O{}^* = \sum_{n=1}^{\infty} \sum_{m=1}^{\infty} \sum_{j=1}^{\infty} \sum_{n'=1}^{\infty} \sum_{m'=1}^{\infty} \sum_{j'=1}^{\infty} O\left(x_j, y_m, z_n\right) O^*\left(x_{j'}, y_{m'}, z_{n'}\right). \tag{12.4}$$

In Eq. 12.4, $E_R E_R{}^*$ is a complete DC term. This term has no object information and simply works as a background noise. The $E_O E_O{}^*$ is the most problematic term because it generates speckle. Speckle is produced by the interference between object waves. Since the reflected beam from the object is composed of reflected beams from points composing the object, a point reflected beam interferes with those from other points. Hence speckle is produced. Mathematically, the $E_O E_O{}^*$ term is divided into two terms; one is the $j = j', m = m', n = n'$ case, which represents a DC term that will work as the background noise. Another is the $j \neq j', m \neq m', n \neq n'$ case, which will be the source of speckles.

$$E_O E_O{}^*|_{DC} = \sum_{n=1}^{\infty} \sum_{m=1}^{\infty} \sum_{j=1}^{\infty} \left|O\left(x_j, y_m, z_n\right)\right|^2$$

$$E_O E_O{}^*|_{Speckle} = \sum_{n=1}^{\infty} \sum_{m=1}^{\infty} \sum_{j=1}^{\infty} \sum_{n'=1}^{\infty} \sum_{m'=1}^{\infty} \sum_{j'=1}^{\infty} O\left(x_j, y_m, z_n\right) O^*\left(x_{j'}, y_{m'}, z_{n'}\right) \left| j \neq j', m \neq m', n \neq n'. \right.$$
$$\tag{12.5}$$

Hence, it is not necessary to calculate Eqs. 12.4 and 12.5 for CGH calculation. Still to be calculated are the last two terms in Eq. 12.3. They are represented as,

$$E_R E_O{}^* = R(x, y, z) \sum_{n=1}^{\infty} \sum_{m=1}^{\infty} \sum_{j=1}^{\infty} O^*\left(x_j, y_m, z_n\right) e^{i\{\varphi_O(x_j, y_m, z_n) - \varphi_r(x, y, z)\}}$$

$$E_O E_R{}^* = R^*(x, y, z) \sum_{n=1}^{\infty} \sum_{m=1}^{\infty} \sum_{j=1}^{\infty} O\left(x_j, y_m, z_n\right) e^{i\{\varphi_r(x, y, z) - \varphi_O(x_j, y_m, z_n)\}} \tag{12.6}$$

In Eq. 12.6, both $E_R E_O{}^*$ and $E_O E_R{}^*$ are complex terms. They have real and imaginary parts. Hence each of them is written as $E_R E_O{}^* = \text{Re}\left(E_R E_O{}^*\right) + \text{Im}\left(E_R E_O{}^*\right)$ and $E_O E_R{}^* = \text{Re}\left(E_O E_R{}^*\right) + \text{Im}\left(E_O E_R{}^*\right)$. Using one of these relationships, amplitude and phase information of an object point on a point on the recording plane can be calculated. This amplitude and phase information is necessary to calculate CGH with gray level information. The CGH with gray level information is good enough to reconstruct the object image with gray level information. In Eq. 12.6, $\varphi_O\left(x_j, y_m, z_n\right)$ is determined as follows.

$$\varphi_O\left(x_j, y_m, z_n\right) = \frac{2\pi}{\lambda}\sqrt{\left(x_j - x\right)^2 + \left(y_m - y\right)^2 + z_n^2}. \tag{12.7}$$

Equation 12.6 is further simplified by summing $E_R E_O{}^*$ and $E_O E_R{}^*$. It is given as,

$$E_R E_O{}^* + E_O E_R{}^* = 2\left|R\left(x, y, z\right)\right|\sum_{n=1}^{\infty}\sum_{m=1}^{\infty}\sum_{j=1}^{\infty}\left|O^*\left(x_j, y_m, z_n\right)\right|$$

$$\cos\left\{\varphi_O\left(x_j, y_m, z_n\right) - \varphi_r\left(x, y, z\right)\right\}. \tag{12.8}$$

In Eq. 12.8, if $\left|O^*\left(x_j, y_m, z_n\right)\right|$ has a constant value, this means that points the making up the object have the same intensity. Equation 12.8 represents the phase-only hologram. It doesn't involve the speckle. If it is not a constant value it represents a gray level hologram. A binary hologram is calculated by assigning a threshold value, then 1 for intensity above the threshold and 0 for below the threshold.

With Eqs. 12.4 and 12.8, Eq. 12.3 is rewritten as,

$$I\left(x, y, z\right) = \left|R\left(x, y, z\right)\right|^2 + \sum_{n=1}^{\infty}\sum_{m=1}^{\infty}\sum_{j=1}^{\infty}\sum_{n'=1}^{\infty}\sum_{m'=1}^{\infty}\sum_{j'=1}^{\infty} O\left(x_j, y_m, z_n\right) O^*\left(x_{j'}, y_{m'}, z_{n'}\right)$$

$$+ 2\left|R\left(x, y, z\right)\right|\sum_{n=1}^{\infty}\sum_{m=1}^{\infty}\sum_{j=1}^{\infty}\left|O^*\left(x_j, y_m, z_n\right)\right|\cos\left\{\varphi_O\left(x_j, y_m, z_n\right) - \varphi_r(x, y, z)\right\}. $$

$$\tag{12.9}$$

Equation 12.9 represents the mathematical description of the fringe pattern recorded on the recording plane for the object represented by $\left|O^*\left(x_j, y_m, z_n\right)\right|$. This equation is for the full parallax hologram calculation generation. For the HPO hologram, both object and recording plane are sampled to the same number of lines in the vertical direction, and then the phase information of the object points on Kth row in the object is recorded on the points in the Kth row of the recording plane. The period of the fringes, δ recorded on recording plane is determined by the following relationship. If the crossing angle between object and reference wave is ξ and the wavelength of the waves λ, the fringe period is given as,

$$\delta = \frac{\lambda}{2\sin\left(\xi/2\right)}. \tag{12.10}$$

This δ value corresponds to two pixels in the recording plane. Hence the crossing angle, ξ, able to be recorded on the recording plane, can be defined by the pixel size on the recording plane. Since this crossing angle is the maximum viewing angle of viewing a point reconstructed in a hologram, it is more desirable to have a recording plane with smaller size pixels.

12.4.3.3 Examples of CGH

The binary hologram of a point on the normal line of the recording plane, which is calculated with Eq. 12.8, is given as in Fig. 12.22. This pattern is compared with the optically obtained Fresnel zone pattern. The patterns look the same but there are periodical appearances of many extra patterns on the CGH, and uneven intensity distribution in the Fresnel zone pattern. The periodic appearance of the same pattern seems to be caused by different diffraction orders, hence the fringe pattern of the point hologram will not be different from the Fresnel zone pattern, and the uneven intensity distribution is caused by the speckle effect. No speckle is seen from the CGH. Since the size of the central rings and the distance between concentric rings in the CGH pattern increases with increasing object point distance from the recording plane, it is also possible to control the object point distance from the recording plane by enlarging or reducing the size of the pattern. Hence, if each point composing an object is replaced by a Fresnel zone pattern, a CGH of the object is obtained. Figure 12.23 shows the hologram made in this way. The object is a letter "A" and it consists of 25 points. This hologram is printed on a transparency and it reconstructs the letter A as shown in Fig. 12.23. In Fig. 12.24, binary and

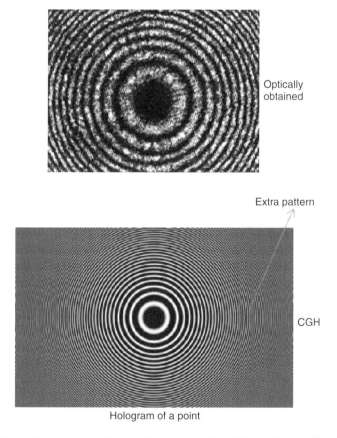

Optically obtained

Extra pattern

CGH

Hologram of a point

Figure 12.22 Fresnel zone pattern. *Source*: Jung-Young Son, Vladmir V. Saveljev, Bahram Javidi, Dae-Sik Kim, and Min-Chul Park 2006. Reproduced with permission from the Optical Society

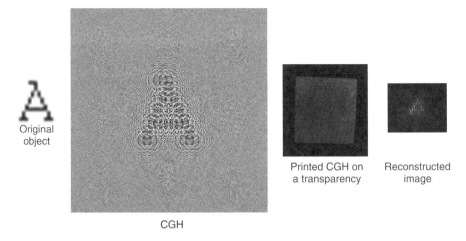

Figure 12.23   Holographic image generation and reconstruction

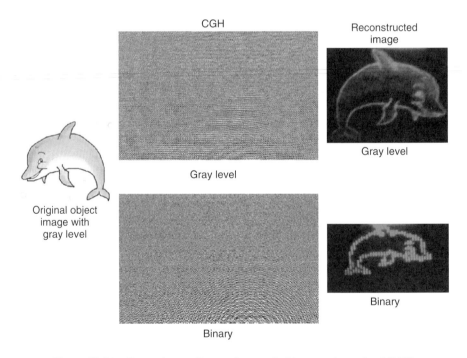

Figure 12.24   Comparisons of image formats for binary and gray level CGH

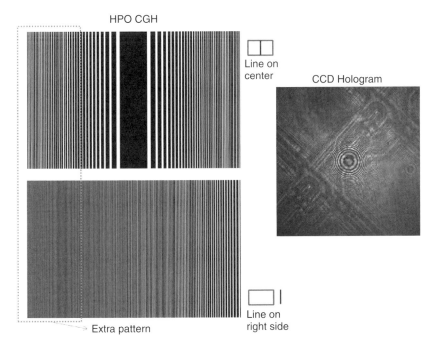

Figure 12.25   HPO CGH and CCD hologram

gray level format CGHs of a dolphin are compared. It appears that the gray level CGH reveals a fabric woven pattern, and the binary, a wavy pattern.

The reconstructed image from the gray level CGH reveals a gray level but the binary reveals only the contour of the object. The gray level CGH can be used for most display chips but the binary is more convenient for DMD, because DMD is a binary device in principle. Figure 12.25 shows two examples of a HPO hologram and CCD hologram in digital holography. Figure 12.25 shows the two cases when a vertical line composed of many points is in the top corner of a triangular prism and when the line is shifted to the outside of the recording plane's right side edge without changing its distance from the plane. The holograms look as if they are composed by the repeated display of the center row image pattern of a Fresnel zone pattern. They also show the presence of extra patterns mentioned with regards to Fig. 12.22. The CCD hologram shows fringe patterns and the object image together. The stereo-hologram is developed to make a natural scene as a hologram. For this purpose, a set of multiview images is taken with a multiview camera array. These images are displayed in the order of the multiview images, on a display device with a laser as the illuminating light source, and each of them are holographically recorded as a thin line with a cylindrical lens or a point with a spherical lens in front of the device. Without these lenses, it is possible to record the hologram by dividing the recording plane into the number of strips, which corresponds to the number of the images in the multiview image set.

The reconstructed image from the hologram will be the multiview image set. Hence the reconstructed images in front of the recording plane will be mixed together and perceived as a 3-D image. An example of a stereo-hologram is shown in Fig. 12.26. Ten multiview images of

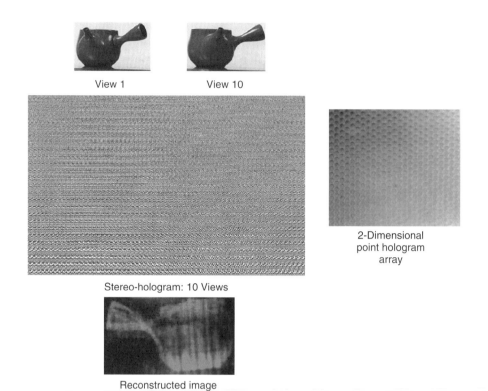

Figure 12.26 An example of a stereo-hologram

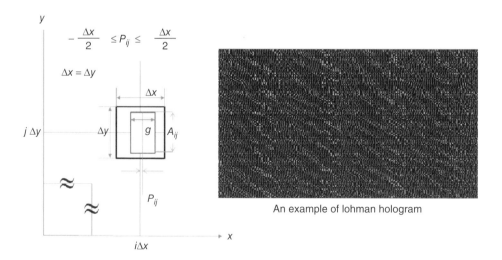

Figure 12.27 Calculation of the Lohmann hologram

a teapot are the object. Figure 12.26 also shows the 2-D point hologram array as in the zebra hologram, Fig. 12.26 shows an example of stereo-hologram on the recording plane.

There are CGHs that are not based on the interference fringe pattern shown so far. One of them is the Lohmann hologram. To design this hologram, a pre-calculated CGH is needed. To design the Lohmann hologram, (1) a recording plane is divided into a 2-D square cell array, which has the same dimensions as the hologram points in the CGH, and (2) a rectangular aperture is created in each cell. The height and relative position of the aperture in the cell is proportional to the amplitude and phase of the CGH point corresponding to the cell, respectively. Figure 12.26 shows the method of designing each cell. The aperture width g is about half of the cell width. The position P_{ij} of the aperture relative to the center of the cell changes in proportional to the phase ϕ of the ij^{th} hologram point, and the height A_{ij} to the amplitude O. Hence $P_{ij} = (\pm\phi/\pi)(\Delta x/2)$ and $A_{ij} = \Delta y \left(O/O_{max}\right)$, where O_{max} is the maximum amplitude value of the CGH. An example of the Lohmann hologram designed in this way is also shown in Fig. 12.27.

References

[1] B. R. Brown and A. W. Lohmann, "Computer generated binary holograms," *IBM J. of Res. Develop.* **13**(3), pp. 160–168, 1969.

[2] Izumi T. (Supervisor), *Fundamentals of 3-D Imaging Techniques* (Japanese Edition), NHK Science and Technology Lab., Ohmsa, Tokyo, 1995.

[3] Son J.-Y. and B. Javidi, "3-Dimensional imaging systems based on multiview images," *IEEE/OSA J. of Display Technology* **1**(1), pp. 125–140, 2005.

[4] Okano F., H. Hoshino, J. Arai, M. Yamada, and I. Yuyama, "Three–dimensional television system based on integral photography," in *Three–Dimensional Television, Video, and Display Technology*, B. Javidi and F. Okano (eds), New–York, USA, Springer, 2002, ch. 4, pp. 101–123.

[5] Son J.-Y., V. V. Saveljev, B. Javidi, and K.-D. Kwak, "A method of building pixel cells with an arbitrary vertex angle," *Optical Engineering* **44**(2), pp. 024003–1–024003–6, 2005.

[6] Okoshi T., *Three Dimensional Imaging Techniques*, New York, Academic Press, 1976.

[7] Son J.-Y., B. Javidi and K.-D. Kwack, "Methods for displaying 3 dimensional images," *Proceedings of the IEEE, Special Issue on: 3–D Technologies for Imaging & Display*, **94**(3), 2006, pp. 502–523.

[8] Son J.-Y., V. V. Saveljev, B. Javidi, D.-S. Kim and M.–C. Park, "Pixel patterns for voxels in a contact–type 3 dimensional imaging system for full–parallax image display," *Appl. Opt.*, **45**(18), pp. 4325–4333, 2006.

[9] Tamura S. and K. Tanaka, "Multilayer 3–D display by multidirectional beam splitter," *App. Opt.*, **21**(20), pp. 3659–3663, 1982.

[10] Downing E., L. Hesselink, J. Ralston, and R. Macfarlane, "A three–color, solid–state, three–dimensional display," *Science*, **2177**, pp. 196–202, 1994.

[11] Langhans K., C. Guill, E. Rieper, K. Oltmann, and D. Bahr, "Solid felix: A static volume 3D–laser display," *SPIE* **5006**, 2003, pp. 161–174.

[12] MacFarlane D. L., "Volumetric three–dimensional display," *App. Opt*, **33**(31), pp. 7453–7457, 1994.

[13] St. Hilaire P., S. A. Benton, M. Lucente, J. Underkoffler, and H. Yoshikawa, "Real–time holographic display: Improvement using a multichannel acousto–optic modulator and holographic optical elements," *SPIE* **1461**, 1991, pp. 254–261.

[14] Shestak S. A. and Jung–Young. Son, "Electroholographic display with sequential viewing zone multiplexing," *SPIE Proc.*, **3293**, pp. 15–22, 1998.

[15] Zachan E., R. Missbach, A. Schwerdtner, and H. Stolle, "Generation, encoding and presentation of content on holographic displays in real time," *Proc. SPIE*, **7690**, pp. 76900E, 2010.

[16] Maeno K., N. Fukaya, O. Nishikawa, K. Sato and T. Honda, "Electro–holographic display using 15–megapixel LCD," *SPIE*, **2652**, 1996, pp. 15–23.

[17] Senoh T., T. Mishina, K. Yamamoto, O. Ryutaro, and T. Kurita, "Viewing–zone–angle–expanded color electronic holography system using ultra–high–definition liquid–crystal displays with undesirable light elimination," *Journal of Display Technology*, **7**(7), pp. 382–390, 2011

[18] Lee W. H., "Sampled Fourier transform hologram generated by computer," *Appl. Opt.*, **9**, 639–643, 1970.

[19] Zebra Imaging Inc.: Klug M. A., C. Newswanger, Q. Huang, and M. E. Holzbach, "Active digital hologram displays," U.S. Patent 7,227,674, June 2007.

[20] Kajiki Y., H. Yoshikawa, and T. Honda, "Hologram like video images by 45–view stereoscopic display," *Proc. SPIE*, **3012**, 1997, pp. 154–166.

[21] Texas Instruments Website. Available at: www.ti.com (accessed December 5, 2013).

[22] Kim S., B. You, H. Choi, B. Berkeley, D. Kim, and N. Kim, "World's first 240 Hz TFT–LCD technology for full–HD LCD–TV and its application to 3D display," *SID 09 DIGEST*, pp. 424–438, 2009.

[23] Saveljev V. V., P. E. Tverdokhleb, and Y. A. Shchepetkin, "Laser system for real–time visualization of three–dimensional objects," *SPIE*, **3402**, 1998, pp. 222–224.

[24] Batchko R. G., "Rotating flat screen fully addressable volume display system," U.S. Pat. 5, 148,310, 1992.

[25] Otsuka R., T. Hoshino, and Y. Horry, "Transport: All–around three–dimensional display system," *SPIE* **5599**, 2004, pp. 56–65.

[26] Son J. Y. and S. A. Shestak, "Live 3D video in a volumetric display," *SPIE*, **4660**, 2002, pp. 171–175.

[27] Slinger C., C. Cameron, and M. Stanley, "Computer–generated holography as a generic display technology," *IEEE Computer*, **38**(8), pp. 46–53, 2005.

[28] Son J.-Y., V. V. Smirnov, V. V. Novoselsky, and Y.-S. Chun, "Designing a multiview 3–D display system based on a spatiotemporal multiplexing," *IDW'98–The Fifth International Display Workshops*, Dec. 7–9, 1998 International Conference Center Kobe, Kobe, Japan, pp. . 783–786.

[29] St.-Hilaire P., M. Lucente, J. D. Sutter, R. Pappu, C. D. Sparrell, and S. A. Benton, "Scaling up the MIT holographic video system," *Proc. SPIE*, **2333**, pp. 374–380, 1994.

[30] Son J. Y., S. Shestak, S. K. Kim, and V. Epikhan, "A multichannel AOM for real time electroholography," *Appl. Opt.* **38**(14), pp. 3101–3104, 1999.

[31] Son J.-Y., V. V. Smirnov, K.-T. Kim, and Y.-S. Chun, "A 16-views TV system based on spatial joining of viewing zones," *SPIE Proc.*, **3957**, pp. 184–190, 2000.

[32] Okano F., H. Hoshino, and I. Yuyama, "Real time pickup method for a three dimensional image based on integral photography," *Appl. Opt.* **36**, pp. 1598–1603, 1997.

[33] Erdmann L. and K. J. Gabriel, "High resolution digital integral photography by use of a scanning microlens array," *Appl. Opt.* **40**, pp. 5592–5599, 2001.

[34] Liao H., M. Iwahara, N. Hata, and T. Dohi, "High Quality Integral Videography By Using A Multi-Projector," *Opt. Exp.*, **12**, pp. 1067–1076, 2004.

[35] Son J.-Y., W.–H. Son, S.-K. Kim, K.–H. Lee, and B. Javidi, "3-D imaging for creating real world like environments," *Proceedings of the IEEE (Invited)*, **101**(1), pp. 190–205, 2013.

[36] Schnars U. and W. P. O. Juptner, "Digital recording and numerical reconstruction of holograms," *Meas. Sci. Technol.*, **13**, pp. R85–R101, 2002.

[37] Mccrickerd J. T. and N. George, "Holographic stereogram from sequential component photographs," *Appl. Phys. Lett.*, **12**(1), pp. 10–12, 1968.

[38] Travis A. R. L., S. R. Lang, J. R. Moore, and N. A. Dodgson, "Time–multiplexed three–dimensional video display," *SID 95 Digest*, pp. 851–852, 1995.

[39] Son J.-Y., V. V. Saveljev, S.-K. Kim and K.-T. Kim, "Comparisons of the perceived image in multiview and IP based 3 dimensional imaging systems," *Japanese J. of App. lied Physics*, **46**(3A), pp. 1057–1059, 2007.

[40] Toshiba website. Website available at: www.toshiba.com/us/tv/3d/47l6200u (accessed December 12, 2013).

[41] Yoshigi M. and M. Sakamoto, "Full–screen high–resolution stereoscopic 3D display using LCD and EL panels," *Proc. SPIE*, **V6399**, pp. 63990Q, 2006.

[42] van Berkel C. and J. A. Clarke, "Characterization and optimization of 3D–LCD module design," *Proc. SPIE*, **3012**, pp. 179–186, 1997.

[43] Schmidt A. and A. Grasnick, "Multi–viewpoint autostereoscopic displays from 4D–vision," *Proc. SPIE*, **4660**, pp. 212–221, 2002.

[44] Nordin G. P., J. H. Kulik, M. Jones, P. Nasiatka, R. G. Lindquist, and S. T. Kowel, "Demonstration of novel three–dimensional autostereoscopic display," *Optics Letters*, **19**, pp. 901–903, 1994.

[45] Toda T., S. Takahashi, and F. Iwata, "3D video system using grating image," *Proc. of SPIE*, **V2406**, pp. 191–198, 1995.

[46] Son J.-Y., V. V. Saveljev, K.-D. Kwack, S.-K. Kim, and M.–C. Park, "Characteristics of pixel arrangements in various rhombuses for full–parallax 3 dimensional image generation," *Appl. Opt.*, **45**(12), pp. 2689–2696, 2006.

[47] Saveljev V. V., J.-Y. Son, B. Javidi, S.-K. Kim, and D.-S. Kim, "A Moiré minimization condition in 3 dimensional image displays," *IEEE/OSA J. of Display Technology*, **1**(2), pp. 347–353, 2005.

[48] Watt A., *3D Computer Graphics*, 3rd Edn, Chap. 13, pp. 370–391, Addison–Wesley, Harlow, 2000.

[49] Halle M. W., "Holographic stereograms as discrete imaging systems," *Proc. SPIE*, **2176**, pp. 73–84, 1994.

[50] Son J.-Y., V. I. Bobrinev, K.–H. Cha, S.K. Kim and M.–C. Park, "LCD based stereoscopic imaging system," *Proc. SPIE* **6311**, pp. 6311021–6311026, 2006.

[51] Blundell B. and A. Schwartz, *Volumetric Three-Dimensional Displays*, John Wiley & Sons, Inc., New York, ISBN, 0471239283, 2000.

[52] Stupp E. H. and M. S. Brenneshaltz, *Projection Displays*, John Wiley & Sons, Ltd, Chichester, ch.14, pp. 330–333, 1999.

[53] Buzak T. S., "A field–sequential discrete–depth–plane three–dimensional display," *SID'85 Digest*, pp. 345–347, 1985.

[54] Endo T., "A cylindrical 3–D video display observable from all directions," *SPIE*, **3957**, pp. 225–233, 2000.

[55] Samui A., "Holographic recording medium," *Recent Patents on Material Science*, **1**(1), pp. 74–94, 2008.

[56] Hariharan P., *Optical Holography: Principles, Techniques and Applications*, Cambridge University Press, New York, 1984.

13

Ray-based and Wavefront-based 3D Representations for Holographic Displays

Masahiro Yamaguchi and Koki Wakunami*
Global Scientific Information and Computing Center, Tokyo Institute of Technology, Japan

13.1 Introduction

A three-dimensional (3D) holography display is capable of reproducing extremely high-quality 3D images by wavefront reconstruction [1–3]. Conventional 3D displays are stereoscopic or multi-view, but more advanced displays based on light-field or light-ray reconstruction have also been investigated actively [4–6]. Then questions arise, what is the advantage of wavefront reconstruction? Is it possible to integrate light-ray-based and wavefront-based systems? This chapter addresses a technology that converts 3D data represented by light-rays into wavefront data and vice versa, and shows how to make use of the advantage of holographic displays. A method for calculating *computer generated holograms* (CGH), which was developed for this purpose, is introduced and experimental results are demonstrated.

13.2 Ray-based and Wavefront-based 3D Displays

Figure 13.1 shows the illustrations of ray-based and wavefront-based 3D displays. In ray-based 3D displays, the light-rays traveling in all directions are reproduced on the display surface in the same manner as those that were reflected or scattered by real objects. Therefore, an observer can see the 3D image as if they were real objects [7]. Integral photography is one of the

*Universal Communication Research Institute, Ultra-Realistic Video Systems Laboratory, National Institute of Information and Communications Technology, Japan

Multi-dimensional Imaging, First Edition. Edited by Bahram Javidi, Enrique Tajahuerce and Pedro Andrés.
© 2014 John Wiley & Sons, Ltd. Published 2014 by John Wiley & Sons, Ltd.

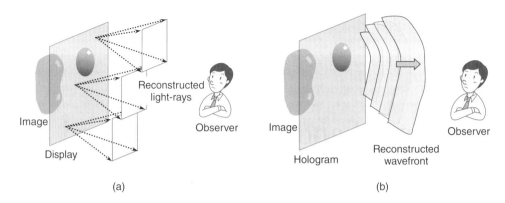

Figure 13.1 (a) Ray-based and (b) wavefront-based 3D displays

ray-based methods with both horizontal and vertical parallax (full-parallax: FP) information. Ray-based techniques have also been developed with *horizontal parallax only* (HPO). Contrary to multi-view or parallax-based methods, a method can be considered ray-based if more than one light ray is incident into the observer's eye pupil. In multi-view 3D displays, the discrepancy in depth perception by parallax and accommodation causes serious eye fatigue, but such discrepancies are diminished in ray-based techniques. In other words, the ray-based display enables 3D image formation by reproducing light rays.

Holography is a technique of wavefront recording and reconstruction as shown in Fig. 13.1(b). From the aspect of a 3D image display, it is important to understand the difference between ray-based and wavefront-based image reproduction. For this purpose, let us consider the wavefront reproduced by a ray-based display.

In references [8,9] the difference between a wavefront reproduced by a ray-based display and holography were discussed. Let us consider displays reproducing a point object as shown in Fig. 13.2. While a wavefront-based display reconstructs a spherical wave as in Fig. 13.2(a), a ray-based display generates an image of the point object by a set of converging rays as in Fig. 13.2(b). Every light ray emitted from a small area on the display plane can be considered a narrow plane wave propagating in different directions. Therefore, it can be said that two components are not reproduced by ray-based displays, whereas they are reproduced by holography; (1) the curvature of every small piece of plane waves, (2) the relative phase differences of the waves propagating in different directions. A component, the inclination of the wavefront depending on the relative location on the hologram plane with respect to the point object, is recorded in both displays.

Since those two components (1) and (2) are not managed in ray-based displays, the reconstructed point image will be broadened even if the sampling of light-rays is done at high density. This means the resolution of ray-based displays is limited compared with holography. It is possible to attach the relative phase difference component to the ray-based reproduction [9], which is called a *phase added stereogram* (PAS), if the optical path length of each light-ray is available. It has been shown that the resolution of the reconstructed image is improved by the PAS technique.

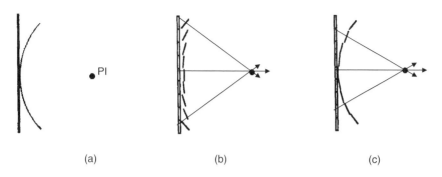

(a)   (b)   (c)

Figure 13.2 The wavefront reproductions by holographic, ray-based, and PAS-based displays for a *point image* (PI)

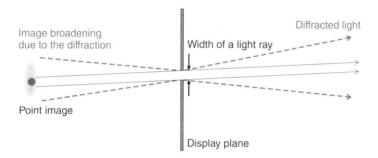

Figure 13.3   Image blur due to diffraction in ray-based displays

A PAS resembles a hologram. If the sampling density of light-rays is enough high, it becomes equivalent to a hologram. This means that ray-based reproductions can also be compared in the wavefront domain. In addition, the phase information of the reconstructed image is not important, but the resolution is the main issue in 3D display applications. Therefore, ray-based displays may be considered low level approximations of wavefront reconstruction, even though the previous two components are not included.

The difference in the wavefront shown in Fig. 13.2 causes a reduction in image resolution. Figure 13.3 shows the reconstruction of a single light-ray. The ray is diffracted on the display plane and the ray broadens at the image plane. If we consider the resolution of the reconstructed image, denoted by δ as the sum of the width of the light ray, d, and the diffraction effect, it is approximately written as

$$\delta = d + \frac{\lambda}{d}|z|, \tag{13.1}$$

where z is the distance of the point object from the display plane, and λ is the wavelength of the light, respectively. If the width of a light ray d expands, the diffraction effect decreases. However, the resolution becomes worse due to the width of the light lay. This is a fundamental limitation of ray-based displays. The resolution can be kept high when z is small. Therefore the ray-based display is suitable for reproducing 3D images near the display plane.

Consider an observer with a visual acuity of 1.0 (in decimals or 20/20), located 500 mm away from the display: the observer can resolve 0.15 mm on the display plane. If we set d = 0.15 mm on the display plane, and let $\lambda = 0.5$ μm and $z = 200$ mm, the second term in Eq. (13.1) becomes 0.67 mm. The degradation in resolution is obviously perceivable by human vision. If we define the specification in the resolution of the reconstructed image as $\delta_l \geq \delta$, then the depth of the image that satisfies this condition is given by

$$|z| \leq \frac{d}{\lambda}\left(\delta_l - d\right),$$ (13.2)

and the right-hand side is maximized when $d = \delta_l/2$ and we have

$$|z| \leq \frac{\delta_l^2}{4\lambda}.$$ (13.3)

From this equation, when $\delta_l = 0.25$ mm, $|z| \leq 31$ mm; the image depth should be substantially smaller. For $\delta_l = 2$ mm, $|z| \leq 2000$ mm, which means that a ray-based display is suitable for larger sizes and lower resolution displays. As the resolution of the display is limited in ray-based methods, a holographic display is required especially when the display size is relatively small and/or the display is viewed close to the display plane.

According to the previous consideration, the main feature of a wavefront-based display is clarified: the reproduction of a deep 3D scene in high resolution, which is impossible with ray-based displays. A possible design for future holographic display is illustrated in Fig. 13.4 [10]. A very deep 3D scene is exhibited as a virtual window and the viewer can see the scene, not the display surface. It is expected that the technologies for holographic 3D displays will develop considering the application shown in Fig. 13.4.

Figure 13.4   The concept of the display that can reproduce very deep and realistic 3D scenes

13.3   Conversion between Ray-based and Wavefront 3D Representations

To make use of the feature of wavefront-based display explained in Section 13.2, the wavefront on the hologram plane must be obtained. Various techniques for the computation of a hologram that simulates the wavefront propagation have been reported to date. However, for realistic display of 3D images, it is necessary to develop methods for representing various phenomena of light interactions. On the other hand, conventional rendering techniques for computer graphics are so advanced and extremely realistic representations are possible, such as occlusion, specular reflections, texture, translucency, and so on. Conventional rendering techniques for computer graphics are based on ray-tracing or light-fields. Thus the introduction of a ray-based rendering technique into wavefront-based displays will provide great benefits. Therefore, we developed a technique for conversion between ray-based and wavefront 3D representations, which enables the application of ray-based rendering techniques in the computation of holograms.

A light-ray represents the direction of energy flow, and is equivalent to a normal vector wavefront in an isotropic medium. Let us consider the wavefront in a small region on a certain plane in 3D space, as shown in Fig. 13.5, then its Fourier transform derives the angular spectrum. The spatial frequency of the angular spectrum corresponds to the direction of the wave propagation, which is parallel to the light ray that passes through the same region. Thus, the intensity of the angular spectrum is considered to be the intensity of light rays.

Under this principle, the conversion from a set of rays to a wavefront is realized in following way; (1) The set of rays, or the light field, like the one shown in Fig. 13.1(a) is sampled on a certain plane as depicted in Fig. 13.6. The plane for light-ray sampling here is called the *ray-sampling plane* (RS plane) [11]. The light-field can be generated by arbitrary rendering

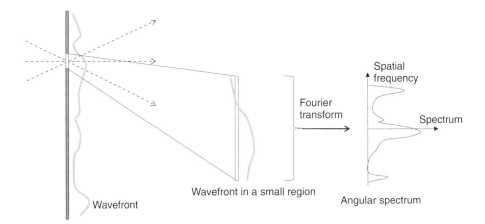

Figure 13.5   The relationship between the light-ray and the wavefront. If considering the wavefront in a small region on a certain plane, its Fourier transform gives the angular spectrum. The spatial frequency of the angular spectrum corresponds to the direction of wave propagation, and the intensity of a particular frequency is considered to be the intensity of the light-ray propagating in the corresponding direction

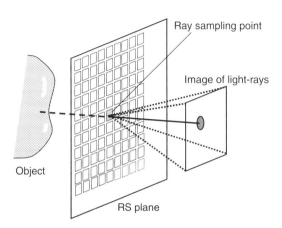

Figure 13.6   Sampling of light rays on the RS plane

techniques for computer graphics. (2) The image of rays that path through a point on the RS plane, which is an image of light rays indicated in Fig. 13.6, is Fourier transformed. Since phase information is not included in the light rays, the phase of each ray should be defined before the Fourier transform. It is possible to assign the phase on the basis of the theory of PAS explained in Section 13.2, or use of random or pseudo-random sequences are other options for phase distribution. If the rays are obtained by inverse conversion from wavefronts to rays, the phase information in the angular spectrum should be retained. The method for assigning phase information to light rays is an issue for further investigation. (3) The Fourier spectrum obtained by the second step is placed at the corresponding location on the RS plane. All the small blocks are converted in the same manner and then the wavefront on the RS plane is obtained.

13.4   Hologram Printer Based on a Full-Parallax Holographic Stereogram

13.4.1   Holographic 3D Printer

A holographic 3D printer was originally proposed on 1990 [7], in which the technique for automatic recording of a *FP holographic stereogram* (HS) was used. In FP-HS, the light ray's path through the hologram plane is recorded at high density; it then enables the observation of distortion-free 3D images from arbitrary locations within the viewing zone. HPO hologram printers have been studied since the 1990s [12–16], and the technology of FP-HS has also been commercialized [18]. The author's group has developed the technology for high-density recording of light rays, and achieved a spatial resolution 50–200 μm and angular resolution $\cong$ 0.3° in full-color FP-HS [18].

13.4.2   Full-Parallax Holographic Stereogram

In recording FP-HS, first a set of images for exposure is prepared using a graphics technique such as ray-tracing or *image-based rendering* (IBR) as shown in Fig. 13.6. The image of

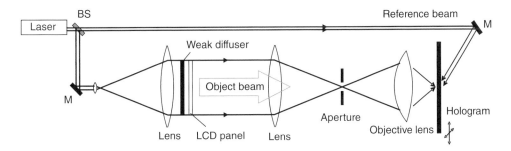

Figure 13.7 The optical system for FP-HS recording. BS: beam splitter, M: mirror, LCD: liquid crystal display. The hologram recording material is set to a XY translation stage to move horizontally and/or vertically after each exposure

light-rays for every block on the RS plane is used as the image for exposure by the optical system shown in Fig. 13.7. The image for exposure is displayed on a spatial light modulator such as liquid crystal display (LCD), which is illuminated by a plane wave from a laser light source, and an objective lens converts the plane wave to converging spherical wave. The holographic recording medium is placed at the focal plane of the objective lens, and the interference pattern with the reference beam that come from the opposite side is recorded in a small area. The small hologram is called elementary hologram (sometimes called a "hogel" [18]). A weak diffuser is used near the LCD panel to homogenize the intensity variation of object beam on the hologram plane [19]. After the exposure of an elementary hologram, the recording medium

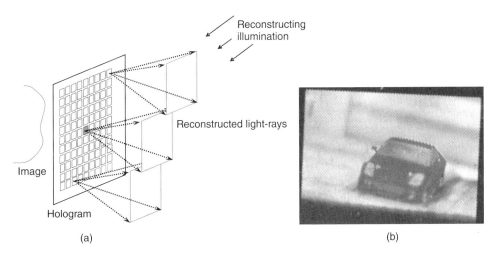

(a)   (b)

Figure 13.8 (Plate 22) Reconstruction of FP-HS. (a) Every elementary hologram reconstructs light rays in all directions. The light field is reproduced as all elementary holograms are illuminated simultaneously. (b) Example of the reconstructed image from full-color FP-HS hologram [20]. *Source*: T. Utsugi and M. Yamaguchi 2013. Reproduced with permission from The Optical Society. *See plate section for the color version*

is moved slightly in the horizontal/vertical direction for the next exposure. By exposing the whole hologram plane, FP-HS is obtained.

In the hologram printer system, a thick holographic medium is used and volume hologram is recorded. Then monochromatic image is reconstructed under white light illumination owing to the wavelength selectivity of the volume hologram. Using red, green, and blue lasers, full-color images have been also recorded. In the reconstruction of hologram, all elementary holograms simultaneously reproduce 2D images, as shown in Fig. 13.8(a). Each elementary hologram reproduces light rays traveling to different directions from the location of the grid point. Then the light field is reconstructed and a viewer can see the reproduced image as if there were real objects. Figure 13.8(b) (Plate 22) shows an example of the reconstructed image.

The optical arrangement shown in Fig. 13.7 is an optical Fourier transform system, and elementary holograms are the Fourier transform holograms of the image for exposure. This means that the wavefront reconstructed from FP-HS is a superposition of sets of wavefronts converted from light rays. Therefore this is a hologram but is also a ray-based 3D display. If the 3D image is located near the hologram plane, the difference between ray-based and wavefront displays can be kept small, and can be regarded an approximation of a real hologram. In such case, FP-HS can be considered as ray-to-wavefront converter.

13.5   Computational Holography Using a Ray-Sampling Plane

13.5.1   Computational Techniques for Electro-Holographic 3D Displays

The main purpose of the ray and wavefront conversion method is to apply it to the computation of holograms for electronic display. To realize electro-holographic 3D display in future, the technology for hologram computation is one of the principal issues as along with the ultra-high resolution display device. There have been various studies on this issue. A basic method for computing CGH is the superposition of spherical waves from a set of point light sources defined on the object surface [3,21]. It is called the *point-source method* (PSM) hereafter. In the PSM the light propagation is correctly simulated and hence high-resolution image can be reconstructed even when the image is distant from the hologram plane. A polygon-based method has also been proposed and the reproduction of excellent quality 3D images has been achieved [22].

Another approach for CGH has been also studied, which can utilize the techniques of conventional computer graphics. It is based on the principle of HS or light ray reproduction. Early work reported the CGH of HS [23], which corresponds to the multi-view 3D display. However, as the resolution of parallax views is increased, this can be deemed a light-field display. A set of parallax images is calculated using rendering technique of computer graphics, and taking the Fourier transform of each parallax image, the wavefront on the hologram plane can be derived as explained in Section 13.3. It is possible to obtain CGH of realistic 3D images by exploiting a variety of conventional rendering techniques. Nevertheless it should be noted that the image resolution is limited if the image location is distant from the hologram plane.

Since the important feature of holographic display is the capability to reproduce a deep 3D scene in high-resolution, both the methods explained earlier may not be satisfactory for this purpose. The method presented in the following sections [11] aims to take the merits of those wavefront-based and ray-based approaches.

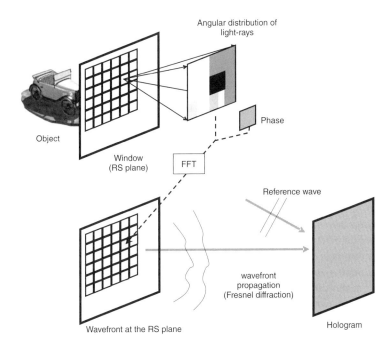

Figure 13.9 The schematic diagram of CGH calculation technique using the RS plane. In the upper figure, the angular distribution of light rays is collected at each sampled point on the RS plane. The ray information is Fourier-transformed after the random phase modulation, yielding the wavefront of a small area on the RS plane (lower figure). After all the ray information is transformed, the obtained wavefront is Fresnel-transformed to derive the wavefront on the hologram plane

13.5.2 Algorithm for CGH Calculation Using a Ray-Sampling Plane

In the proposed method for CGH calculation, a rectangular window is defined near the object location, and the images of light-rays are derived as already shown in Fig. 13.6, where the window is equivalent to the RS plane. As depicted in Fig. 13.9, the light field is sampled on the RS plane. However, the RS plane does not represent the hologram plane in contrast to Section 13.4. Now we intend to reproduce a deep 3D scene. By locating the RS plane near the object, the degradation due to ray sampling and the diffraction effect can be kept satisfactory small, and it is possible to avoid degradation of resolution. The conversion from the light field to wavefront described in Section 13.3 is applied on the RS plane; taking the Fourier transform of the angular distribution of light-ray intensity, after a random phase pattern is attached to the image so as to homogenize the intensity variation on the RS plane. Then the wavefront on the RS plane is obtained. Next, the wavefront propagation from the RS plane to the final hologram is simulated by Fresnel transform and the hologram fringe is obtained by the interference with a reference wave.

The features of the proposed method are summarized as follows:

- High-resolution sampling is possible even for the objects distant from the display plane, since light rays are sampled near the objects.

- Image blur or artifacts due to diffraction do not affect the reconstructed image because the propagation from the RS plane to the hologram is calculated based on diffraction theory.
- If the RS plane is defined in parallel to the hologram plane, the distance of wave propagation is constant, and high-speed computation of Fresnel diffraction can be implemented by using FFT (*Fast Fourier Transform*).

13.5.3   Comparison with Ray-based Techniques

First the reconstructed image of the proposed method (RS) was compared with that of the ray-based method (R-CGH hereafter) by numerical simulation [11]. In this case a 2D object was defined and located 10 and 200 mm behind the hologram plane as shown in Fig. 13.10. In the proposed method, RS plane is defined at 5 mm front of each object plane, while in R-CGH the rays are sampled at the CGH plane. The number of sampling points is 128×128, the number of angular sampling was 32×32 in both R-CGH and proposed method, and the total number of pixels of CGH was 4096×4096. To simulate reconstruction, an imaging lens with a 7 mm pupil size that mimics the human eye was defined and the reconstructed image calculated by scalar diffraction theory.

Figure 13.11 shows the simulated reconstructed images. It can be confirmed that a high-resolution image was reproduced by the proposed method, while the image reconstructed by ray-based hologram was blurred to large extent.

13.5.4   Optical Reconstruction

In the experiment, a CGH was calculated by the proposed technique and recorded in a holographic recording material using a CGH printer developed for this purpose [11]. For the object shown in Fig. 13.12(a), a car, the images of various different views were calculated by using an off-the-shelf rendering software. Then the images of angular ray-distribution were calculated based on the IBR technique. The object is located 200 mm behind the hologram plane and the image size was about 50×50 mm. The number of ray-sampling points was 768×768, the size of a ray-sampling point was $64 \times 64\,\mu m$, and the number of rays in each ray-sampling point was 32×32. The total number of pixels of the CGH was $24\,576 \times 24\,576$ pixels. The

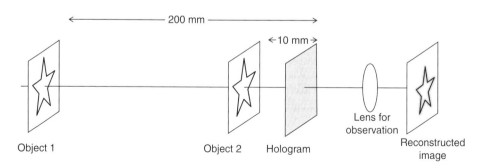

Figure 13.10   Geometry for simulation

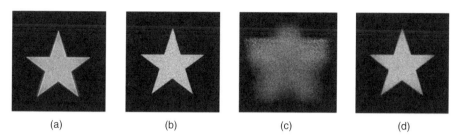

Figure 13.11 Simulation result. Reconstructed images of object 2 by (a) R-CGH and (b) proposed method. Reconstructed images of object 1 by (c) R-CGH and (d) proposed method. The object and hologram are both in 8×8 mm

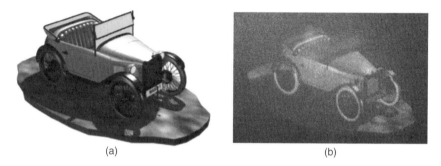

Figure 13.12 (a) The object rendered by computer graphics software. (b) Optical reconstruction of the CGH generated by proposed computation technique using the RS plane

calculated CGH pattern was exposed onto a holographic recording material using the fringe printer described previously. The image reconstructed by a plane wave generated from a DPSS (*Diode-Pumped Solid State*) green laser of 532 nm is shown in Fig. 13.12(b). As the pixel pitch of the recorded CGH was about 2 μm, the viewing angle is limited in ±3.6°. It can be observed that the surface reflection property is exhibited by rendering graphics.

13.6 Occlusion Culling for Computational Holography Using the Ray-Sampling Plane

13.6.1 Algorithm for Occlusion Culling Using the Ray-Sampling Plane

Occlusion processing is one of the important issues in the calculation of holograms [22,24–26]. In ray-based rendering, the hidden surface removal can be done by a z-buffer or ray-tracing. However, occlusion culling has not been established in the simulation of wave propagation. Strictly speaking, the wave equation should be solved with boundary conditions given by the object arrangement, but it is not practical because computation becomes very complicated. A method for exact occlusion culling was reported in [25] where the wave propagation was calculated step-by-step for every depth where occlusion took place. But

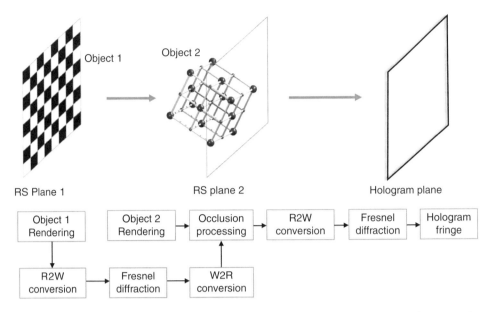

Figure 13.13   Schematic diagram of the proposed method incorporating mutual occlusion processing. R2W and W2R represent ray-to-wavefront and wavefront-to-ray transformations, respectively

such a method requires a lot of computation and most of the previous work on occlusion processing was based on ray tracing. In reference [25], the wavefront from the hidden surface was removed on the hologram plane. In this case, the diffraction of the light emitted from the hologram was taken into account, but the diffraction by the interrupting object was not. If the hologram reconstructs a very deep scene, the diffraction by the interrupting object cannot be ignored. A simple method for this purpose, called the silhouette method, was also proposed, in which the occlusion mask was defined near the interrupting object [22]. Although the calculation is simple, some errors can appear due to the discrepancy between depths of the interrupting object and the silhouette mask.

In the proposed method using the RS plane, self-occlusion, some surfaces are hidden by other surfaces of the same object, and this can be directly achieved by the rendering technique. Another type of occlusion, mutual occlusion between different objects, should be dealt with by defining an RS plane for each object located at different depths, as shown in Fig. 13.13. In the occlusion culling method using the RS plane [26], the conversion from wavefront to ray, as well as the conversion from ray to wavefront, is employed as follows.

The wavefront propagation from the RS plane 1 to the RS plane 2 is first calculated by Fresnel transformation to derive the wavefront on the RS plane 2, and is converted to the ray-information using inverse Fourier transform (W2R) on the second RS plane. Then occlusion processing is realized in the light-ray domain; the rays from the RS plane 1 are overwritten by those from the object 2. If there is no ray from object 2 at a certain location on the RS plane 2 traveling to a certain direction, then the ray from the background object is used. The

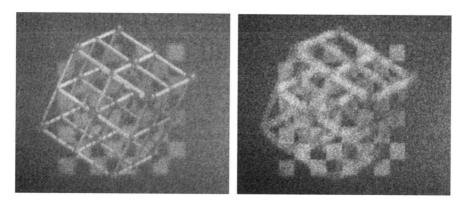

Figure 13.14   Results of optical reconstruction of the hologram calculated by the proposed method. The camera focus was adjusted to the front object in the left and the background object in the right

ray-information after occlusion processing is converted to wavefront (R2W) again, and the Fresnel transformation gives the wavefront on the hologram plane.

13.6.2   *Experiment on Occlusion Culling Using the Ray-Sampling Plane*

To prove the capability of occlusion culling using the RS plane, an experiment was carried out. Two objects, front and background as shown in Fig. 13.13, were used in this experiment. RS planes were defined at 100 and 150 mm behind the hologram plane, the background object was located on the RS plane. The hologram size and resolution were 50×50 mm and $16\,384 \times 16\,384$ pixels, respectively. Figure 13.14 shows the reconstructed images with changing the focusing position. Both the self and mutual occlusions were represented correctly and focusing on each object was possible.

13.7   Scanning Vertical Camera Array for Computational Holography

13.7.1   *Acquisition of a High-Density Light Field*

The holographic 3D printer introduced in Section 13.4 and the CGH using RS plane presented in Sections 13.5 and 13.6 are both generated from the ray-based 3D data. The images presented in Fig. 13.14 were synthesized by artificial computer graphics but it is also expected they can produce holograms of real pictured objects. In the HPO case, shooting images for HPO 3D is not as difficult as using horizontal camera motion or a horizontal camera array. On the other hand, capturing FP images requires much effort, and large-scale systems that use a huge number of cameras have been reported for light-field acquisition [3,27,28]. It is also possible to apply model-based rendering using 3D measurement and texture mapping, but the reproduction of high-fidelity realistic images, capturing and modeling of angular dependent reflections have turned out to be complicated, and the problems due to occlusion often affect

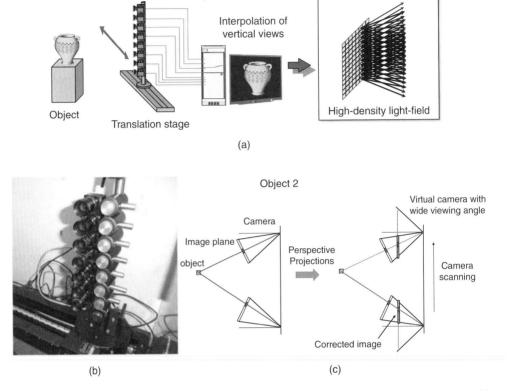

Figure 13.15 (a) Scanning vertical camera array system for the acquisition of high-density light field information. (b) Experimental of vertical camera array. (c) Geometry for keystone distortion correction

3D images as well. In this chapter, a relatively simple system for capturing the FP light field of still images is presented with a scanning vertical camera array [29].

13.7.2 Scanning Vertical Camera Array

In the proposed system, a vertical camera array is scanned horizontally as shown in Fig. 13.15(a), and vertical parallax information is interpolated from the captured data. Then only a small number of cameras are required, which allows relatively simple and compact implementation. Figure 13.15(b) shows the vertical array of seven cameras controlled by a single PC, which was used in the following experiment.

If light-rays are captured by horizontal scanning of a camera array, the angular range of rays is limited by the angle of view of the camera. In order to acquire wider angular range, the camera array is rotated with the horizontal motion such that the camera is always pointed at the object as shown in Fig. 13.15(c). In this converged camera motion, the captured images have keystone distortion. For the distortion correction, the image of a checkerboard pattern was captured in the experiment, and the parameters of keystone distortion were obtained in advance. Then perspective projection was applied to derive the images that were virtually captured by the cameras arranged parallel to the camera motion axis. Figure 13.15(c) shows the keystone distortion correction. The distortion correction had been done in advance of the following steps.

13.7.3   Vertical Interpolation

To obtain high-density light-ray information, the rays passing through the gap between cameras are interpolated. In the system shown in Fig. 13.15(a), since the horizontal parallax information is captured at high-resolution, the depth of an object can be estimated at each pixel without much difficulty. With the depth information at each pixel, vertical views between cameras are interpolated by a *depth image-based rendering* (DIBR) technique [30].

The depth data are calculated for every pixel of captured images after the distortion correction. Stereo matching or *epipolar plane image* (EPI) analysis for depth estimation is first applied in the horizontal direction for this purpose. In horizontal matching, only a limited range of camera positions around the target camera is used as shown in Fig. 13.16(a). This is for the purpose to avoid mismatching due to possible object motion or specular reflection on the object surface because their effect increases if a long sequence of horizontal camera motion is used. But when the range of camera positions used in matching is small, the depth resolution becomes low. Therefore, from horizontal camera motion, the depth of each pixel is roughly estimated. Then another stereo matching is performed using the images captured by the cameras at different heights, for fine estimation.

The vertical interpolation is performed basing on the geometry shown in Fig. 13.16(b). Consider the interpolation between upper (camera 1) and lower (camera 2) cameras. The distance between lower and upper cameras is Δy in the vertical direction, and the height of the new

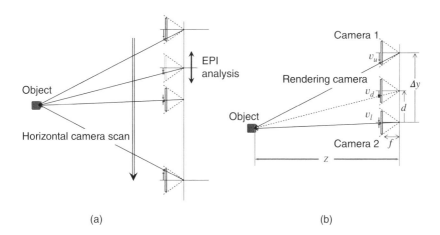

(a)   (b)

Figure 13.16   The geometry for depth estimation and vertical-parallax interpolation. (a) *Top view*: After the distortion correction, virtual cameras with wide viewing fields are moved in the horizontal direction. (b) *Side view*. The image of the rendering camera (new camera) is interpolated from the images captured by the upper (camera 1) and lower (camera 2) cameras

camera to be interpolated is d from the lower camera. f denotes the distance between the image plane and the projection center.

Let the depth of a point photographed at v_u on the upper camera image be z, then the same point should be projected at v_l in the lower camera image, given by

$$v_l = v_u + \frac{f}{z}\Delta y, \tag{13.4}$$

if occlusion does not affect the corresponding point. So, the estimated depth of the pixel at v_l of lower camera image should be almost equal to that of the pixel at v_u of upper camera image. Then, in the image of a new camera, $f_d(u, v)$, the pixel value is given by the linear interpolation of upper and lower images as,

$$f_d\left(u, v_u + \frac{f}{z}d\right) = \frac{d}{\Delta y}f_u(u, v_u) + \left(1 - \frac{d}{\Delta y}\right)f_l\left(u, v_u + \frac{f}{z}\Delta y\right), \tag{13.5}$$

where $f_l(u, v)$ and $f_u(u, v)$ denote the pixel values of lower and upper camera image, respectively. If no corresponding points are found in a vertical camera pair, the interpolation using either upper or lower images is applied, or the result of horizontal matching is used.

13.7.4   Synthesis of Ray Images

Next, the light-ray images are synthesized from a set of interpolated camera images. Figure 13.17 shows the geometry for generating the ray image at kp_e of the RS plane, where $k = 0, 1, \ldots K-1$, K (is the number of RS points in horizontal dimension) is the index for

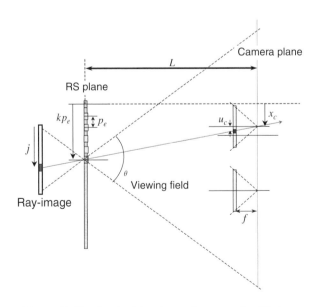

Figure 13.17   Synthesizing the ray image on the RS plane

a RS point and p_e is the sampling pitch of light rays on the RS plane. Only the example of horizontal direction is shown, where the same process is applied to both horizontal and vertical directions after the previous vertical interpolation.

The j-th pixel of the ray-image is picked-up from the image of the camera at x_c, as

$$x_c = kp_e - \left(\frac{j}{N-1} - \frac{1}{2} \right) 2L \tan \frac{\theta}{2}, \tag{13.6}$$

where $j = 0, 1, \ \ldots \ N-1$, N is the number of pixels of the ray-image in horizontal dimension, L is the distance of the camera plane from the RS plane, and θ denotes the angle of viewing field. The location of the pixel in this camera image, u_c, is given by

$$u_c = - \left(\frac{j}{N-1} - \frac{1}{2} \right) 2f \tan \frac{\theta}{2}. \tag{13.7}$$

then the corresponding pixel can be found from the camera images. f is a parameter that determines the magnification of the camera image.

A set of the ray images is generated by the earlier image transform process, which is based on IBR. Note that the position of the RS plane should be set near the object, as discussed in Section 13.5, so as to avoid reducing resolution far from the RS plane.

Figure 13.18 The results of distortion correction. *Upper row*: captured images, *Lower row*: corrected images corresponding to upper images

13.7.5 Experiment on Full-Parallax Image Generation

In the experiment, a FP 3D image captured by the proposed system was employed to synthesize holographic 3D images. The experimental system consists of seven compact CCD cameras (480 × 640 pixels, Point Gray Research, Flea) (see Fig. 13.15b) connected to PC through IEEE1394 interface, and a translation stage (Sigma, SGSP601200 (X)), which is also controlled by the same PC. The vertical intervals of cameras are 60 mm. The system is rather simple and concise, for a single PC can control the entire system. Scanning 576 mm in the horizontal direction and 577 × 7 images were captured in the experiment in 40 s, where the maximum angle of camera rotation was about 60°. The distortion correction described in Section 13.7.2 was applied as shown in Fig. 13.18.

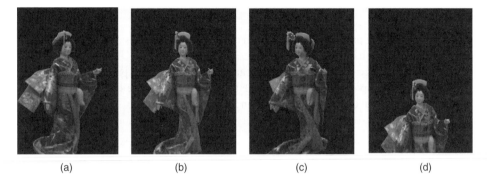

    (a)   (b)   (c)   (d)

Figure 13.19 Examples of camera images after distortion correction. Captured from (a) left, (b) center, (c) right, and (d) upper center

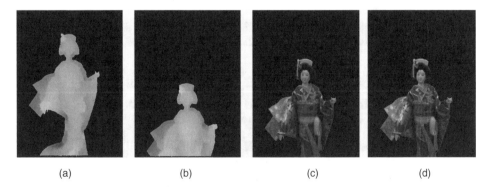

    (a)   (b)   (c)   (d)

Figure 13.20 Estimated depth images for (a) Figure 13.19(b) and (b) Figure 13.19(d). (c) The result of vertical interpolation. (d) The image captured by real camera at the position of interpolated camera of (c)

The captured images after the distortion correction are shown in Fig. 13.19. Depth estimation was done by stereo matching analysis between the images of camera intervals at 6 mm, following stereo matching using a vertical camera pair. Block matching using SAD (*Sum of Absolute Difference*) with 25×25 pixel windows was used. In vertical matching the search area was only ± 5 pixels around the pixel estimated by horizontal matching. The estimated depth images are shown in Fig. 13.20(a) and (b). Figure 13.20(c) shows the interpolated image, where Fig. 13.20(d) represents the image captured by the real camera placed in the same position. The accuracy of the interpolated image was satisfactorily high, since both horizontal and vertical parallax information were effectively employed. The PSNR (*Peak Signal-to-Noise Ratio*) of the interpolated image was 28.64 dB, whereas it was 27.05 dB in the case when only vertical stereo matching was used and 27.16 dB for horizontal EPI analysis only; about 1.5–1.6 dB improvement was achieved.

The total number of interpolated images was 577 (H) $\times$ 361 (V) at 1 mm intervals. It is possible to generate free-viewpoint images using the light-field data acquired by the proposed system, though this is limited to still image applications. Figure 13.21 shows three different view images generated using the IBR technique. Excellent quality images were synthesized from high-density light-field data captured by the scanning system. It should be noted that the specular reflections are also well reproduced, which considerably contributes to the realism of the images.

In the next experiment, CGHs were recorded using the captured light-field data. An RS plane was defined near the object, and the CGH plane was defined about 200 mm apart from the RS plane (Fig. 13.22a). The number of RS points was 256×256 and the resolution of each ray-image was 64×64 pixels. The total number of pixels in RS plane and CGH was both $16\,384 \times 16\,384$ pixels. The sampling intervals and the sizes of RS plane and CGH were the same; 2.1 µm and 34.4×34.4 mm, respectively. The calculated CGH pattern was conveyed to the output using the CGH printer presented in Section 13.5. The images reconstructed by the 532 nm green laser are shown in Fig. 13.22(b) and (c). It can be confirmed that the images are fairly reproduced by CGH, although the quality is limited because the hologram size is small and the resolution of the CGH printer is not high enough.

Figure 13.21   Free-viewpoint images generated from the light-field data captured by the proposed system. Left, close center, and upper right views

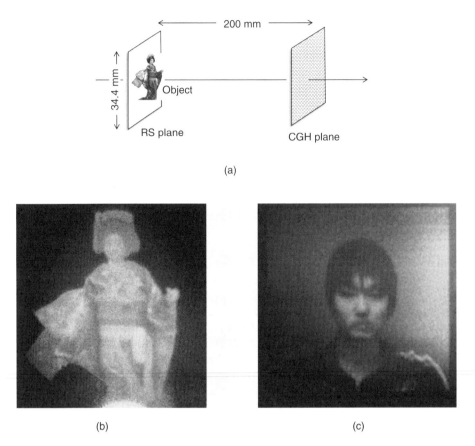

(a)

(b)   (c)

Figure 13.22 (a) The geometry for CGH calculation. The sizes of RS plane and CGH plane were both 34.4 mm. (b, c) Reconstructed images of the CGH. (b) Japanese doll, and (c) human portrait

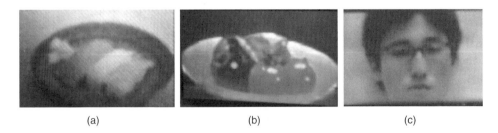

(a)   (b)   (c)

Figure 13.23 (Plate 23) Reconstructed images from the FP HS recorded from the captured light-field data. (a) Sushi, (b) vegetables, and (c) human face. *See plate section for the color version*

The light-field data captured by the proposed FP 3D image scanner were also employed for the hologram printer explained in Section 13.4. The light-ray data on the RS plane were directly printed as HS by the hologram printer. Examples of reconstructed images are shown in Fig. 13.23 (Plate 23). Holographic images of food, vegetables, and human portraits are reproduced with proper surface reflection characteristics.

13.8   Conclusion and Future Issues

In this chapter, the features of light-field displays and wavefront displays are discussed and the technique for converting ray-based and wavefront-based 3D data is introduced. The limitation of ray-based 3D display is clarified: the resolution of a 3D image becomes lower for images far from the display plane. Thus the important feature of a holographic display is the capability to reproduce deep 3D scenes at high resolution. To take the advantage of holography, a new method is proposed for the computation of hologram using the RS plane.

The proposed method exploits the conversion between the light rays and wavefront, and it becomes possible to employ the advanced rendering techniques in the computation of holograms. It also enables improved occlusion processing and surface shading for the display of deep 3D images at high resolution. Moreover, a system for capturing high-density light-field information is presented. The scanning vertical camera array technique proposed here allows for a small-scale system, controlled by a single PC, collecting high-resolution FP 3D images. The angular-dependent characteristics of 3D objects, that is, specular or glossy surfaces, can be reproduced from captured images owing to the application of the IBR technique and high-resolution parallax data, especially in horizontal direction. High-resolution light-field data captured systems are applied to 3D displays by CGH and HS.

The technology of converting light-rays and wavefronts also enables the integration of 3D imaging systems based on stereoscopic, ray-based, light-field, and holographic methods. Realization of extremely high-quality and innovative holographic display is expected, along with the advancement in device technologies, as well as a system for light-field 3D imaging by integrating ray- and wavefront-based 3D information.

Acknowledgments

This work was partly supported by the JSPS Grant-in-Aid for Scientific Research #17300032, and Toppan Printing Co. The authors acknowledge Hiroaki Yamashita, Takeru Utsugi, Mamoru Inaniwa, in Tokyo Institute of Technology, and Shingo Maruyama, Toppan Printing Co., for providing experimental data.

References

[1] Leith, E.N. and Upatnieks, J., "Wavefront reconstruction with diffused illumination and three-dimensional objects," *J. Opt. Soc. Am.* Vol. **54**, 1295–1301 (1964).

[2] Bjelkhagen, H.I. and Mirlis, E., "Color holography to produce highly realistic three-dimensional images," *Appl. Opt.* Vol. **47**, No. 4, A123–A133 (2008).

[3] St-Hilaire, P., Benton, S. A., Lucente, M. E., Jepsen, M. L., Kollin, J., Yoshikawa, H., and Under-koffler, J. S., "Electronic display system for computational holography," *Proc. SPIE* **1212**, 174 (1990).

[4] Levoy, M. and Hanrahan, P., "Light field rendering," *Computer Graphics (Proc. SIGGRAPH'96)*, 31–42 (1996).

[5] Jones, A., McDowall, I., Yamada, H., Bolas, M., and Debevec, P., "Rendering for an interactive 360° light field display," *ACM Transactions on Graphics* Vol. **26**, No. 3, *Proc. ACM SIGGRAPH 2007*, Article No. 40 (2007).

[6] Wetzstein, G., Lanman, D., Hirsch, M., Heidrich, W., and Raskar, R., "Compressive Light Field Displays," *IEEE Computer Graphics and Applications*, Vol. **32**, No. 5, 6–11 (2012).

[7] Yamaguchi, M., Ohyama, N., and Honda, T., "Holographic 3-D printer," *Proc. SPIE*, Vol. **1212**, 84–90 (1990).

[8] Yamaguchi, M., Ohyama, N., and Honda, T., "Imaging characteristics of holographic stereogram," *Japanese J. of Optics*, (Kogaku), Vol. **22**, No. 11, 714–720 (1993) (in Japanese).

[9] Yamaguchi, M., Hoshino, H., Honda, T., and Ohyama, N., "Phase added stereogram: calculation of hologram using computer graphics technique," *Proc. SPIE*, Vol. **1914**, 25–31 (1993).

[10] Yamaguchi, M., "Ray-based and wavefront-based holographic displays for high-density light-field reproduction," *Proc. SPIE*, Vol. **8043**, 804306 (2011).

[11] Wakunami, K. and Yamaguchi, M., "Calculation for computer generated hologram using ray-sampling plane," *Opt. Express*, Vol. **19**, No. 10, 9086–9101 (2011).

[12] Halle, M., Benton, S. A., Klug, M. A., and Underkoffler, J., "The Ultragram: A generalized holographic stereogram," *Proc. SPIE*, Vol. **1461**, 142–155 (1991).

[13] Yamaguchi, M., Ohyama, N., and Honda, T., "Holographic three-dimensional printer: new method," *Appl. Opt.* Vol. **31**, 217–222 (1992).

[14] Spierings, W. C. and Nuland, E. van, "Development of an office holoprinter II", *Proc. SPIE* Vol. **1667**, 52 (1992).

[15] Bains, S., "The rise and rise of the holographic printer," *OE Reports, SPIE*, May (1996).

[16] Shirakura, A., Kihara, N., and Baba, S., "Instant Holographic Portrait Printing System," *Proc. SPIE*, Vol. **3293** (1998).

[17] Klug, M. A., Klein, A., Plesniak, W. J., Kropp, A. B., and Chen, B., "Optics for full-parallax holographic stereograms," *Proc. SPIE*, Vol. **3011**, 78–88 (1997).

[18] Maruyama, S., Ono, Y., and Yamaguchi, M., "High-density recording of full- color full-parallax holographic stereogram," *Proc. of SPIE*, Vol. **6912**, 69120N-1-10 (2008).

[19] Yamaguchi, M., Endoh, H., Honda, T., and Ohyama, N., "High-quality recording of a full-parallax holographic stereogram with a digital diffuser," *Opt. Lett.* Vol. **19**, No. 2, 135–137 (1994).

[20] Utsugi, T. and Yamaguchi M., "Reduction of the recorded speckle noise in holographic 3D printer," *Opt. Express*, Vol. **21**, No. 1, 662–674 (2013).

[21] Waters, J. P., "Holographic image synthesis utilizing theoretical methods," *Appl. Phys. Lett.* Vol. **9**, 405–407 (1966).

[22] Matsushima, K. and Nakahara, S., "Extremely high-definition full-parallax computer-generated hologram created by the polygon-based method," *Appl. Opt.* Vol. **48**, H54–H63 (2009).

[23] Yatagai, T., "Stereoscopic approach to 3-D display using computer-generated holograms," *Appl. Opt.* Vol. **15**, 2722–2729 (1976).

[24] Underkoffler, J. S., "Occlusion processing and smooth surface shading for fully computed synthetic holography," *Proc. SPIE*, Vol. **3011**, 53–60 (1997).

[25] Matsushima, K., "Exact hidden-surface removal in digitally synthetic full-parallax holograms," *Proc. SPIE*, Vol. **5742**, 25–32 (2005).

[26] Wakunami, K., Yamashita, H., and Yamaguchi, M., "Occlusion culling for computer generated hologram based on ray-wavefront conversion," Submitted to *Optics Express* Vol. **21**, No. 19, 21811–21822 (2013).

[27] Wilburn, B., Joshi, N., Vaish, V., Talvala, E.-V., Antunez, E., Barth, A., *et al.*, "High performance imaging using large camera arrays," *ACM Trans. Graphics*, Vol. **24**, No. 3, 765–776 (2005).

[28] Brewin, M., Forman, M., and Davies, N. A., "Electronic capture and display of full-parallax 3D images," *Proc. SPIE*, Vol. **2409**, 118–124 (1995).

[29] Yamaguchi, M., Kojima, R., and Ono, Y., "Full-parallax 3D image scanning for holoprinter," *Nicograph International* (2008).

[30] Zhang, L., "Stereoscopic image generation based on depth images for 3D TV," *IEEE Trans. Broadcasting*, Vol. **51**, No. 2, 191–199 (2005).

14

Rigorous Diffraction Theory for 360° Computer-Generated Holograms

Toyohiko Yatagai, Yusuke Sando and Boaz Jessie Jackin
Center for Optical Research and Education, Utsunomiya University, Japan

14.1 Introduction

In computer-generated holography (CGH) [1] the fast Fourier transform (FFT) algorithm is commonly used to reduce the calculation time. Based on FFT, various types of algorithms for calculating diffraction have been developed [2]. Yoshikawa *et al.* have proposed a fast calculation method for large sized holograms by interpolation [3].

Most algorithms for a *computer-generated hologram* (CGH) using FFT are effective only under the condition that both the input and observation surfaces are finite planes parallel to each other. Some authors proposed fast calculation methods using FFT that can be applied to the case where the input plane is not parallel to the observation plane [4, 5]. These methods are very useful for the calculation of reconstructed images observed from different points of view [6]. However, since the observation surfaces, in any of the methods, are assumed to be planes, very high-resolution display devices are necessary to enlarge the viewing angles. Although such devices are available, the reconstructed images cannot be observed from the opposite side of a hologram.

A remarkable technique developed to achieve 360° field of view is 360° holography [7]. To synthesize a 360° hologram on a computer, a numerical simulation of diffraction on the non planar observation surfaces is required. Rosen synthesized a CGH on a spherical observation surface [8]. However, this method does not yield a 360° field of view because the origin of the object does not correspond to the center of the observation sphere, and also considerable computing time is required, since the FFT algorithm cannot be applied in this method.

Multi-dimensional Imaging, First Edition. Edited by Bahram Javidi, Enrique Tajahuerce and Pedro Andrés.
© 2014 John Wiley & Sons, Ltd. Published 2014 by John Wiley & Sons, Ltd.

To date, methods that enable the fast calculation of diffraction on spherical or cylindrical observation surfaces have been proposed based on the convolution theorem using the FFT algorithm. In this method, both the object and observation surfaces are cylindrical and concentric about the cylindrical axis [9, 10]. These methods are valid only in specific cases of geometrical configurations.

At first, we propose a simple but rigorous equation that describes the relationship between the diffracted wavefront of a 3D object and its 3D Fourier spectrum. In the present method, an exact solution of the diffraction integral is given by the Green function [11]. This principle gives us an intuitive understanding of calculation processes for various diffraction situation. To verify this method and confirm its effectiveness, CGHs with full view-angles are reconstructed using simulated experiments.

Alternatively, fast computation solutions for spherical computer generated hologram employing PSF (convolution method) was proposed by Tachiki et al. [12]. Here, the object and hologram were assumed to be concentric spherical surfaces in order to achieve shift invariance and hence, enable fast calculation. However, though the object was assumed to be a concentric spherical surface with the hologram surface, shift invariance does not exist due to unequal sampling points on a spherical grid (i.e., the grid points are more crowded at the poles). To facilitate fast calculation using FFT, an approximation to the convolution integral was proposed, which forced the PSF to be spatially invariant. Hence this calculation produced errors that were quantified in the same report. To solve the problem, we start the Helmholtz equation again, considering a boundary value problem in spherical coordinates. The solution will define the transfer function and the spectral decomposition of the wave field on the spherical surface. Using the transfer function and the wave spectrum we can develop a spectral propagation formula (for spherical surfaces in spherical coordinates) analogous to the angular spectrum formula (for plane surfaces in Cartesian coordinates).

This chapter explains how such theories for spherical CGH are devised.

14.2 Three-Dimensional Object and Its Diffracted Wavefront

Consider the geometry of a 3D object and an observation space as shown in Fig. 14.1. The coordinate system in the object is (x_o, y_o, z_o) and that of the observation space (x, y, z).

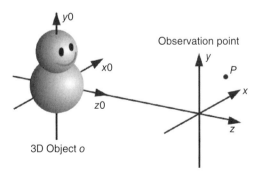

Figure 14.1 Geometry of 3-D object coordinates (x_o, y_o, z_o) and observation coordinates (x, y, z). x_o- and y_o-axes are parallel to the x- and y-axes, respectively, and the z_o-axis is the same as z-axis

The complex amplitude $u(r)$ of a wave in a homogeneous field at a position $r(x, y, z)$ satisfies the following Helmholtz equation

$$(\nabla^2 + k_0^2)u(r) = 0, \tag{14.1}$$

where $k_0 = 2\pi/\lambda$ is a constant wavenumber in a homogeneous field and λ denotes the wavelength.

The diffracted wavefront from the object $o(r)$ satisfies the following equation,

$$(\nabla^2 + k_0^2)u(r) = -o(r) \tag{14.2}$$

To solve Eq. (14.2), we employ the Green function. Green function is a solution of

$$(\nabla^2 + k_0^2)g(r, r') = -\delta(r - r'), \tag{14.3}$$

and the Green function is given by

$$g(r, r') = \frac{\exp(ik_0|r - r'|)}{4\pi|r - r'|}. \tag{14.4}$$

Therefore, the solution of Eq. (14.2) is given by

$$u(r) = \int o(r')g(r - r')\mathrm{d}r. \tag{14.5}$$

Equation (14.5) is a convolution integral and is rewritten by

$$u(x, y, z) = \mathcal{F}^{-1}[O(k_x, k_y, k_z) \cdot G(k_x, k_y, k_z)], \tag{14.6}$$

where

$$O(k_x, k_y, k_z) = \mathcal{F}[o(x, u, z)] \tag{14.7}$$

$$G(k_x, k_y, k_z) = \mathcal{F}[g(x, u, z)] \tag{14.8}$$

and $\mathcal{F}[\dots]$ denotes the 3D Fourier transform operator and $\mathcal{F}^{-1}[\dots]$ its inverse transform operator.

To obtain $G(k_x, k_y, k_z) = \mathcal{F}[g(x, y, z)]$, we Fourier-transform both sides of Eq. (14.3)

$$[-(k_x^2 + k_y^2 + k_z^2) + k_0^2]G(k_x, k_y, k_z) = -1. \tag{14.9}$$

So we have

$$G(k_x, k_y, k_z) = \frac{1}{k_x^2 + k_y^2 + k_z^2 - k_0^2}. \tag{14.10}$$

We use spatial frequencies (u,v,w) and since $k_x = 2\pi u, k_y = 2\pi v, k_z = 2\pi w$, we have

$$G(u, v, w) = \frac{1}{4\pi^2(u^2 + v^2 + w^2 - 1/\lambda^2)}. \tag{14.11}$$

Finally, from Eqs. (14.6) and (14.11), we have

$$u(x, y, z) = \frac{1}{4\pi^2} \iiint \frac{O(u, v, w)}{u^2 + v^2 + w^2 - 1/\lambda^2} \cdot$$
$$\times \exp[i2\pi(ux + vy + wz)] du dv dw \tag{14.12}$$

Next, we will find the integral with respect to w. Integral (14.12) has a singularity for

$$w_{\pm} = \pm\sqrt{1/\lambda^2 - u^2 - v^2} \tag{14.13}$$

By using contour integration, the integral with respect to w along the path shown in Fig. 14.2 is performed, where a complex value is defined by $\zeta = w + i\eta$. The residues at the singularities $w_{\pm}$ are given by

$$A_{\pm} = \lim_{\zeta \to w_{\pm}} (\zeta - w_{\pm}) \frac{O(u, v, \zeta)}{\zeta^2 + u^2 + v^2 - 1/\lambda^2} \exp(i2\pi\zeta z)$$
$$= \pm \frac{i2\pi O(u, v, \pm\sqrt{1/\lambda^2 - u^2 - v^2})}{2\sqrt{1/\lambda^2 - u^2 - v^2}}$$
$$\times \exp\left(\pm i2\pi\sqrt{1/\lambda^2 - u^2 - v^2}z\right). \tag{14.14}$$

So we have

$$u(x, y) = \frac{i}{4\pi} \iint \frac{O(u, v, \sqrt{1/\lambda^2 - u^2 - v^2})}{\sqrt{1/\lambda^2 - u^2 - v^2}}$$
$$\times \exp\left[i2\pi(ux + vy + \sqrt{1/\lambda^2 - u^2 - v^2}z)\right] du dv$$
$$- \frac{i}{4\pi} \iint \frac{O(u, v, -\sqrt{1/\lambda^2 - u^2 - v^2})}{\sqrt{1/\lambda^2 - u^2 - v^2}}$$
$$\times \exp\left[i2\pi(ux + vy - \sqrt{1/\lambda^2 - u^2 - v^2}z)\right] du dv. \tag{14.15}$$

Since only the forward propagating wave is considered, only the singularity w_+ is employed to calculate the contour integral. Suppose the observation plane is located at a plane $z = R$ and

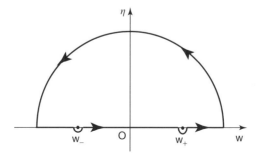

Figure 14.2   Singularities ($w_{\pm}$) and integration path in the complex plane $\zeta = w + i\eta$

we have

$$u(x, y)|_{z=R} = \frac{i}{4\pi} \iint \frac{O(u, v, \sqrt{1/\lambda^2 - u^2 - v^2})}{\sqrt{1/\lambda^2 - u^2 - v^2}}$$

$$\times \exp[i2\pi(ux + vy + R\sqrt{1/\lambda^2 - u^2 - v^2})] du dv. \qquad (14.16)$$

Finally, by 2D inverse Fourier transforming Eq. (14.16), the spectrum of the observed complex amplitude at $z = R$ is given by

$$U(u, v)|_{z=R} = \frac{i}{4\pi} \frac{O(u, v, \sqrt{1/\lambda^2 - u^2 - v^2})}{\sqrt{1/\lambda^2 - u^2 - v^2}}$$

$$\times \exp(i2\pi R\sqrt{1/\lambda^2 - u^2 - v^2}). \qquad (14.17)$$

In Eq. (14.17), $O(u, v, \sqrt{1/\lambda^2 - u^2 - v^2})$ implies a hemispherical surface component with the diameter of $1/\lambda$ in the 3D spectrum $O(u, v, w)$ of the 3D object $o(x, y, z)$. This means that the 2D spectrum $U(u, v)$ of the diffracted wave is given by the hemispherical surface component $O(u, v, \sqrt{1/\lambda^2 - u^2 - v^2})$ of the 3D spectrum with the weight of $i/(4\pi\sqrt{1/\lambda^2 - u^2 - v^2})$ multiplied with a phase component $\exp(i2\pi R\sqrt{1/\lambda^2 - u^2 - v^2})$.

Next, the correspondence relation between the diffracted wavefront and the hemispherical Fourier spectrum equivalent to it is illustrated in Fig. 14.3. Here, the diffracted wavefront to the $+z_0$ direction corresponds to the hemispherical spectrum with the $+w$ axis of symmetry. Similarly, the directions $-z_0$, $+y_0$, and $-y_0$ correspond to the directions $-w$, $+v$, and $-v$, respectively. Thus, it is possible to express the wavefront diffracting in all directions only by the hemispherical Fourier spectrum with the radius $1/\lambda$.

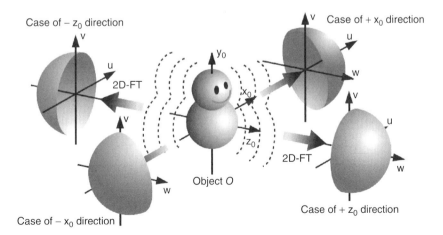

Figure 14.3 Three-dimensional hemispherical Fourier spectrum of diffracted wavefronts in each direction

14.2.1   Diffracted Waves with Full View Angles

As an example of the fast calculation method of diffraction, the diffracted images in all directions are demonstrated. Once the 3D Fourier spectrum is obtained, the diffracted images to the arbitrary direction can be readily calculated by extracting the hemispherical spectrum, multiplying the weight factor and the phase components, and running the 2D inverse FFT. To verify this method, we have simulated the diffraction under the configuration where the 2D asterism image shown in Fig. 14.4 is distributed cylindrically as shown in Fig. 14.5.

θ is the angle between the observation direction and the z axis and the diffraction distance R is set to the radius of the cylinder of the object. The results calculated with the observation angle θ varying are shown in Fig. 14.6.

As shown in Fig. 14.6, only the object corresponding to the angle θ is focused on the center of the observation plane, while both ends of each image are out of focus. These results are very natural considering the configuration shown in Fig. 14.5 and the validity of the derived principle has been verified at once.

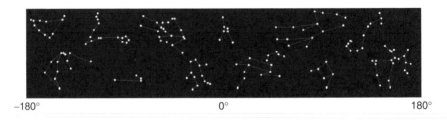

$-180°$ $0°$ $180°$

Figure 14.4   Asterism map for the object

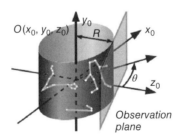

Figure 14.5   Schematic for the simulation. Perspective view of the observation plane

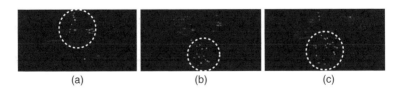

(a)   (b)   (c)

Figure 14.6   Diffracted images observed at $\theta = -109°$ (a), $-23°$ (b), and $152°$ (c)

14.3   Point-Spread Function Approach for Spherical Holography

14.3.1   Spherical Object and Spherical Hologram

As far as we know, diffraction to a spherical surface cannot be directly expressed as a convolution. The pixel resolution for a grid on a spherical surface in spherical coordinates is not constant, so the PSF is not spatially invariant. To facilitate fast calculation, an approximation to the integral is proposed, which forces the PSF to be spatially invariant and allows the diffraction integral to be expressed as a convolution.

In this method, to obtain a convolution form, the object surface must also be spherical and concentric with the hologram surface. The geometries of the object and CGH are shown in Fig. 14.7, where a spherical coordinate system is used. Note that the azimuthal axis is defined as ϕ, and the latitudinal axis is defined as θ. The object and hologram surfaces are denoted as $f_o(r_o, \phi_o, \theta_o)$ and $f_h(r_h, \phi_h, \theta_h)$. In this paper we limit the discussion to the case of a two-spherical dimensional object with hologram radius r_h constant, so we can rewrite the distributions as $f_o(\phi_o, \theta_o)$ and $f_h(\phi_h, \theta_h)$. A CGH of a 3D object can be created by taking the superposition of multiple holograms, f_h, computed with successive object radii to an arbitrary resolution. In this discussion we also assume that the object surface radius is smaller than the hologram radius, although this is not a necessary condition mathematically.

Treating the object as a composite of many point light sources, the hologram amplitude distribution $f_h(\phi_h, \theta_h)$ is given as the integral of the spherical wavefronts emerging from all the point light components on the object surface:

$$f_h(\phi_h, \theta_h) = C \iint \frac{f_o(\phi_o, \theta_o) \exp(ikd)}{d} \, d\phi_o \, d\theta_o, \tag{14.18}$$

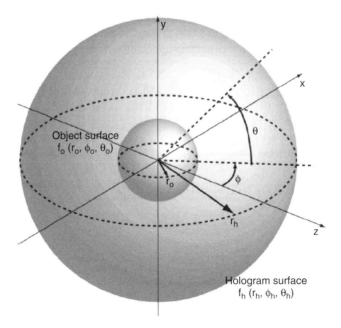

Figure 14.7   Spherical coordinate system

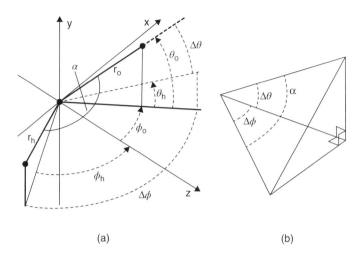

(a)   (b)

Figure 14.8   Geometry of the angular approximation (a) and geometry to solve the angle α(b)

where k denotes the wavenumber of the incident light, C denotes a constant, and d, which is the distance between a point on the object and a point on the hologram, is given by

$$d = \{r_o^2 + r_h^2 - 2r_o r_h[\sin(\theta_h)\sin(\theta_o) + \cos(\theta_h)\cos(\theta_o)\cos(\phi_h - \phi_o)]\}^{1/2}. \tag{14.19}$$

To allow the integral to be expressed as a convolution, the distance formula must be expressed as a function of $(\theta_h - \theta_o)$ and $(\phi_h - \phi_o)$.

We propose an approximation to the distance using the geometry of Fig. 14.8(a). The angle α between the two points can be approximated from the angles $\Delta\phi$ and $\Delta\theta$ through the law of cosines using the geometry of Fig. 14.8(b). Then the distance can be calculated using the lengths r_o and r_h, and the angle α with another application of the law of cosines. The resulting equation for distance is

$$d = \{r_o^2 + r_h^2 - 2r_o r_h[\cos(\theta_h - \theta_o)\cos(\phi_h - \phi_o)]\}^{1/2}. \tag{14.20}$$

The approximated PSF can then be defined as

$$h(\phi, \theta) = \frac{\exp\{ik[r_o^2 + r_h^2 - 2r_o r_h \cos(\theta)\cos(\phi)]^{1/2}\}}{[r_o^2 + r_h^2 - 2r_o r_h \cos(\theta)\cos(\phi)]^{1/2}}. \tag{14.21}$$

We substitute the PSF into Eq. (14.18) and obtain

$$f_h(\phi_h, \theta_h) = C \iint f_o(\phi_o, \theta_o)h(\phi_h - \phi_o, \theta_h - \theta_o)d\phi_o d\theta_o, \tag{14.22}$$

which takes the form of a convolution integral,

$$f_h = Cf_o * h, \tag{14.23}$$

where $*$ denotes the convolution operation. Thus, the calculation of the diffracted wavefront on the spherical hologram surface can be performed using the FFT, according to the convolution theorem.

14.3.2   *Approximation Error*

We examine the error analytically and numerically. The exact formula for the distance between a point on the object and a point on the hologram is given by Eq. (14.2). If we insert a $\cos(\phi_h - \phi_o)$ term next to the sine terms,

$$d = \{r_o^2 + r_h^2 - 2r_o r_h[\sin(\theta_h)\sin(\theta_o)\cos(\phi_h - \phi_o) + \cos(\theta_h)\cos(\theta_o)\cos(\phi_h - \phi_o)]\}^{1/2} \tag{14.24}$$

the equation reduces to the approximation form of Eq. (14.20). It is clear that the error is minimal when the $\cos(\phi_h - \phi_o)$ term approaches a value of one, or $(\phi_h - \phi_o)$ approaches zero. Thus, contributions to the hologram from object points far in the azimuthal axis have larger errors. In addition, when θ_o approaches zero, the approximated distance equation becomes

$$d = \{r_o^2 + r_h^2 - 2r_o r_h[\cos(\theta_h)\cos(\phi_h - \phi_o)]\}^{1/2}, \tag{14.25}$$

which is equivalent to the exact distance. Thus, the contributions to the hologram from object points near $\theta_o = 0$ have a smaller error. The same is true of θ_h. That is, points near $\theta_h = 0$ on the hologram have a smaller error, regardless of the area of the object image observed. Also, as r_o approaches zero, the error decreases. As can be seen from Eq. (14.14), as r_o becomes very small compared with r_h, the r_o^2 and $-2r_o r_h \cos(\phi_h - \phi_o)\cos(\theta_h - \theta_o)$ terms become very small compared with the r_h^2 term, and the distance approaches that given by the exact equation.

14.3.3   *Computer Simulation for Spherical Holography*

To demonstrate the application of the fast calculation method to holograms of more realistic object distributions, a hologram of the image in Fig. 14.9(a) was generated with the $r_o = 1$ cm

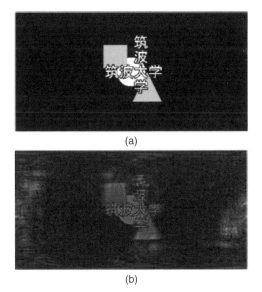

(a)

(b)

Figure 14.9   Object (a) and reconstructed image (b)

and $r_h = 10$ cm, with a wavelength of 300 μm. The reconstructed image on the same spherical surface is shown in Fig. 14.9(b), reconstructed from the hologram region of $|\phi_h| < \pi/4$ and $|\theta_h| < \pi/3$. We note that a reconstruction from a hologram computed through direct calculation of the diffraction integral would yield a perfect reconstruction of the original image in Fig. 14.9(a). Although there is distortion throughout the image, particularly as θ increases, the image is clearly recognizable. This implies that images can be reconstructed from holograms created using this method to a reasonable degree of practical usefulness.

14.4   Rigorous Point-Spread Function Approach

We will describe points on the sphere by their latitude $\theta \in [-\pi/2, \pi/2]$ and longitude $\phi \in [0, 2\pi]$ as shown in Fig. 14.7. Throughout this work we will be dealing with square-integrable functions that span a space $L^2(S)$ on the unit sphere S. Fast computation algorithms take advantage of the cyclic and periodic properties of the transformation kernel for fast calculations. The cyclic and periodic properties exists only if the system is shift invariant between the transformation planes for that operation. Which means that the object and hologram surface should be shift invariant in order to devise a fast computation scheme. To achieve this the object and hologram surface are chosen to be concentric spherical surfaces, so that they can remain shift invariant in rotation along ϕ and θ directions. But these surfaces are defined on a spherical grid, where the sampling points are more dense at the poles than at the equator. Hence, shift invariance is not satisfied. However, in the hologram generation process, the hologram and object are band-limited functions on $L^2(S)$ space. The band-limited functions on $L^2(S)$ space have a very useful and important property that *any rotated version of a band-limited function is also a band-limited function with the same bandwidth* [13–15]. Thus, they are refereed to as having *uniform resolution* at all points on the sphere, meaning that they are shift invariant. The triangular truncation and Gauss–Legendre quadrature method that occurs in the transformation operation is responsible for this property (explained in the next section). Hence the system shown in Fig. 14.7 does have a shift invariance relationship between object and hologram surfaces, and confirms the possibility of a fast computation formula. With this assurance we proceed to develop the fast calculation method for computer generated spherical holography starting from the basic equations of electromagnetism.

An electromagnetic field is defined by Maxwell's equations and its propagation by the Helmholtz wave equation. Hence for any particular system the complex amplitude of a propagating wave at any instance of time and anywhere in space can be found by solving the wave equation, applying its constraints and conditions. Accordingly, for the system shown in Fig. 14.7 the solution can be derived starting from the wave equation as follows. The time independent vector wave equation $u(r, \theta, \phi)$ is expressed the by the Helmholtz equation as

$$\nabla^2 u + k^2 u = 0, \tag{14.26}$$

where r is the radius of the spherical surface of interest and θ and ϕ represent the azimuthal and polar angles in the surface, respectively. The Laplacian operator ∇^2 in spherical coordinates is defined as

$$\nabla^2 = \frac{1}{r^2} \frac{\partial}{\partial r} \left(r^2 \frac{\partial}{\partial r} \right) + \frac{1}{r^2 \sin\theta} \frac{\partial}{\partial \theta} \left(\sin\theta \frac{\partial}{\partial \theta} \right) + \frac{1}{r^2 \sin^2\theta} \frac{\partial^2}{\partial \phi^2}. \tag{14.27}$$

The scalar wave equation given in spherical coordinates becomes

$$\frac{1}{r^2}\frac{\partial}{\partial r}(r^2\frac{\partial u}{\partial r}) + \frac{1}{r^2\sin\theta}\frac{\partial}{\partial\theta}(\sin\theta\frac{\partial u}{\partial\theta}) + \frac{1}{r^2\sin^2\theta}\frac{\partial^2 u}{\partial\phi^2} + k^2 u = 0. \tag{14.28}$$

The solution of Eq. (14.28) can be found by separation of variables [16–18], which can be expressed as shown next

$$u(r,\theta,\phi,t) = R(r)\Theta(\theta)\Phi(\phi). \tag{14.29}$$

The separable variables obey the following four ordinary differential equations:

$$\frac{d^2\Phi}{d\phi^2} + m^2\Phi = 0 \tag{14.30}$$

$$\frac{1}{\sin\theta}\frac{d}{d\theta}(\sin\theta\frac{d\Theta}{d\theta}) + [n(n+1) - \frac{m^2}{\sin^2\theta}]\Theta = 0 \tag{14.31}$$

$$\frac{1}{r^2}\frac{d}{dr}(r^2\frac{dR}{dr}) + k^2 R - \frac{n(n+1)}{r^2}R = 0. \tag{14.32}$$

The solution to all these equations are derived in by Arfken [18]. Only the final results are used in this research work. The solution to azimuthal Eq. (14.30) is

$$\Phi(\phi) = \Phi_1 e^{im\phi} + \Phi_2 e^{-im\phi}, \tag{14.33}$$

where m must be an integer so that there is continuity and periodicity in $\Phi(\phi)$. Φ_1 and Φ_2 are constants.

The solution to polar Eq. (14.31) is

$$\Theta(\theta) = \Theta_1 P_n^m(\cos\theta) + \Theta_2 Q_n^m(\cos\theta) \tag{14.34}$$

where, P_n^m and Q_n^m are the associated Legendre polynomials of first and second kind, respectively, and Θ_1 and Θ_2 are constants. Q_n^m is not finite at the poles where $\cos(\theta) = \pm 1$, so this solution is discarded ($\Theta_2 = 0$).

For the radial differential equation Eq. (14.32) the solutions are

$$R(r) = R_1 h_n^{(1)}(kr) + R_2 h_n^{(2)}(kr) \tag{14.35}$$

where, $h_n^{(1)}$ and $h_n^{(2)}$ are the spherical Hankel functions of the first and second kind, respectively. Since we are only interested in the outgoing wave, we can neglect the second term ($R_2 = 0$).

The angle functions in Eq. (14.33) and Eq. (14.34) are conveniently combined into a single function called a *spherical harmonic* Y_n^m [17, 18] defined by

$$Y_n^m(\theta,\phi) \equiv (-1)^m \sqrt{\frac{(2n+1)(n-m)!}{4\pi(n+m)!}} P_n^m \cos(\theta) e^{im\phi} \tag{14.36}$$

where the quantity

$$\overline{P}_n^m = \sqrt{\frac{(2n+1)(n-m)!}{4\pi(n+m)!}} P_n^m(\cos\theta) \tag{14.37}$$

is known as the orthonormalized associated Legendre polynomial. The term $(-1)^m$ is called the Condon–Shortley phase. Hence the spherical harmonics can also be represented in short form as shown next (neglecting the Condon–Shortley phase)

$$Y_n^m(\theta, \phi) = \overline{P}_n^{-m}(\cos \theta)e^{im\phi}. \tag{14.38}$$

Combining all these equations, the traveling wave solution to Eq. (14.28) can be represented as

$$u(r, \theta, \phi, \omega) = \sum_{n=0}^{\infty} \sum_{m=-n}^{n} A_{mn}(\omega)h_n(kr)Y_n^m(\theta, \phi). \tag{14.39}$$

The radiated field is completely defined when the coefficient A_{mn} is determined. This is achieved by using the orthonormal property of spherical harmonics. Assume that the wave field $u(r, \theta, \phi)$ is known on a sphere of radius $r = a$. We also drop the time variable (which is not important) for simplicity. Now multiplying each side of Eq. (14.39) (evaluated at $r = a$) by $Y_n^m(\theta, \phi)^*$ and integrating over the sphere, gives

$$A_{mn} = \frac{1}{h_n(ka)} \int_{-\frac{\pi}{2}}^{\frac{\pi}{2}} \int_0^{2\pi} u(a, \theta, \phi)Y_n^m(\theta, \phi)^* \sin(\theta)d\theta d\phi \tag{14.40}$$

where, $d\Omega = \sin(\theta)d\theta d\phi'$, is the solid angle for integrating on a sphere. Inserting the expression for A_{mn} back into Eq. (14.39), we get,

$$u(r, \theta, \phi) = \sum_{n=0}^{\infty} \sum_{m=-n}^{n} Y_n^m(\theta, \phi) \left(\left[\int_{-\frac{\pi}{2}}^{\frac{\pi}{2}} \int_0^{2\pi} u(a, \theta', \phi')Y_n^m(\theta', \phi')^* d\Omega' \right] \frac{h_n(kr)}{h_n(ka)} \right). \tag{14.41}$$

Hence the wavefield at any spherical surface $u(r, \theta, \phi)$ can be calculated knowing the wavefield at $u(a, \theta', \phi')$.

For easy understanding and interpretation of Eq. (14.41), it is compared with the well-known angular spectrum of plane waves equation [19], which is given by

$$u(x, y, z) = \frac{1}{4\pi^2} \int_{-\infty}^{\infty} \int_{-\infty}^{\infty} e^{i(k_x x + k_y y)}dk_x dk_y$$

$$\left(\left[\int_{-\infty}^{\infty} \int_{-\infty}^{\infty} u(x', y', 0)e^{-i(k_x x' + k_y y')}dx'dy' \right] e^{ik_z z} \right). \tag{14.42}$$

It is well known that, in Eq. (14.42), the quantity within square brackets represents the source wave field decomposed into spectrum of plane waves in (k_x, k_y). The propagation of the spectrum is defined by the quantity $e^{ik_z z}$, which is known as the transfer function. The propagated spectrum is recomposed into the wave field at destination by the integral over k_x, k_y. Thus this equation defines wave propagation from one plane surface $u(x', y', 0)$ to another surface $u(x, y, z)$. When comparing Eq. (14.41) with Eq. (14.42), the following can be deduced:

- The wavefield in (θ, ϕ) at a radiating spherical surface of radius $r = a$ is decomposed into its wave spectra in (m, n) defined by

$$U_{mn}(a) = \int u(a, \theta, \phi)Y_n^m(\theta, \phi)^* d\Omega. \tag{14.43}$$

- The decomposed wave components (spectra) are expressed by Eq. (14.36), which is composed of a traveling wave component in ϕ, and defined by $e^{im\phi}$ and a standing wave component in θ given by $P_n^m(\cos\theta)$.
- The decomposed wave components in this system can be named spherical wave components in analogy to plane wave components for planar systems. Similarly Eq. (14.43) can be termed a spherical wave spectrum as analogous to the angular spectrum of plane waves.
- The wavenumbers k_x and k_y are imitated by the quantities m/a and n/a. Hence, we can refer to the spherical wave spectrum as a k-space spectrum, due to this analogy.
- Equation (14.43) can be viewed as a forward Fourier transform using $Y_n^m(\theta, \phi)$ as the basis function. In other words, the spectral space for spherical system is spanned by spherical harmonic functions ($Y_m^n(\theta, \phi)$). This is also termed *spherical harmonic transform* (SHT).
- The propagation of a spherical wave spectrum from one spherical surface of radius a to another of radius r is given by

$$U_{mn}(r) = \frac{h_n(kr)}{h_n(ka)} U_{mn}(a). \tag{14.44}$$

- Hence the quantity $h_n(kr)/h_n(ka)$ can be referred to the transfer function (TF) for a spherical system, as opposed to the quantity $e^{ik_z z}$ for a planar system.
- The inverse spherical harmonic transform (ISHT) that recomposes the wave field back from the spectrum is given by

$$u(r, \theta, \phi) = \sum_{n=0}^{\infty} \sum_{m=-n}^{n} U_{nm}(r) Y_n^m(\theta, \phi). \tag{14.45}$$

The transfer function is crucial since it completely defines propagation and hence it is worth discussing some of its properties. During propagation, amplitude and phase of the spectral components change with distance as defined by the transfer function. What is most important is the rate of change of phase of the transfer function, which determines the sampling requirements. Accordingly the plot shown in Fig. 14.10 reveals that the phase change increases with increasing orders "n" of the transfer function (the plot was generated for wavelength 100 μm,

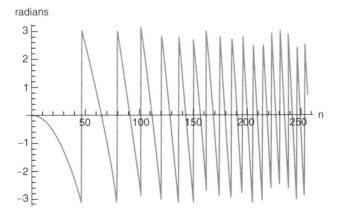

Figure 14.10   Plot of phase (in radians) of the transfer function for increasing order (n)

radius 10 and 0.5 cm, and up to 256 orders). Hence, sampling requirements will be satisfied if highest order "n" of the transfer function is sampled according to Nyquist criteria. It can also be understood that the rate of phase change becomes higher as the distance of propagation increases. This will demand a lot of sampling and also increase numerical errors. It is worth noting here that the spherical Hankel functions are asymptotic in nature. In the far field, the spherical Hankel functions can be approximated by their asymptotic expressions, which is as shown in Eq. (14.46).

$$h_n^{(1)}(x) = (-i)^{n+1} \frac{e^{ix}}{ix}. \qquad (14.46)$$

This could yield a formula analogous to the far field Fresnel diffraction formula. However, it requires systematic development of theory with proper analysis of approximations and is not discussed in this chapter. But this can be considered as future work to the proposed method. Therefore, the devised formula, which is analogous to the angular spectrum of plane waves (AS) formula, defines wave propagation between spherical surfaces. The numerical procedure to implement the devised formula is discussed in the next section.

14.4.1 Numerical Computation

The numerical computation of angular spectrum (AS) method heavily depends on the FFT operations for which a lot of tools and methods are available. But the numerical computation of the proposed method heavily depends on the SHT operations. Fast computation was guaranteed from the theory and geometry of the system. Now a numerical procedure that takes advantage of this is required. Fortunately, lot of fast computation numerical methods have been reported for SHT and this research work could make use of it. Since FFT and numerical computation of AS method are well understood, this section intends to introduce SHT and numerical computation of the proposed method in close analogy to the former.

Continuing with the comparison from the previous section, the numerical computation of wave propagation for planar and spherical systems according to Eq. (14.41) and Eq. (14.42) can be represented respectively as

$$u(r, \theta, \phi) = \text{ISHT}[\text{SHT}(u(a, \theta, \phi)) \times TF_s] \qquad (14.47)$$

$$u(x, y, z) = \text{IFFT}[\text{FFT}(u(x, y, 0)) \times TF_c] \qquad (14.48)$$

where, FFT[...] and IFFT[...] denote the forward and inverse fast Fourier transform operations, while SHT[...] and ISHT[...] denote the forward and inverse spherical harmonic transform operations. TF_s and TF_c are the transfer functions for spherical and planar wave propagations as defined by Eq. (14.41) and Eq. (14.42), respectively. Comparison reveals that the computation method (for spherical systems) is analogous to the angular spectrum of plane waves method (for planar systems) except for the fact that the Fourier transform is replaced by the spherical harmonic transform. Hence evaluation of SHT becomes key, the others being only basic mathematical evaluations. SHT operations are fundamentally different from the FFT operations. In FFT operation the transform is done with a sinusoid and its higher harmonics as the basis function, whereas in the case of SHT, spherical harmonics defined by Eq. (14.36) are the basis. This also implies that since the transformation kernel for the FFT is a sinusoid, it is an operation defined on a circle. But spherical harmonics are functions on a sphere and hence the transformation is an operation defined on a sphere. Spherical harmonic transforms have been studied extensively, and fast computation algorithms and optimization methods have been proposed. A method that requires only $O(N^2 \log N)$ operations for N grid points, as opposed to the standard N^3 operations, was proposed by Chien *et al.* [13]. They imposed truncations on the spectral components, and used a fast multipole method and fast Fourier transform for evaluation. Their method is called the "spectral truncation method" and the errors due to truncation were well within acceptable limits. Later on, Healy *et al.* [14] achieved the same using $O(N(\log N))^2$ operations. They took advantage of the recursive properties of the associated Legendre polynomial for fast computation. This method was found to be more efficient and hence was chosen for numerical evaluation of the spherical harmonic transforms in this work. A brief outline of the numerical evaluation is presented. For more details please refer to Healy [14] and Driscoll [15].

Though FFT and SHT are fundamentally different, they both are variable separable. Which means a 2D FFT is computed by separating the transformation kernel into its variables and evaluating it as 1D column transform followed by a 1D row transformation. Similarly the spherical harmonic transform kernel Y_n^m given by Eq. (14.38) is also variable separable and can be separated into a ϕ component and θ component. Then it is evaluated as a 1D transform along the ϕ direction followed by another 1D transform along the θ direction. Accordingly, the transformation given by Eq. (14.43) can be represented as shown next

$$U_{mn}(r) = \int_{-\pi/2}^{\pi/2} \left(\int_{-\pi}^{\pi} u(r,\theta,\phi)e^{im\phi}d\phi \right) \overline{P}_n^m(\cos\theta)d\theta. \tag{14.49}$$

First, the quantity within the round brackets alone is to be computed, which is nothing but a Fourier transform operation. The Fourier coefficients $U^m(\theta)$ are evaluated for $m = -N, ..., N$ as shown next

$$U_m(\theta) = \int_{-\pi}^{\pi} u(r,\theta,\phi)e^{im\phi}d\theta \tag{14.50}$$

$$= \frac{1}{I}\sum_{i=1}^{I} u(r,\theta,\phi_i)e^{im\phi_i} \tag{14.51}$$

where $\phi_i = 2\pi i/I$ for $i = 1,...,I$. The equispaced longitudes ϕ_i enables the use of fast Fourier transform.

Second, the Legendre transform of the Fourier coefficients $U_m(\theta)$ is to be evaluated for $|m| \le n \le N$. This is done using the Gaussian–Legendre quadrature as shown next,

$$U_{nm} = \int_{-\pi/2}^{\pi/2} U_m(\theta_j)\overline{P}_n^m(\cos\theta)\sin\theta d\theta \tag{14.52}$$

$$= \sum_{j=|m|}^{N} U_m(\theta_j)\overline{P}_n^m(\cos\theta_j)w_j, \tag{14.53}$$

where θ_j and w_j are, respectively, the Gauss nodes and weights, and are calculated using the Fourier–Newton method as described by Swarztrauber [16]. The Gauss–Legendre quadrature replaces the integral with the sum. The fact that the summation runs only from $|m|$ to N is referred to as triangular truncation. The use of the Gauss–Legendre quadrature method redistributes θ into Gaussian nodes θ_j. This along with the triangular truncation are responsible for uniform resolution on the latitudinal points. Again, the triangular truncation, along with the recurrence property of the Legendre polynomial, helps to achieve fast computation.

In a similar way the inverse spherical harmonic transform can be represented as

$$u(\theta, \phi) = \sum_{m=-N}^{N} \left(\sum_{n=|m|}^{N} U_{nm} P_n^m (\cos \theta) \right) e^{im\phi}. \tag{14.54}$$

The inverse transform also follows the same procedure and is computed in two steps but in the reverse order (i.e., Legendre transform first and Fourier transform next) as shown here

$$U_m(\theta) = \sum_{n=|m|}^{N} U_{nm} \overline{P}_n^m (\cos \theta) \tag{14.55}$$

$$u(\theta, \phi) = \sum_{m=-N}^{N} U_m(\theta) e^{im\phi}. \tag{14.56}$$

Thus using this numerical procedure fast computation of wave propagation in spherical computer generated holograms is achieved, which uses only $O(N(\log N)^2)$ operations for SHT computation.

There are a lot of software tools available on the Internet to do the SHT operation. Most of them are tuned and dedicated for geophysical processes, which only requires real SHT but holography requires complex SHT. However, the package of SHTools by Wieczorek [20] could do a complex SHT operation in the Fortran language. Though not tested by us, this is the best one available that suits the work related to this paper. Hence, we recommend using this if it is required to quickly reproduce the work in this paper. The next section describes the testing and verification of this numerical procedure using simulations.

14.4.2 Simulation Results on Rigorous Theory

The system considered for simulation experiments is shown in Fig. 14.7. The object $(O(a, \theta, \phi))$ is a spherical surface of radius 1 cm and the hologram $(H(r, \theta, \phi))$ is another concentric spherical surface of radius 10 cm. The reference is considered to be a virtual source emitting spherical waves from the center, that is, the wavefield, due to reference has same phase and amplitude throughout the hologram plane. This is similar to using a plane reference wave with normal incidence in plane holography.

14.4.3 Verification through Comparison

Since this the first occurrence of such a formula in computer generated holography first it is required to test it to see if it obeys the basic diffraction laws. For this, the proposed method by

expecting it to reproduce the already known diffraction results. To achieve this, the proposed method is made to generate already reported diffraction patterns for spherical surfaces. For this the diffraction pattern reported by Tachiki *et al.* [12] for spherical surfaces is used as a reference. Accordingly, a simple object was chosen, which is a spherical surface with two irradiating points at $\phi = -\pi/2$ and $\phi = \pi/2$, as shown in Fig. 14.11. The object and hologram are composed of 256 pixels in the longitudinal (north-south) direction and 512 pixels in the latitudinal (east-west) direction. The wavelength was chosen to be $\lambda = 100\,\mu\text{m}$, in order to reduce sampling requirements and visibility of fringes. The procedure for numerical generation of hologram using the proposed method is expressed in an abstract form as shown next.

$$AmplitudeHologram = |\text{ISHT}[\text{SHT}(Object) \times TF]$$

$$+ \text{ISHT}[\text{SHT}(Reference) \times TF]|^2.$$

Then a hologram for the same object was simulated using the the well-known direct integration formula defined in Eq. (14.57).

$$H(r, \theta, \phi) = \iint \frac{O(\theta', \phi') \exp\ (ikL)}{L} dx dy \qquad (14.57)$$

where,

$$L = \sqrt{r^2 + a^2 - 2ra[\sin(\theta)\sin(\theta') + \cos(\theta)\cos(\theta')\cos(\phi - \phi')]} \qquad (14.58)$$

$$Hologram = |H_{object}(r, \theta, \phi) + H_{reference}(r, \theta, \phi)|^2.$$

The simulation results are shown in Fig. 14.12. The pattern generated by the proposed method matches the one generated by the direct integration method. However, the distribution of brightness and contrast across the pattern is constant for the direct integration method, while it decreases gradually from the center for our proposed method. This inconsistency can be explained as follows. The direct integration formula Eq. (14.57) is the Rayleigh–Sommerfeld diffraction formula [12] without the obliquity factor. The obliquity factor is the cosine of the angle between the normal of the radiating surface to the direction of the observation point. This is responsible for the distribution of light intensity based on the angle (i.e., more bright at the center, gradually decreases outward, and no radiation backwards). However, the spectral method, which is the solution to the boundary value problem of the wave equation, incorporates the obliquity factor, and hence the brightness and contrast varies radially. Moreover, the

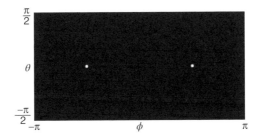

Figure 14.11   Two point source object

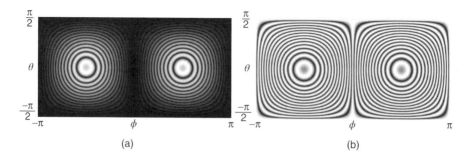

Figure 14.12   Computed hologram (intensity) using (a) proposed method and (b) direct integration

obliquity factor does not alter the phase of the traveling wave, which in turn does not affect the interference pattern, and hence guarantees a fair comparison.

14.4.4   Hologram Generation

Since the theory is developed in the context of computer generated holography, it is mandatory to verify its applicability to the same. For this, it was decided to perform spherical hologram generation and then reconstruction from the hologram on the computer using our proposed method. The object was assumed to be a single spherical surface with some images inscribed on it. The spherical object for which the hologram was to be made is shown in Fig. 14.13. The object and hologram were composed of 256×512 pixels. The wavelength for simulation was chosen to be $\lambda = 30\,\mu$m. Again, here the wavelength was assumed to be large in order to reduce sampling requirements. Now, using the developed formula, wave propagation was simulated from the object surface to the hologram surface. Since the reference was assumed to be a spherical wave emanating from the center, it contained the same phase and amplitude at the hologram surface. So we have the complex amplitudes of the object and reference as a matrix of complex numbers at the hologram surface. By adding these two complex amplitudes and calculating the intensity, the hologram is produced and is shown in Fig. 14.14.

Figure 14.13   Object

Figure 14.14 Hologram (intensity)

Figure 14.15 Reconstruction

From this hologram the object was reconstructed back onto the original spherical surface using Eq. 14.41. Reconstruction with the original reference will only produce a virtual image at the location of the object. In order to obtain a real reconstructed image at the original object location, the hologram should be illuminated (or reconstructed) using the conjugate of the reference. This means that we are attempting to reconstruct a real image on the spherical surface where the object was earlier located. The conjugate of the reference was produced by taking the complex conjugate of the reference wave field matrix. Accordingly, the numerical procedure of the for reconstruction is expressed in an abstract form as shown next.

$$Reconstruction = |ISHT[SHT(Hologram \times Conjugate[Reference]) \times TF]|^2.$$

The reconstructed real image is shown in Fig. 14.15. The reconstruction matches exactly with the object chosen. The reconstruction is crisp and is also free from any noise. As mentioned earlier, the object and hologram are square integrable band-limited functions on a closed surface. Hence, a rotated (shifted in theta or phi) version of the object or hologram will produce a rotated version of the reconstruction. The wave propagation calculation requires $O(N(\log N))^2$ operations for N sampling points, and hence it is a fast computation formula. The calculations were executed using a scripted language-python on a Dell Precision T7400 machine with 12 GB of RAM memory. A calculation time for the direct integration, convolution, and spectral methods are 3730 s, 0.057 s, and 0.039 s, respectively. This means that the proposed method took the least time for calculation, and hence is the fastest.

14.5 Conclusion

We have revealed the information inherent to the diffracted wavefront by the derivation of the relation with the 3D Fourier spectrum based on the Helmholtz equation and its Green function. The diffracted wavefront propagating in all directions can be expressed completely by just the spherical Fourier spectrum. Moreover, as an example of the application of this basic principle, the simulation of the diffracted images to all directions has been demonstrated and the validity of this principle has been also verified. This principle gives us the basis of the diffraction calculation based on the Fourier spectrum and it is flexibly applicable to other various diffractions, such as the rotation of wavefronts, cylindrical observation, and so on.

References

[1] Lohmann A. W. and D. P. Paris, "Binary Fraunhofer holograms, generated by computer," *Appl. Opt.,* **6**, 1739–1748 (1967).

[2] Leseberg D., "Sizable Fresnel-type hologram generated by computer," *J. Opt. Soc. Amer.,* **6**, 229–233 (1989).

[3] Yoshikawa N., M. Itoh, and T. Yatagai, "Interpolation of reconstructed image in Fourier transform computer-generated hologram," *Opt. Commun.,* **119**, 33–40 (1995).

[4] Leseberg D. and C. Frére, "Computer-generated holograms of 3-D objects composed of tilted planar segments," *Appl. Opt.,* **27**, 3020–3024 (1988).

[5] Matsushima K., H. Schimmel, and F. Wyrowski, "Fast calculation method for optical diffraction on tilted planes by use of the angular spectrum of plane waves," *J. Opt. Soc. Am. A,* **20**, 1755–1762 (2003).

[6] Yu L., Y. An, and L. Cai, "Numerical reconstruction of digital holograms with variable viewing angles," *Optics Express,* **10**, 1250–1257 (2002).

[7] Soares O. D. D. and J. C. A. Fernandes, "Cylindrical hologram of 360° field of view," *Appl. Opt.,* **21**, 3194–3196 (1982).

[8] Rosen J., "Computer-generated holograms of images reconstructed on curved surfaces," *Appl. Opt.,* **38**, 6136–6140 (1999).

[9] Sando Y., M. Itoh and T. Yatagai, "Fast calculation method for cylindrical computer-generated holograms," *Opt. Express,* **13**, 1418 (2005).

[10] Jackin B. J. and T. Yatagai, "Fast calculation method for computer-generated cylindrical hologram based on wave propagation in spectral domain," *Opt. Express,* **18**, 25546–25555 (2010).

[11] Kak A. C. and M. Slaney, *Principles of Computerized Tomographic Imaging* (IEEE, New York, 1988).

[12] Tachiki M. L., Y. Sando, M. Itoh, and T. Yatagai, "Fast calculation method for spherical computer-generated holograms," *Appl. Opt.,* **45**, 3527–3533 (2006).

[13] Chien R. J. and B. K. Alpert. "A fast spherical filter with uniform resolution." *J. Comput. Phys.,* **136**, 580–584 (1997).

[14] Healy Jr D. M., D. Rockmore, P. J. Kostelec, and S. Moore. "FFTs for the 2-sphere–improvements and variations" *J. Fourier. Anal. Appl.* **9**, 341–385 (1998).

[15] Driscoll J. R. and D. M. Healy. "Computing Fourier transforms and convolutions on the sphere," *Adv. Appl. Math.,* **15**, 201–250 (1994).

[16] Swarztrauber P. N. "On computing the points and weights for Gauss–Legendre quadrature," *SIAM. J. Sci. Computing*, **24**, 945–954 (2002).

[17] Lebedev N.N. *Special Functions and their Applications*. Prentice Hall (1965).

[18] Arfken G.B. *Mathematical Method for Physicisst*. Academic Press, p. 702 (2001).

[19] Goodman J.W., *Introduction to Fourier Optics*, 3rd edn. Roberts and Company Publishers, (2004).

[20] Wieczorek M. SHTools. Website, available at: URL http://shtools.ipgp.fr/ (accessed December 6, 2013).

Part Four

Spectral and Polarimetric Imaging

15

High-Speed 3D Spectral Imaging with Stimulated Raman Scattering

Yasuyuki Ozeki[1] and Kazuyoshi Itoh[1,2]
[1] *Graduate School of Engineering, Department of Material & Life Science, Osaka University, Japan*
[2] *Science Technology Entrepreneurship Laboratory (e-square), Osaka University, Japan*

15.1   Introduction

Ultrafast laser pulses with durations from femtoseconds to picoseconds can have high peak intensities while the average power is kept low. When such pulses are tightly focused on materials, various types of nonlinear-optical effects can occur such as harmonic generation, wave mixing, and stimulated scattering. Compared with linear optical effects such as refraction and absorption, nonlinear-optical effects lead to rich light-matter interactions. One of the important applications of such nonlinear optical effects is nonlinear optical microscopy, which exploits nonlinear optical effects for microscopic imaging. Depending on the nonlinear optical effect involved, nonlinear optical microscopy provides various contrast mechanisms. A common advantage of nonlinear optical microscopy is that it has 3D resolution because nonlinear optical interaction is confined to the vicinity of focal volume where the laser intensity is the highest.

Nowadays, biology researchers use *two-photon excited fluorescence* (TPEF) microscopy, which was developed in 1990 [1]. In TPEF microscopy, samples are labeled with fluorescent molecules and proteins and then irradiated by focused femtosecond pulses at an optical frequency of ω. When the fluorescent molecules and/or proteins are transparent at a frequency of ω and absorptive at a frequency of 2ω, they can simultaneously absorb two photons to cause electronic transition to an excited state, and then emit fluorescence. This is called TPEF. TPEF is especially advantageous for deep tissue imaging because we can use near-infrared laser pulses, which is less likely to be scattered in tissue compared with visible light used for single photon excitation. Therefore, the intrinsic 3D resolution of TPEF is maintained even in scattering samples. In contrast, in previous single photon fluorescence microscopy, where a confocal pinhole is required for obtaining 3D resolution, light scattering is detrimental because scattered fluorescence can no longer pass through the confocal pinhole.

Other forms of nonlinear-optical microscopy have been developed based on second harmonic generation [2], third harmonic generation [3], four-wave mixing including coherent anti-Stokes Raman scattering (CARS) [4,5], stimulated parametric emission (SPE) [6], two-photon absorption [7], stimulated Raman scattering (SRS) [8–10], and stimulated emission [11]. Since these nonlinear optical interactions do not rely on fluorescence, they allow label-free imaging of non-fluorescent molecules, even without labeling or staining.

Of these techniques, SRS microscopy has various advantages [8–10]. For example, SRS gives chemical contrast based on vibrational spectroscopy. The contrast is quantitative, that is, the signal is proportional to the density of molecules of interest. Furthermore, the signal is so intense that fast imaging at up to video rate is possible [12]. These features have been proved by several demonstrations of label-free biomedical imaging with SRS [12–20]. One of the important research trends in SRS microscopy is spectral imaging where SRS images at various Raman shifts are acquired [21–30]. Such spectral data can be analyzed to differentiate tiny spectral features, leading to further improvement of molecular specificity.

This chapter reviews the current status of SRS microscopy, and introduces high-speed SRS spectral microscopy developed by the authors [28]. Note that detailed principles and theoretical treatments of SRS microscopy are available in recent review papers [31] and a book [32].

This chapter is organized as follows. Section 15.2 summarizes the principle and advantages of SRS microscopy. Section 15.3 explains how spectral imaging is accomplished in SRS microscopy. Section 15.4 describes the principle, operation, and imaging results of our high-speed SRS spectral microscopy. Section 15.5 summarizes this chapter.

15.2   Principles and Advantages of SRS Microscopy

This section describes the principle of SRS microscopy, and then explains the advantages of SRS microscopy over existing Raman microscopy techniques.

15.2.1   Operation Principles

Figure 15.1 schematically shows the basic configuration of SRS microscopy. Two-color pulsed laser beams, which we call pump and Stokes beams, are prepared at optical frequencies of ω_p and ω_s ($\omega_p > \omega_s$), respectively. One of the laser beams is intensity-modulated beforehand. Then these pulses are combined and tightly focused by an objective lens. When the optical frequency difference between the two-color pulses (i.e., $\omega_p - \omega_s$) matches the frequency of Raman-active molecular vibrations (ω_R) at the focus, SRS occurs. Through SRS, pump pulses at ω_p are attenuated whereas Stokes pulses at ω_p are amplified. As a result, the intensity modulation of the pulses at one color is transferred to the pulses at the other color. In order to detect this modulation transfer caused by SRS, the pulses are collimated by another lens. After the intensity-modulated laser pulses are removed by an optical filter, the laser pulses with the transferred modulation are led to a photodiode. The modulation in the photocurrent is detected by a lock-in amplifier, which can measure the tiny electric signal at the modulation frequency as an SRS signal. For imaging, the focus position is scanned by a laser scanner (not shown in Fig. 15.1).

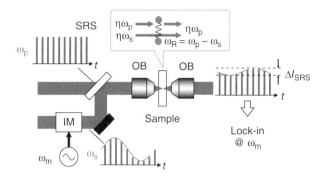

Figure 15.1 Basic configuration of SRS microscopy. IM: intensity modulator. OB: objective lens

Note that, in Fig. 15.1, Stokes pulses at ω_s (i.e., lower frequency and longer wavelength) are modulated, and the modulation transfer to pump pulses at ω_p is detected. In principle, one can modulate pump pulses and detect the modulation transfer to Stokes pulses. Care has to be taken to choose a proper configuration because the optical pulses for lock-in detection should be quiet (i.e., have quite a low noise property) so that small modulation transfer can be detected without being affected by laser intensity noise. Ideally, the noise of laser pulses should be as small as shot noise, which is the theoretical limit set by the Poissonian distribution of the number of photons. Another important issue is the wavelength dependence of sensitivity of photodetectors. For example, Si photodiodes can receive optical pulses only at shorter wavelengths than $1.0\,\mu m$.

We also note that the lock-in detection of SRS signal was originally employed in SRS spectroscopy [33–35]. Therefore, the novelty of SRS microscopy lies in the tight focusing of laser beams, which leads to high spatial resolution in 3D, and in the imaging, either by laser scanning or stage scanning. Nevertheless, application of SRS spectroscopy to microscopy leads to various advantages, as will be described shortly.

15.2.2 Comparison with Previous Raman Microscopy Techniques

Here we compare SRS microscopy with previous Raman microscopy techniques such as spontaneous Raman microscopy [36] and CARS microscopy, which are schematically shown in Figs. 15.2(a) and (b), respectively. We also refer to the energy diagrams shown in Figs. 15.3(a)–(c) (later), which correspond to spontaneous Raman scattering, CARS, and SRS, respectively.

As shown in Fig. 15.2(a), spontaneous Raman scattering microscopy uses a continuous-wave laser beam at an optical frequency of ω_p. The beam is focused on a sample, and spontaneous Raman scattering from the focus at an optical frequency of ω_s is collected and detected by a spectrometer. Imaging is carried out either by scanning the laser beam or the sample stage. As shown in Fig. 15.3(a), spontaneous Raman scattering involves the absorption of a photon at ω_p and spontaneous emission of a photon at ω_s, leading to transition to the vibrationally excited state with a vibrational frequency of $\omega_R = \omega_p - \omega_s$. Since molecules have several resonance frequencies of molecular vibrations, spontaneous Raman scattering can simultaneously occur at several frequencies, ω_s. The spectrometer can spectrally resolve them, providing us with

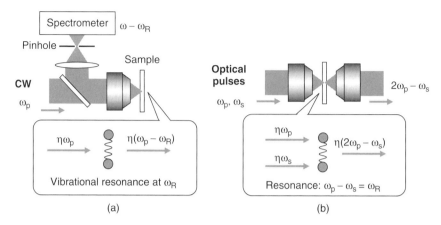

Figure 15.2 Schematic of previous Raman microscopy techniques. (a) Spontaneous Raman scattering microscopy. (b) CARS microscopy

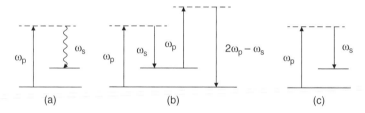

Figure 15.3 Energy diagrams of representative Raman processes. (a) Spontaneous Raman scattering. (b) Coherent anti-Stokes Raman scattering (CARS). (c) Stimulated Raman scattering

rich spectral information of Raman-active molecular vibrations at the focus in the sample. Such spectral information allows us to discriminate between various kinds of molecules in biological samples. A critical issue of spontaneous Raman scattering microscopy, however, is that the imaging speed is slow because the signal is quite weak. Typical pixel dwell time is in the order of 1 s, and total imaging time ranges from several tens of minutes to several hours, depending on the number of pixels.

The issue of slow imaging speed in spontaneous Raman scattering can be mitigated by CARS microscopy [4,5]. In CARS microscopy, two color pulsed laser beams at ω_p and ω_s are focused on a sample as shown in Fig. 15.2(b). Such two-color beams can excite molecular vibrations at $\omega_p - \omega_s$. The molecular vibrations are probed by the interaction between the vibrations and one of the laser beams at ω_p, leading to the generation of a CARS signal at $2\omega_p - \omega_s$. The CARS signal is strong when the sample molecules have a vibrational resonance at $\omega_R = \omega_p - \omega_s$. Imaging is accomplished by scanning the focus position with a laser scanner placed in front of the focusing lens. Since the CARS signal is stronger than the spontaneous Raman scattering by several orders of magnitude, CARS microscopy allows high-speed imaging. Indeed, several groups have reported video-rate operations of CARS microscopy [37–39]. Such high-speed imaging capability with molecular vibrational information with CARS is attractive for label-free imaging.

However, CARS microscopy has suffered from several limitations in terms of molecular specificity, difficulty in signal interpretation, distortion of CARS spectrum, and so on. Such limitations originate from the onset of so-called nonresonant background generated at the same frequency as CARS (i.e., $2\omega_p - \omega_s$). The nonresonant background comes from the third-order nonlinear electronic response [40,41].

Theoretically, the nonresonant background issue is well understood. The effect of CARS and nonresonant background can be described by third-order nonlinear susceptibility $\chi^{(3)}$ [31,32], which relates the Fourier components among the nonlinear polarization $P^{(3)}$ and the electric fields E of pump and Stokes beam by

$$P^{(3)}(2\omega_p - \omega_s) = \chi^{(3)}(\omega_p, \omega_p, -\omega_s)E(\omega_p)E(\omega_p)E^*(\omega_s). \tag{15.1}$$

Then the oscillation of polarization leads to the radiation of CARS signal. Therefore, the intensity of CARS radiation is proportional to $|P^{(3)}|^2$ and $|\chi^{(3)}|^2$. Here, $\chi^{(3)}$ is contributed to by Raman response χ_R and electronic response χ_{NR}. If we assume that Raman resonance exists at ω_R, χ_R has a complex Lorenzian line shape. If the molecules are electronically nonresonant, that is, molecules are transparent at ω_p and $2\omega_p$, χ_{NR} is real. Therefore, $\chi^{(3)}$ can be described by

$$\chi^{(3)} = \chi_{NR} + \chi_R$$

$$= \chi_{NR} + \frac{A}{\omega_R{}^2 - (\omega_p - \omega_s)^2 - 2i\Gamma(\omega_p - \omega_s)}, \tag{15.2}$$

where A is a proportional constant and Γ is the damping factor of molecular vibration. Equation 15.2 is plotted in Fig. 15.4. You can see that $|\chi^{(3)}|^2$ has an offset and a distorted line shape, whose peak is shifted to lower frequency, due to the interference between χ_{NR} and χ_R. This explains the spectral distortion and the background in CARS microscopy. On the other hand, Im $\chi^{(3)}$ has a single peak at ω_R and is free from the nonresonant background.

In SRS microscopy, the signal is known to be proportional to Im $\chi^{(3)}$. Therefore, compared with the previous Raman imaging techniques, SRS microscopy has various advantages as follows.

1. Compared with spontaneous Raman scattering, the sensitivity of SRS is higher by several orders of magnitude. This allows video-rate operation of SRS microscopy, similar to CARS.
2. Interpretation of SRS signal is easy because the SRS spectrum is the same as corresponding spontaneous Raman scattering.

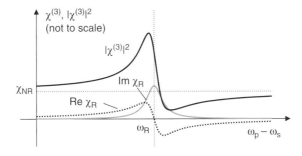

Figure 15.4 Theoretical curves of $\chi^{(3)}$ plotted as functions of difference frequency of two-color light

3. The signal is quantitative because the signal is proportional to the density of molecules of interest. In contrast, in previous Raman microscopy techniques, the quantitative measurement is troublesome due to unwanted fluorescence and stray light in spontaneous Raman scattering, and to the nonresonant background in CARS.

These advantages have been demonstrated by several applications of SRS microscopy such as imaging of lipids [8,3,16,29,30] and nucleic acids [19] in cells, lipids and proteins in tissues [20,28], drugs absorbed in skin [8,17], foods [18], and monitoring of delignification in biofuel production [15].

Note that developments in spontaneous Raman scattering microscopy and CARS microscopy are still in progress. Line-scanning geometry in spontaneous Raman scattering microscopy [42] reduces the acquisition time down to several minutes. In multiplex CARS microscopy, where broadband pulses are utilized for the simultaneous acquisition of the CARS spectrum, spectral distortion can be numerically compensated for, giving a quantitative image contrast [43–45]. Future optimization will further enhance the features of each technique.

15.2.3   Artifacts in SRS Microscopy

As described in Section 15.2.2, SRS signal is proportional to the density of molecules of interest. Therefore, an SRS image has a smaller amount of artifacts compared to CARS microscopy, where the signal is not proportional to the density of molecules. Nevertheless, it would be fair to point out that SRS microscopy is also susceptible to image artifacts. The origins of the artifacts include two-photon absorption (TPA) and cross-phase modulation (XPM).

TPA occurs when the sum frequency of two-color pulses (i.e., $\omega_p + \omega_s$) matches the absorption frequency of molecules, leading to the simultaneous absorption of both photons. Since TPA also causes modulation transfer in SRS microscopy, TPA results in the artifact. In typical SRS microscopy, where near-infrared pulses at 0.8 and 1.0 µm are used, the sum frequency corresponds to 0.45 µm (i.e., visible blue region). Such an absorption wavelength is common in some kinds of biomolecules such as porphyrin. Recently, a method for discriminating between SRS and TPA has been demonstrated [46] where three-color pulsed laser beams are used and the phase of one color of the pulsed beams is modulated. As a result, the Raman excitation can be modulated, allowing us to detect only SRS, even in the presence of TPA.

XPM is another important source of artifacts. XPM slightly changes the phase front of the focused beam depending on the spatial intensity distribution of the beam of the other color. When the numerical aperture (NA) of the collecting lens is insufficient for receiving the entire laser beam that comes from the focus, the change in the phase front leads to the change in the beam intensity. This results in the artifact as an offset in the lock-in signal. The XPM artifact can be suppressed by using a collecting lens with a NA higher than the focusing lens [8].

15.2.4   Physical Background

Here, we describe the physical background of SRS and CARS because it allows you to intuitively understand the reason why SRS has the various advantages described in Section 15.2.2. In the quantum-mechanical picture, CARS and SRS are different phenomena. As shown in Fig. 15.3(b), CARS is a parametric process, which does not leave photon energy in sample molecules, while SRS is the transition of molecular states from the ground state to the vibrationally excited state as shown in Fig. 15.3(c). On the other hand, in a classical picture described later, SRS and CARS can be viewed as the same phenomenon.

When two-color laser pulses at ω_p and ω_s are combined temporally, the temporal intensity waveform oscillates at $\omega_p - \omega_s$ due to the beating effect. When molecules are irradiated by such beating beams, the light can apply a force at the same frequency as the intensity beat, leading the forced oscillation of molecular vibrations. As described in basic physics textbooks, the forced oscillator oscillates at the same frequency as the force. Importantly, the oscillation amplitude and phase are dependent on the frequency of the force. When the frequency of the force is much lower than the resonance frequency of the oscillator, the oscillation is in phase with the force. When the frequency of the force is increased slowly and it matches the resonance frequency, the oscillation has maximum amplitude, and its phase is lagged by 90°. If the frequency is further increased, the oscillation amplitude is decreased and out-of-phase. The same phenomenon occurs in the Raman-active molecular vibrations and the force given by two-color laser beams. Then the molecular vibrations lead to the refractive index modulation of molecules. Finally, the refractive index modulation leads to the optical phase modulation of the excitation laser beams. As a result of the phase modulation at $\omega_1 - \omega_2$, frequency sidebands are generated at different frequencies, which are separated from the original laser beams by $\omega_1 - \omega_2$, that is, $\omega_1 \pm (\omega_1 - \omega_2)$ and $\omega_2 \pm (\omega_1 - \omega_2)$. The amplitudes and phases of these sidebands are dependent on how phase modulation is applied, that is, the amplitude and phase of molecular vibrations. Note that the electronic nonlinear response also leads to the phase modulation through optical Kerr effect, and the modulation is in-phase to the optical beat.

As the result of optical phase modulation, the following effects occur: (1) CARS appears at $2\omega_p - \omega_s$ as an upper sideband generated from ω_p. (2) The upper sideband at ω_p generated from ω_s destructively interferes with the original laser beam at ω_p. (3) The lower sideband at ω_s generated from ω_p constructively interferes with the original laser beam at ω_s. In this way, CARS measures the intensity of one of the sidebands, whereas SRS measures the interference between another sideband and the excitation beam. Since the interference is phase sensitive, SRS is unsusceptible to the nonresonant background. On the other hand, since intensity measurement is independent of the optical phase, CARS is susceptible to the nonresonant background.

The classical picture described previously allows us to compare the signal-to-noise ratios (SNRs) of SRS and CARS. Although there are various noise sources in SRS and CARS microscopy, the ultimate sensitivity is limited by the shot noise, which stems from the fact that the number of photons obeys the Poissonian distribution. When the number of photons is N, its standard deviation is $N^{1/2}$. On the other hand, in the quantum optics point of view, shot noise can be considered as the onset of vacuum fluctuation with a spectral density of $\hbar\omega/2$ [47]. Therefore, both in the intensity measurement and interference measurement, the shot-noise-limited SNR is approximately given by the ratio between the signal field and the vacuum field. Therefore, the SNR in CARS and SRS is limited by the ratio between the phase modulation sideband and the vacuum field. This is why CARS and SRS have similar signal-to-noise ratios at the theoretical limit. Since the optical pulses used in SRS and CARS are similar, the optical damages caused by the excitation beams are also similar.

In order to achieve shot-noise-limited SNR in SRS microscopy, however, care has to be taken as follows:

1. The one color of optical pulses used for the lock-in detection has to have a low-noise property. This is why mode-locked solid-state lasers and optical parametric oscillators have been used in SRS microscopy.

2. The frequency for intensity modulation and lock-in detection has to be high so that the low-frequency intensity fluctuation of lasers does not affect the SNR of lock-in signal.
3. In the photodiode circuit, electronic filters should be carefully implemented to extract the frequency components of lock-in signal and to reject unwanted strong signals at the repetition rate. Otherwise, electronic amplifiers can be easily saturated and lead to the degradation of SNR.

Another important issue in achieving shot-noise limited SNR arises when fiber lasers are used for realizing compact and practical systems. Since the optical power of fiber laser oscillators is as low as several milliwatts, it is necessary to use optical amplifiers. Through optical amplification process, amplified spontaneous emission (ASE) noise is added to the laser pulses. Since ASE is much stronger than shot noise, ASE limits the SNR in SRS microscopy. In order to reject the effect of ASE, one may use the balanced detection technique, where the laser beam is split into the signal and the reference beams, and the former is sent to an SRS microscope while the latter directly detected by a photodiode to measure the intensity noise due to ASE. As a result, we can subtract the effect of ASE from the SRS signal. In balanced detection it is crucial to maintain the balance in the subtraction. This is accomplished by a special electric amplifier with variable gain [48]. Alternatively, we proposed the co-linear balanced detection (CBD) technique [49]. In CBD, pump beam is split into signal and reference beams, and recombined with a delay difference. When the beams are detected, the photocurrents originated from the signal and reference beams add up destructively at a certain frequency because the time delay is equivalent to the frequency-dependent phase shift. Such a frequency region with suppressed photocurrent noise can be used for low-noise lock-in detection of SRS signal. The CBD technique can maintain the balance between the signal and reference because both beams co-linearly pass through the microscope.

15.3   Spectral Imaging with SRS

As described in the previous sections, SRS microscopy has attractive features for label-free biological imaging. Nevertheless, the molecular specificity in typical SRS microscopy is still limited because SRS signal reflects molecular vibrations at a single frequency determined by the frequencies of laser pulses, while the spectral characteristics of biological molecules can overlap. If spectral information (i.e., SRS images at various vibrational frequencies) can be acquired in SRS microscopy, it can be analyzed in detail to discriminate tiny spectral features of different molecules.

Figure 15.5 summarizes possible techniques for SRS spectral imaging. In Fig. 15.5(a), the wavelength of one of the two-color lasers is scanned while the images are successively acquired. This simple technique, however, may suffer from the slow imaging speed limited by the time required for tuning the laser wavelength. Typical tuning time in wavelength-tunable lasers is in the order of 1 s. In Fig. 15.5(b), narrow-band laser pulses are used along with broadband pulses, in which SRS spectra appear through the SRS processes at various vibrational frequencies. This technique is called multiplex inverse Raman spectroscopy or femtosecond stimulated Raman spectroscopy [50]. An important technical challenge in this technique is that spectrometer has to have a high dynamic range so that it can detect tiny intensity changes due to SRS, while the detector arrays in spectrometers are typically small, and easily saturated. Therefore, it seems difficult to achieve shot-noise-limited SNR in this

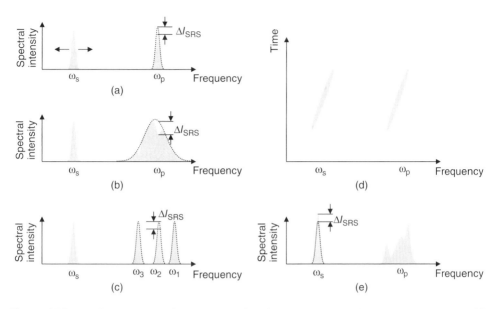

Figure 15.5 Possible methods of SRS spectral imaging. (a) Laser wavelength scanning. (b) Multiplex inverse Raman scattering spectroscopy. (c) Multicolor SRS. (d) Spectral focusing SRS. (e) Tailored spectrum SRS

technique at the moment. In Fig. 15.5(c), multicolor pulses with discrete spectrum are used [23]. Such multicolor pulses can be generated through the spectral filtering of broadband laser pulses with a 4-f pulse shaper. Spectral information can be acquired simultaneously by using multiple photodetectors, which measure SRS effects occurred in different pulses. An alternative technique demonstrated in [22] uses an acousto-optic tunable filter, which can modulate the intensities of different spectral portions of broadband pulses at different modulation frequencies. Then the modulation transfer to the narrowband pulses at various lock-in frequencies are detected by a photodetector followed by an array of lock-in amplifiers. Figure 15.5(d) shows the spectral focusing technique [24–26], where two-color broadband pulses with frequency chirp (i.e., the instantaneous frequency of pulses changes along the time) are used. If the frequency chirp of one color matches that of the other color, the different frequencies between them are kept over the entire pulse duration. Therefore, such pulses can excite molecular vibrations at a single frequency over a long time, leading to a spectral resolution higher than the spectral widths of the excitation pulses. Furthermore, the difference frequency can be controlled by changing the relative delay between the two-color pulses, allowing us to obtain spectral information about SRS. Nevertheless, it is technically challenging to perfectly adjust the frequency chirps of both colors, and to calibrate the difference frequency because it is affected by the group velocity dispersion of optics components (i.e., the frequency dependent time delay in the light propagation.) Another approach for improving molecular specificity is shown in Fig. 15.5(e), where spectrally shaped broadband pulses along with narrowband pulses are used [21]. This technique allows us to acquire spectral correlation between the sample molecules and the excitation spectrum.

In principle, these techniques have similar signal-to-noise ratios in the shot-noise limit under the same pulse repetition rate, the average power, and the pixel dwell time. In practice, however, broadband pulses without frequency chirp have high peak powers, which can cause optical damage on biological samples. Furthermore, when a spectrometer is used, it seems difficult to achieve the shot-noise-limited sensitivity as mentioned previously. By considering these issues, we recently developed a high-speed wavelength-tunable pulsed laser [27] to realize high-speed SRS spectral imaging [28], which will be described in the next section.

15.4   High-Speed Spectral Imaging

The authors developed a high-speed SRS spectral imaging system, which allowed us to acquire spectral information quickly, leading to improved molecular specificity. This system can visualize biological samples, without labeling, at a frame rate of >30 frames/s, while the wavenumber is controlled in a frame-by-frame manner. This section describes the principle of the wavelength-tunable pulse source, the configuration of the microscopy system, processing of spectral images with independent component analysis, and the results of visualization of biological samples.

15.4.1   High-Speed Wavelength-Tunable Laser

In order to realize high-speed spectral imaging with SRS microscopy, wavelength-tunable pulsed lasers have to satisfy several requirements as follows:

1. *Wide wavelength tunability*. For example, in order to access whole CH-stretching region ($2800-3100\,\text{cm}^{-1}$), the tunability has to be $>300\,\text{cm}^{-1}$.
2. *High-speed wavelength tuning capability*. In order to realize frame-by-frame wavenumber tunability, the tuning time should be on the order of ~ 1 ms.
3. *Narrow spectral width*. In order to have the high spectral resolution in SRS spectrum, the spectral width should be on the order of $3-5\,\text{cm}^{-1}$, which corresponds to the picosecond, transform-limited pulses.
4. *High power*. For video-rate SRS imaging, the average optical power at the laser output has to be $>100\,\text{mW}$.
5. *Constant delay*. The delay should be kept even when the wavelength is scanned so that the two-color pulses can overlap in time.

We developed a wavelength-tunable laser source that satisfies these requirements [27]. Figure 15.6 shows the schematic of the laser source. Broadband pulses from an Yb-fiber laser (YbF) are spectrally filtered by a tunable optical filter (TBPF), and then amplified by polarization-maintaining Yb fiber amplifiers. In the TBPF, optical pulses are reflected by the GM, and its mirror plane is imaged by 4-f relay lenses on a reflection grating in the Littrow configuration. The diffracted beam goes back through the relay lenses and GM. To pick up the output beam, the incident beam on the GM is vertically misaligned. As a result, the output beam propagates on a slightly different axis from the incident beam, while the output beam is reflected at the same location as the incident beam on the GM. Thus, the output beam can be picked up by a mirror and is launched to a fiber collimator, which acts as a single-mode spatial filter that extracts the Littrow wavelength component. When the angle of the GM is changed, the incident angle on the grating also changes. Therefore, we can tune the transmission wavelength of the TBPF with the GM. The response time of the TBPF is governed by the

Tunable bandpass filter

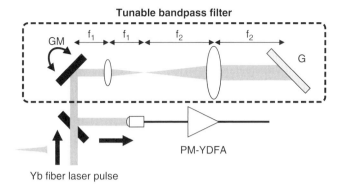

Figure 15.6 Schematic of the wavelength-tunable pulse source using a high-speed tunable bandpass filter [27]. f_1: 50 mm, f_2: 100 mm. GS: galvanomirror scanner. G: reflection grating with 1200 grooves/mm. PM-YDFA: polarization-maintaining Yb-doped fiber amplifier. *Source*: Y. Ozeki, W. Umemura, K. Sumimura, N. Nishizawa, K. Fukui, and K. Itoh 2011. Reproduced with permission from The Optical Society

speed of the GM and it is in the order of milliseconds. Hence, it would be possible to tune the wavelength in a frame-by-frame manner even at the standard video rate. Furthermore, the path length and the group delay of this filter are almost independent of the transmission wavelength because the grating plane is imaged on the GM, as mentioned previously (Fermat's principle). This property is important for SRS microscopy, where two-color pulses have to coincide in time even when the wavelength is varied.

Figure 15.7 shows the output spectra of the developed pulse source. The wavelength was tunable over ~32 nm, which corresponds to ~300 cm^{-1}. The half-width half maximum of each spectrum was approximately ~3 cm^{-1}, and the average power was ~120 mW.

Note that it is important to use polarization-maintaining fiber amplifiers. Otherwise, the polarization states of the amplified pulses are dependent on the wavelength due to the polarization mode dispersion in optical components in the amplifiers.

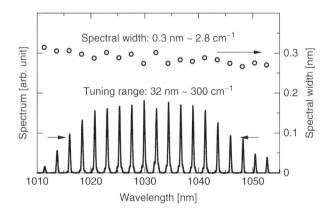

Figure 15.7 Spectra of wavelength-tunable optical pulses [28]. *Source*: Y. Ozeki, W. Umemura, Y. Otsuka, S. Satoh, H. Hashimoto, K. Sumimura, N. Nishizawa, K. Fukui, and K. Itoh 2012. Reproduced with permission from Nature Publishing

15.4.2   Experimental Setup

Figure 15.8 shows the schematic of the high-speed SRS spectral microscopy system. The YbF was synchronized to a Ti:sapphire laser (TiS) (Coherent, Mira 900D), which generated 76 MHz trains of 4 ps pulses at a wavelength of 790 nm. Note that the repetition rate of the YbF was set to half that of the TiS. This allowed us to increase the lock-in frequency to the maximum one (i.e., 38 MHz) as shown in Fig. 15.9(b), without relying on external optical modulation,

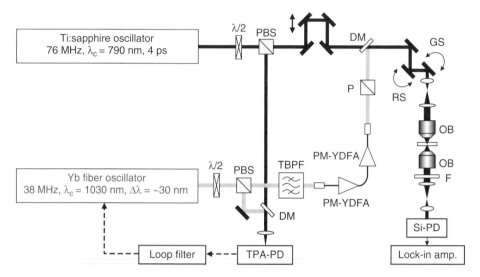

Figure 15.8   Schematic of high-speed SRS spectral microscopy. PBS: polarization beam splitter. DM: dichroic mirror. TPA-PD: two-photon absorption photodiode. RS: resonant galvanometer scanner. GS: Galvanometer scanner. OB: objective lens. F: short-pass optical filter

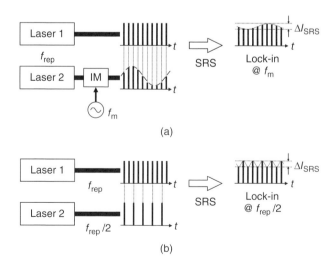

Figure 15.9   SRS microscopy using (a) an intensity modulator and (b) two-color, subharmonically synchronized pulses. IM: intensity modulator

which are typically used in SRS microscopy as shown in Fig. 15.9(a). The mechanism of synchronization was as follows [51,52]. The pulses from both lasers were tapped and introduced to a two-photon absorption photodiode (TPA-PD) as shown in Fig. 15.8. The resultant cross-correlation signal between TiS and YbF was fed back to YbF to control the repetition rate of the YbF by an intracavity electro-optic modulator and a piezo-driven delay stage. After the synchronization was achieved, the spectrally filtered YbF pulses were combined with TiS pulses by a dichroic mirror, led to a resonant galvanometer scanner, a standard galvanometer scanner and a beam expander, and focused on a sample by an objective lens (60×, NA 1.2, water). The transmitted pulses were collected by another objective lens (60×, NA 1.2). After rejecting YbF pulses with an optical short-pass filter, the TiS pulses were detected by a Si photodiode. Its photocurrent was sent to a homemade lock-in amplifier to obtain the SRS signal, which was led to a frame grabber. As a result, we were able to acquire SRS images with 500 × 480 pixels at a frame rate of 30.8 frames/s while the wavelength scanner was controlled in a frame-by-frame manner.

15.4.3   Observation of Polymer Beads

Figure 15.10 shows the results of spectral imaging of polymer beads. We observed ∼5-μm poly(methyl methacrylate) (PMMA) beads and 6-μm polystyrene (PS) beads in water. We acquired 91 spectral images at wavenumbers ranging from 2800–3100 cm^{-1} within 3 s. We can see that SRS images at 2913, 2946, and 3053 cm^{-1} shown in Figs. 15.10(a)–(c) have different contrasts, depending on the wavenumber. Figure 15.10(d) shows the spectra taken at different locations indicated by the arrows in Fig. 15.10(b) and 1.10(c). Obviously, these spectra have different shapes and indicate the high signal-to-noise ratio of our imaging system.

15.4.4   Spectral Analysis

In order to analyze the spectral data, we used principal component analysis (PCA) followed by a modified version of independent component analysis (ICA) [53]. PCA allows one to extract characteristic features with a smaller number of dimensions. Then ICA is used for the blind separation of independent sources. Ordinary ICA assumes that the data given are linear combination of independent signal sources. Blind separation is possible through the iterative calculations, which typically look for the spectral bases whose linear projections maximize the difference of fourth-order moment (kurtosis) of probability density from that of Gaussian distribution because independent sources have maximum non-Gaussianity. Although ICA has been applied to the analysis of Raman spectral imaging, we found that modification is needed because the ordinary ICA assumes that the average values of sources and those of signals are zero, and the results of ICA tend to give bipolar values. In contrast, Raman spectra and Raman images always have positive values. To cope with this, we modified the ICA algorithm so that it maximizes the third order moment (skewness) of probability density. As a result, the modified ICA enables blind separation of positive sources and tends to give images with positive values [28].

Figure 15.11 demonstrates the source separation through the SRS spectral imaging and the modified ICA [47]. Figure 15.11(a) shows the spectral bases obtained by ICA. Through a simple inverse matrix calculation, we can calculate independent component (IC) spectra shown in Fig. 15.11(b), which are in good agreement with those of PMMA and PS shown in Fig. 15.9(d).

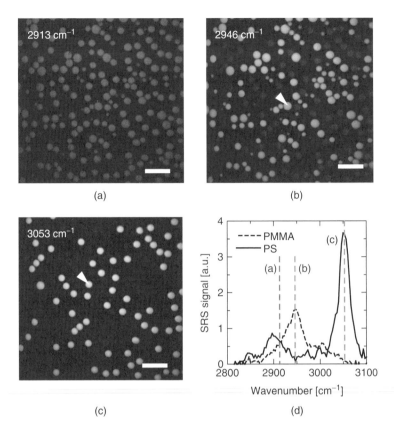

Figure 15.10 Spectral imaging of poly(methyl methacrylate) (PMMA) and polystyrene (PS) beads. (a–c): SRS images taken at 2913, 2946, and 3053 cm^{-1}, respectively. (d): SRS spectra taken at the locations of arrows in (b) and (c). Scale bar: 20 µm

From the IC spectra, we can speculate that the first and second independent component (IC) images in Figs 15.11(c) and 15.11(d) correspond to the distributions of PMMA and PS, respectively. In this way, ICA allows blind source separation for SRS spectral images, and IC spectra can be used for the assignment of IC images.

Note that in [29], the technique of multivariate curve resolution was successfully applied to blind source separation in SRS spectral imaging. Future development of feature extraction and blind source separation will be important in SRS spectral microscopy.

15.4.5 Tissue Imaging

Figure 15.12 (Plate 24) shows the imaging results of a rat liver observed with the developed system. The tissue was cryo-sectioned into a nominal thickness of 100 µm, and preserved between cover glasses with phosphate buffered saline (pH 7.4). In order to have a good signal-to-noise ratio, we repeated the acquisition of 91 spectral images at wavenumbers from 2800 to 3100 cm^{-1} 10 times. Nevertheless, the total acquisition time is less than 30 s.

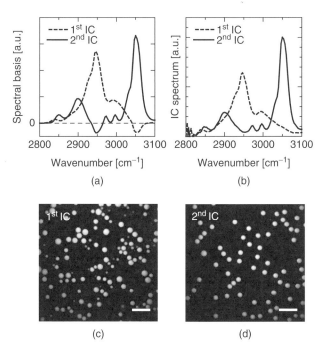

Figure 15.11 Analysis of spectral images of polymer beads by ICA. (a) Spectral basis of IC images. (b) IC spectra. (c) first IC image. (d) second IC image. Scale bar: 20 μm

Figures 15.12(a)–(c) show the first, second, and third IC images, respectively. These images correspond to the distributions of (a) lipids and cytoplasm, (b) water-rich regions, and (c) fibrous texture and nuclei, respectively. Their corresponding spectra shown in Fig. 15.12(d) indicate that the spectral difference is quite small. All the IC spectra have a CH_3 stretching mode at 2930 cm^{-1}, CH_2 stretching vibration in 2850 cm^{-1}, and the tails of OH stretching vibrations centered on ~3400 cm^{-1}. However, their ratios of vibrational modes are different: The first IC has a prominent CH_2 stretching mode, the second IC is dominated by the OH stretching mode, and the third IC has a strong CH_3 stretching mode. By combining these images and inverting the contrast [28], we could obtain the multicolor image shown in Fig. 15.12(e). Various structures in the liver tissue such as lipid droplets (A), cytoplasm (B), fibrous texture (C), nucleus (D), and the water-rich region (E), can be seen with different pseudo-colors, and their morphological shapes and locations are clearly visualized, which would be useful for pathological diagnosis. It is interesting to compare the IC spectra with the original SRS spectra shown in Fig. 15.12(f). Obviously, IC spectra have the much higher signal-to-noise ratios because ICA can extract spectral features from enormous numbers of pixels in a statistical manner.

It is also possible to conduct optical 3D sectioning in SRS spectral imaging. Figure 15.13 (Plate 25) shows the eight sections of multicolor images of intestinal villi in the mouse at different z positions separated by 5.6 μm. For each section, 91 spectral images at wavenumbers from 2800–3100 cm^{-1} were taken. The total acquisition time was 24 s. Figure 15.13(i) indicates that the first IC has a large amount of CH_2 stretching vibrations at 2850 cm^{-1}, and

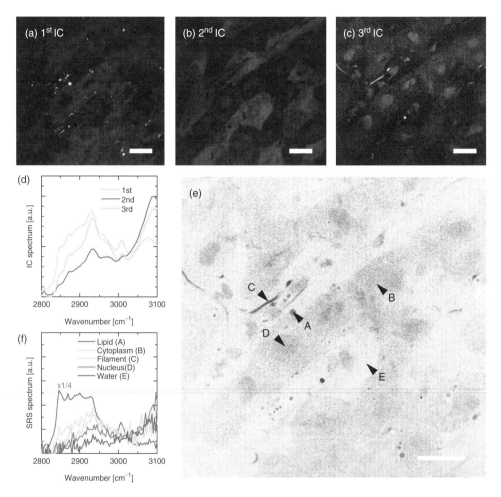

Figure 15.12 (Plate 24)   Spectral imaging of a rat liver tissue [28]. 91 images at wavenumbers from 2800–3100 cm^{-1} were taken and averaged over 10 times. The total acquisition time was <30 s. The spectral images were analyzed by using 5 ICs. (a) First IC image reflecting the distribution of lipid-rich region. (b) Second IC image reflecting the distribution of water-rich regions. (c) Third IC image reflecting the distribution of protein-rich region. (d) IC spectra. (e) Multicolor image produced by combining images (a–c) and inverting the contrast. (a)–(e) are explained in the text. (f) SRS spectra in locations indicated by arrows in (e). Scale bar: 20 μm. *Source*: Y. Ozeki, W. Umemura, Y. Otsuka, S. Satoh, H. Hashimoto, K. Sumimura, N. Nishizawa, K. Fukui, and K. Itoh 2012. Reproduced with permission from Nature Publishing. *See plate section for the color version*

the fourth IC is a mixture of CH$_3$ stretching mode at 2930 cm^{-1} and the tails of OH stretching modes. Multicolor images shown in Figs. 15.13(a)–(h) are produced by combining the first (cytoplasm, cyan) and fourth (nuclei, yellow) IC images and inverting the contrast. We can clearly see the morphology of the tissue such as cell nuclei and the cytoplasm. Note that, based on the conventional staining procedure, several thin slices are prepared, stained, and separately observed with a microscope. In contrast, our technique enables quick, label-free

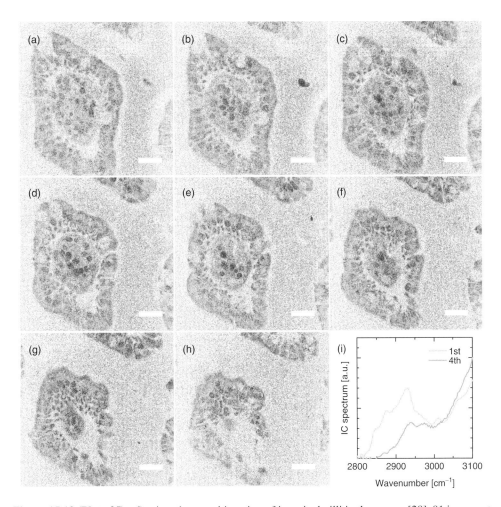

Figure 15.13 (Plate 25) Sectioned spectral imaging of intestinal villi in the mouse [28]. 91 images at wavenumbers from 2800–3100 cm⁻¹ were taken by changing the z position by 5.6 μm. The total acquisition time was 24 s. The spectral images were analyzed by using 4 ICs. The first IC (cytoplasm) and the fourth IC (nuclei) images were colored cyan and yellow, respectively, and then combined and the contrast was inverted. (a–h) Sectioned multicolor images. (f. Spectra of the first and fourth ICs. Scale bar: 20 μm. *Source*: Y. Ozeki, W. Umemura, Y. Otsuka, S. Satoh, H. Hashimoto, K. Sumimura, N. Nishizawa, K. Fukui, and K. Itoh 2012. Reproduced with permission from Nature Publishing. *See plate section for the color version*

observation of unstained samples with optical sectioning capability. This would be powerful for the medical diagnosis of tissues.

15.5 Summary

We have introduced SRS microscopy, which allows high-speed label-free biomedical imaging, even in 3D. We also discussed several ways of SRS spectral imaging for improving molecular specificity. Then we introduced our spectral imaging system using the high-speed wavelength-tunable pulsed laser. This system can acquire >30 spectral images in a second, allowing high-speed spectral imaging with high molecular specificity. Combined with the spectral analysis method based on ICA, we can observe unstained tissues quickly, which will be useful for medical applications. The future developments include wider wavelength tunability and the development of practical systems based on fiber lasers.

Acknowledgments

We are grateful to the joint research of SRS spectral microscopy with Professor K. Fukui (Osaka University), Professor N. Nishizawa (Nagoya University), Dr K. Sumimura, W. Umemura (Osaka University), Dr H. Hashimoto, Dr Y. Otsuka, and Dr S. Sato (Canon Inc.).

References

[1] Denk W., J. H. Strickler, and W. W. Webb: *Science* **248**, 73 (1990).

[2] Campagnola P. J., M.-D. Wei, A. Lewis, and L. M. Loew: *Biophys. J.* **77**, 3341 (1999).

[3] Barad Y., H. Eisenberg, M. Horowitz, and Y. Silberberg: *Appl. Phys. Lett.* **70**, 922 (1997).

[4] Zumbusch A., G. R. Holtom, and X. S. Xie: *Phys. Rev. Lett.* **82**, 4142 (1999).

[5] Hashimoto M., T. Araki, and S. Kawata: *Opt. Lett.* **25**, 1768 (2000).

[6] Isobe K., S. Kataoka, R. Murase, W. Watanabe, T. Higashi, S. Kawakami, *et al.*: *Opt. Express* **14**, 786 (2006).

[7] Fu D., T. Ye, T. E. Matthews, B. J. Chen, G. Yurtserver, and W. S. Warren: *Opt. Lett.* **32**, 2641 (2007).

[8] Freudiger C. W., W. Min, B. G. Saar, S. Lu, G. R. Holtom, C. He, *et al.*: *Science* **322**, 1857 (2008).

[9] Nandakumar P., A. Kovalev, and A. Volkmer: *N. J. Phys.* **11**, 033026 (2009).

[10] Ozeki Y., F. Dake, S. Kajiyama, K. Fukui, and K. Itoh: *Opt. Express* **17**, 3651 (2009).

[11] Min W., S. Lu, S. Chong, R. Roy, G. R. Holtom, and X. S. Xie: *Nature* **461**, 1105 (2009).

[12] Saar B. G., C. W. Freudiger, J. Reichman, C. M. Stanley, G. R. Holtom, and X. S. Xie: *Science* **330**, 1368 (2010).

[13] Slipchenko M. N., T. T. Le, H. Chen, and J.-X. Cheng: *J. Phys. Chem. B* **113**, 7681 (2009).

[14] Slipchenko M. N., H. Chen, D. R. Ely, Y. Jung, M. T. Carvajal and J.-X. Cheng: *Analyst* **135**, 2613 (2010).

[15] Saar B. G., Y. Zeng, C. W. Freudiger, Y. -S. Liu, M. E. Himmel, X. S. Xie, and S.-Y. Ding: *Angew. Chem. Int. Ed.* **122**, 5608 (2010).

[16] Wang M. C., W. Min, C. W. Freudiger, G. Ruvkun, and X. S. Xie: *Nature Meth.* **8**, 135 (2011).

[17] Saar B. G., L. R. Contreras-Rojas, X. S. Xie, and R. H. Guy: *Mol. Pharmaceutics* **8**, 969 (2011).

[18] Roeffaers M. B. J., X. Zhang, C. W. Freudiger, B. G. Saar, M. van Ruijven, G. van Dalen, *et al.*: *J. Biomed. Opt.* **16**, 021118 (2011).

[19] Zhang X., M. B. J. Roeffaers, S. Basu, J. R. Daniele, D. Fu, C. W. Freudiger, *et al.*: *ChemPhysChem.* **13**, 1054 (2012).

[20] Freudiger C. W. , R. Pfannl, D. A. Orringer, B. G. Saar, M. Ji, Q. Zeng, *et al.*: *Lab. Invest.* **92**, 1492 (2012).

[21] Freudiger C. W., W. Min, G. R. Holtom, B. Xu, M. Dantus, and X. S. Xie: *Nature Photon.* **5**, 103 (2011).

[22] Fu D., F.-K. Lu, X. Zhang, C. Freudiger, D. R. Pernik, G. Holtom, and X. S. Xie: *J. Am. Chem. Soc.* **134**, 3623 (2012).

[23] Lu F.-K., M. Ji, D. Fu, X. Ni, C. W. Freudiger, G. Holtom, and X. S. Xie: *Mol. Phys.* **110**, 1927 (2012).

[24] Andresen E. R., P. Berto, and H. Rigneault: *Opt. Lett.* **36**, 2387 (2011).

[25] Beier H. T., G. D. Noojin, and B. A. Rockwell: *Opt. Express* **19**, 18885 (2011).

[26] Fu D., G. Holtom, C. Freudiger, X. Zhang, and X. S. Xie: *J. Phys. Chem. B* DOI: 10.1021/jp308938t.

[27] Ozeki Y., W. Umemura, K. Sumimura, N. Nishizawa, K. Fukui, and K. Itoh: *Opt. Lett.* **37**, 431–433 (2011).

[28] Ozeki Y., W. Umemura, Y. Otsuka, S. Satoh, H. Hashimoto, K. Sumimura, *et al.*: *Nature Photon.* **6**, 845 (2012).

[29] Zhang D., P. Wang, M. N. Slipchenko, D. Ben-Amotz, A. M. Weiner, and J. –X. Cheng: *Anal. Chem.* **85**, 98 (2013).

[30] Kong L., M. Ji, G. R. Holtom, D. Fu, C. W. Freudiger, and X. S. Xie: *Opt. Lett.* **38**, 145 (2013).

[31] Min W., C. W. Freudiger, S. Lu, and X. S. Xie: *Annu. Rev. Phys. Chem.* **62**, 507 (2011).

[32] Cheng J.-X. and X. S. Xie: *Coherent Raman Scattering Microscopy*, CRC Press (2012).

[33] Levine B., C. V. Shank, and J. P. Heritage: *IEEE J. Quantum Electron.* **15**, 118 (1979).

[34] Eesley G. L.: *Coherent Raman Spectroscopy*, Pergamon Press, Oxford (1981).

[35] Levenson M. D. and S. Kano: *Introduction to Nonlinear Laser Spectroscopy*, Academic Press (1989).

[36] Puppels G. J., F. F. M. De Mul, C. Otto, J. Greve, M. Robert-Nicoud, D. J. Arndt-Jovin, and T. M. Jovin: *Nature* **347**, 301 (1990).

[37] Evans C. L., E. O. Potma, M. Puoris'haag, D. Côté, C. P. Lin, and X. S. Xie: *Proc. Natl. Acad. Sci. U.S.A.* **102**, 16807–16812 (2005).

[38] Heinrich C., A. Hofer, A. Ritsch, C. Ciardi, S. Bernet, and M. Ritsch-Marte: *Opt. Express* **16**, 2699–2708 (2008).

[39] Minamikawa T., M. Hashimoto, K. Fujita, S. Kawata, and T. Araki: *Opt. Express* **17**, 9526–9536 (2009).

[40] Volkmer A.: *J. Phys. D: Appl. Phys.* **38**, R59 (2005).

[41] Cheng J.-X., A. Volkmer, and X. S. Xie: *J. Opt. Soc. Am. B* **19**, 1363 (2002).

[42] Hamada K., K. Fujita, N. I. Smith, M. Kobayashi, Y. Inouye, and S. Kawata: *J. Biomed. Opt.* **13**, 044027 (2008).

[43] Kano H., and H. Hamaguchi: *Appl. Phys. Lett.* **85**, 4298 (2004).

[44] Kee T. W., M. T. Cicerone: *Opt. Lett.* **29**, 2701 (2004).

[45] Cicerone M. T., K. A. Aamer, Y. J. Lee, and E. Vartiainen, *J. Raman Spectrosc.* **43**, 637 (2012).

[46] Garbacik E. T., J. P. Korterik, C. Otto, S. Mukamel, J. L. Herek, and H. L. Offerhaus: *Phys. Rev. Lett.* **107**, 253902 (2011).

[47] Yariv A. and P. Yeh, *Photonics: Optical Electronics in Modern Communications*, Oxford University Press (2006).

[48] Yang W., C. W. Freudiger, G. R. Holtom, and X. S. Xie: *Photonics West*, paper 8588–8580 (2013).

[49] Nose K., Y. Ozeki, T. Kishi, K. Sumimura, N. Nishizawa, K. Fukui, *et al.*: *Opt. Express* **20**, 13958 (2012).

[50] Ploetz E., S. Laimgruber, S. Berner, W. Zinth and P. Gilch: *Appl. Phys. B* **87**, 389 (2007).

[51] Ozeki Y., Y. Kitagawa, K. Sumimura, N. Nishizawa, W. Umemura, S. Kajiyama, *et al.*: *Opt. Express* **18**, 13708 (2010).

[52] Umemura W., K. Fujita, Y. Ozeki, K. Goto, K. Sumimura, N. Nishizawa, *et al.*: *Jpn. J. App. Phys.* **51**, 022702 (2012).

[53] Hyvärinen A., J. Karhunen, and E. Oja, *Independent Component Analysis*, John Wiley & Sons, Inc., New York (2001).

16

Spectropolarimetric Imaging Techniques with Compressive Sensing

Fernando Soldevila[1], Esther Irles[1], Vicente Durán[1,2], Pere Clemente[1,3], Mercedes Fernández-Alonso[1,2], Enrique Tajahuerce[1,2] and Jesús Lancis[1,2]

[1]*GROC·UJI, Departament de Física, Universitat Jaume I, Spain*
[2]*Institut de Noves Tecnologies de la Imatge (INIT), Universitat Jaume I, Spain*
[3]*Servei Central d'Instrumentació Científica, Universitat Jaume I, Spain*

16.1   Chapter Overview

The information that the human eye can provide is limited. Although we are able to see in a wide range of distances, under different light conditions, and in a relatively broad spectral range, in many applications it is necessary to acquire information far beyond the limits imposed by the human eye. To this end, a great variety of image techniques have been developed [1]. As an archetypical example, microscopy, which is essential in fields like biology or medicine, provides a tool for obtaining high-resolution images of very close objects [2]. Many of these imaging techniques share a common feature: they measure the intensity of the light coming from the scene under consideration. However, it is sometimes required to measure other physical quantities, like the phase of the optical field, its spectral content, or its polarization state. The spectral content of a sample is normally used to obtain information about its material components. Polarization, that is, the knowledge of the vector nature of light, gives information about surface features such as shape, shading, and roughness of an object [3]. Advanced imaging techniques make it possible to acquire multi-dimensional images, which provide information not only about the spatial distribution of intensity but also about the previously mentioned primary physical quantities associated with an optical field.

Multi-dimensional Imaging, First Edition. Edited by Bahram Javidi, Enrique Tajahuerce and Pedro Andrés.
© 2014 John Wiley & Sons, Ltd. Published 2014 by John Wiley & Sons, Ltd.

In general, the measurement of multi-dimensional images involves the acquisition of a huge amount of information, which causes both storage and transmission difficulties [4]. In addition, techniques such as multispectral or hyperspectral imaging, require a sequential acquisition of images in the spectral domain, leading to a dramatic increase in measurement time. A recent approach to hyperspectral and polarimetric imaging is based on the use, respectively, of miniaturized spectral and polarimetric filters [5,6] that are incorporated to each pixel of the sensor, which allows acquiring multi-dimensional images in one shot. However, the development of such systems implies the use of high-end micro-optical components.

In this chapter, we describe several single-pixel multi-dimensional imaging systems based on *compressive sensing* (CS), a new sampling paradigm that has revolutionized data acquisition protocols, enabling us to start the signal compression at the measurement stage. In Section 16.2 we show how single-pixel imaging techniques work and how CS can boost their performance. In Sections 16.3 to 16.5 we describe single-pixel architectures that use off-the-shelf components in the fields of polarimetry, multispectral imaging, and spectropolarimetry.

16.2   Single-Pixel Imaging and Compressive Sensing

The operation principle of single-pixel imaging can be briefly described as follows. Let us consider a sample object, whose N-pixel image is arranged in an $N \times 1$ column vector, $\mathbf{x}$. That image can be expressed in terms of a basis of functions, $\mathbf{\Psi} = \{\mathbf{\Psi}_\ell\}$ ($\ell=1, \ldots, N$). In mathematical terms, $\mathbf{x} = \mathbf{\Psi} \cdot \mathbf{s}$, where $\mathbf{\Psi}$ is a $N \times N$ matrix that has the vectors $\{\mathbf{\Psi}_\ell\}$ as columns and $\mathbf{s}$ is the $N \times 1$ vector which contains the expansion coefficients of $\mathbf{x}$ in the chosen basis. Single-pixel cameras exploit the fact that those coefficients can be measured by using detectors with no spatial resolution. The acquisition process is governed by a spatial light modulator (SLM), which generates a set of patterns directly related to the selected basis. The irradiance corresponding to the inner product between the patterns and the object provides the coefficients of the image expansion.

In recent years, the introduction of CS has dramatically improved the performance of these single-pixel architectures. CS exploits the fact that natural images tend to be sparse, that is, only a small set of the expansion coefficients is nonzero when a suitable basis is chosen [7]. In this way, images can be retrieved without measuring all the projections of the object on the chosen basis. The mathematical formulation behind CS ensures that the object under study, $\mathbf{x}$, can be reconstructed from just a random subset of the expansion coefficients that make up $\mathbf{s}$. To this end, we randomly choose M different functions of the basis ($M < N$) and measure the projections of the object. This process can be expressed in matrix form as

$$\mathbf{y} = \mathbf{\Phi} \cdot \mathbf{x} = \mathbf{\Phi}(\mathbf{\Psi} \cdot \mathbf{s}) = \mathbf{\Theta} \cdot \mathbf{s}, \tag{16.1}$$

where $\mathbf{y}$ is a $M \times 1$ vector which contains the measured projections and $\mathbf{\Phi}$ is a $M \times N$ matrix called sensing matrix. Each row of $\mathbf{\Phi}$ is a function of $\mathbf{\Psi}$ chosen randomly, and the product of $\mathbf{\Phi}$ and $\mathbf{\Psi}$ gives the matrix $\mathbf{\Theta}$ acting on $\mathbf{s}$. If the chosen basis is orthonormal, every row of $\mathbf{\Theta}$ randomly selects a unique element of $\mathbf{s}$. As $M < N$, the underdetermined matrix relation obtained after the measurement process is resolved through an off-line algorithm. The best approach to recover the object is based on the minimization of the ℓ_1-norm of $\mathbf{s}$ subjected to the constrain given by Eq. (16.1), that is, that the solution given by the algorithm has to be compatible with the performed measurements. In this case, the proposed reconstruction $\mathbf{x}^*$ is given by the optimization program

$$\min_{\mathbf{x}^*} \| \mathbf{\Psi}^{-1} \mathbf{x}^* \|_{\ell_1} \text{ subject to } \mathbf{\Phi} \mathbf{x}^* = \mathbf{y}. \tag{16.2}$$

In the experiments described in this paper, the chosen basis is a family of binary intensity patterns derived from the Walsh–Hadamard basis. This basis has proved to be suitable for single-pixel architectures due to being easily implemented on a SLM. A Walsh–Hadamard matrix of order N ($\mathbf{H}_N$) is a $N \times N$ matrix with ± 1 entries that satisfies $\mathbf{H}_N^T \mathbf{H}_N = N \cdot \mathbf{I}_N$, where $\mathbf{I}_N$ is the identity matrix and $\mathbf{H}_N^T$ denotes transposed matrix. Walsh–Hadamard matrices form an orthonormal basis that was first proposed in image coding and transmission techniques [8]. By shifting and rescaling the different $\mathbf{H}_N$, it is possible to generate binary waveforms taking values 0 or 1 that can be simply encoded onto the SLM as an intensity modulation.

A suitable SLM for a single-pixel camera is a display formed by voltage-controlled liquid-crystal (LC) cells, such as those found in video projection systems [9]. Another option is a digital micromirror device (DMD), composed of an array of micromirrors that can rotate between two positions. In this way, only selected portions of the incoming light beam are reflected in a given direction [10]. Both devices are used in the different optical systems described in the following sections. Regarding detection, in general, a photodiode is used as single-pixel camera, which measures the irradiance of the light coming from an object for each pattern generated by the SLM. In the optical systems described in this chapter other single-pixel detectors, such as a beam polarimeter or a fiber spectrometer, are used.

16.3 Single-Pixel Polarimetric Imaging

Polarimetric imaging (PI) has the aim of measuring spatially resolved polarization properties of a light field, an object, or an optical system [11]. These properties are usually the Stokes parameters of light (passive imaging polarimeters) or the Mueller matrix that characterizes a sample or a system (active imaging polarimeters). The use of PI includes a great variety of optical applications, like scene analysis, target detection [3], polarization-sensitive microscopy [12], or segmentation of rough surfaces [13], among others. Polarimetric techniques have been used in the field of biomedical imaging for enhanced visualization of biological samples at different depths [14], as well as *in vivo* detection and diagnosis of cancerous tumors in tissues [15,16]. PI can be also combined with optical coherence tomography [17] and ophthalmic adaptive optics [18].

In this chapter, we describe how the concept of single-pixel imaging by CS has been extended to the design of a passive polarimetric camera. In particular, we describe a PI system able to measure spatially resolved Stokes parameters by means of a commercial beam polarimeter [19]. This commercial beam polarimeter is designed for free-space and fiber-based measurements, and provides the state of polarization (SOP) of an optical beam as a whole; that is, without spatial resolution. The PI system exhibits high dynamic range (up to 70 dB), broad wavelength range, and high accuracy on the Poincaré sphere, thanks to the use of the beam polarimeter. This fact simplifies the design and optimization procedures of current polarimetric cameras based on pixelated image sensors [16,20]. A programmable SLM is at the heart of this imaging polarimeter. This modulator controls the time-multiplexed acquisition process required by a single-pixel imaging scheme. The amount of acquired data is minimized by the application of a CS algorithm, which implies a proper selection of light patterns generated by the SLM, in accordance with the theory briefly described in Section 16.2.

A *Stokes polarimeter* (SP) is a device that measures the irradiance of a light beam whose SOP is modulated by a *polarization state analyzer* (PSA). In the commercial SP used here, which is sketched in Fig. 16.1, the PSA is formed by two voltage-controlled liquid-crystal variable retarders (LCVR$_1$ and LCVR$_2$) and a polarizing beam splitter (PBS). Two photodiodes (PD$_1$ and PD$_2$) are respectively located at the output ports of the PBS. The sum of the signals of PD$_1$ and PD$_2$ gives the total irradiance I_0 impinging onto the SP, despite of slight (and measurable) losses. The application of the Stokes–Mueller formalism allows obtaining the SOP of the input light, which is given by the Stokes vector $(I_0, S_1, S_2, S_3)^T$. If the retardances of LCVR$_1$ and LCVR$_2$ are δ_1 and δ_2, respectively, the irradiance I_{PD} measured by one photodiode is given by

$$I_{PD}(\delta_1, \delta_2) = m_{00}(\delta_1, \delta_2)I_0 + \sum_{i=1}^{3} m_{0i}(\delta_1, \delta_2)S_i. \tag{16.3}$$

In this expression $m_{0k}(k = 0, \ldots, 3)$ are the voltage-dependent elements of the first row of the PSA Mueller matrix. A proper calibration process, usually performed by the manufacturer, can be used to determine these elements. The description of such a process is out of the scope of the present study [21]. By a sequential reconfiguration of the PSA, the SOP of the incoming light is derived through the measurement of at least three values of I_{PD}, together with the irradiance I_0. In commercial devices, the LCVRs perform a wide retardance sweep. In this way, the input SOP is obtained through a least-squares fitting routine to minimize measurement errors [22]. The quantities registered by the SP are usually the normalized version of the Stokes parameters, $\sigma_i = S_i/I_0 (i = 1, \ldots, 3)$. It should be noted that the retardance of the LCVRs strongly depends on the light frequency, so the device calibration is valid for a given wavelength and must be repeated if the light source spectrum is changed.

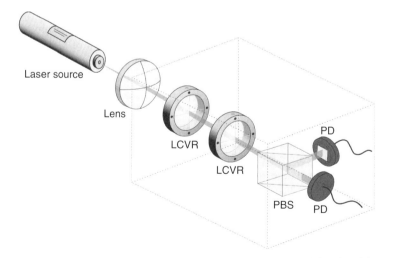

Figure 16.1   Scheme of the Stokes polarimeter acting as a single-pixel detector

A polarimetric detector with no spatial structure, such as the SP detector in Fig. 16.1, can be adapted to perform PI with the aid of the single-pixel architecture discussed on Section 16.2. The idea is simple: the problem of measuring a spatial-dependent Stokes vector is equivalent to resolving three times the CS algorithm of single-pixel imaging. This is possible because the linearity of Eq. (16.3) implies that each Stokes parameter S_i^{SP} provided by the SP is the sum of the values taken by S_i at each point of the input light beam. As a consequence, the measurement process expressed by Eq. (16.1) can be separately applied to each Stokes parameter whose spatial distribution (described by an N-pixel matrix) is recovered using $M < N$ polarimetric measurements. A layout of the PI system is depicted in Fig. 16.2. A collimated (unpolarized) laser beam passes through an LC-SLM, programmed to generate a set of intensity patterns. Just after the modulator there is a polarization object (PO), which produces a space-variant Stokes vector. As an LC-SLM is a polarization-dependent device; it is sandwiched between properly oriented linear polarizers (P_1 and P_2), so the object is illuminated with linearly polarized light. The light emerging from the object is guided to the SP by means of an afocal optical system, like an inverted beam expander, which fits the beam width to the typically small entrance window of the SP. This coupling optic ensures that all the light emerging from the object is collected by the SP and it preserves the normal incidence, which contributes to the optimal performance of the polarimeter.

The light source used in this experiment was an He-Ne laser emitting at 632.8 nm. The LC-SLM was a transmissive twisted nematic LCD (TNLCD) with SVGA resolution (800 $\times$ 600 pixels) and a pixel pitch of 32 µm. To reach an intensity modulation regime, the LC-SLM was sandwiched between two linear polarizers, respectively oriented parallel and normal to the input molecular director of the TNLCD, which was previously determined by a polarimetric technique [23]. In this configuration, the LC-SLM worked as a spatial intensity modulator. Pixels were individually addressed by sending gray-level images to the TNLCD. Each gray level corresponded to a transmitted intensity level, ranging from the dark state (extinction) to the bright state (maximum transmission).

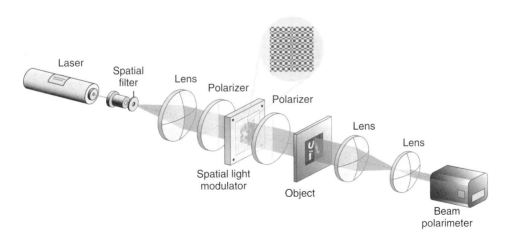

Figure 16.2 Experimental setup for the polarimetric single-pixel camera. One of the binary intensity patterns displayed by the SLM is also shown

Table 16.1   Technical specifications of the polarimetric camera

Wavelength	632.8 nm
Image resolution	64 × 64 pixels
Compression ratio	3:1
Pixel pitch	64 µm

In order to perform CS, the Walsh–Hadamard functions were chosen as the reconstruction basis Ψ. This election was particularly useful because the intensity patterns $\{\Phi_m\}$ generated by the TNLCD were binary masks (see Fig. 16.2). The corresponding images addressed to the display had a resolution of 64×64 cells, and the cell pitch was 64 µm. The number of binary patterns displayed onto the TNLCD was 1225, which represents $\sim 30\%$ of the Nyquist criterion. Custom software written with LabVIEW was used to synchronize the SP with the modulator. These technical specifications are summarized in Table 16.1. For each realization, the values of the Stokes parameters $\{S_i^{SP}\}(i = 1, \ldots, 3)$, as well as the signals of PD_1 and PD_2, are measured. The maximum measurement rate of the SP (10 Stokes vectors per second) was the speed limiting factor, since the refreshing frequency of the TNLCD was 60 Hz.

The selected object, shown in Fig. 16.3(a), was a cellophane film, acting as linear retarder, attached to an amplitude mask, which reproduces the logotype of the university UJI. Linearly

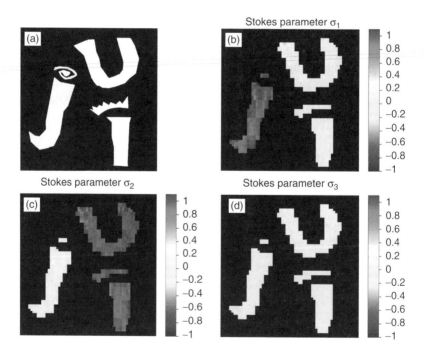

Figure 16.3   (a) High-resolution representation of the object under study, which is an amplitude mask representing the logo of the university UJI, with a cellophane film over the letter in gray shading (was yellow). Pseudocolored pictures showing the distribution of the Stokes parameters are shown in (b), (c), and (d). *Source*: V. Durán, P. Clemente, M. Fernández-Alonso, E. Tajahuerce, and J. Lancis 2012, Figure 3. Reproduced with permission from The Optical Society

polarized light emerging from the polarizer P_2 illuminated the object. An inhomogeneous polarization distribution was generated by covering just the capital letter J with cellophane. With the introduction of this element, the polarization of the light coming from this letter was approximately rotated by the cellophane film. The parameters of the polarization ellipse (the azimuth α and the ellipticity e) of the light passing through the J were measured by the SP (blocking the light emerging from the other part of the object). For this measurement, the TNLCD was configured in its bright state. The results were $\alpha_J = 8.62°$ and $e_J = -0.07$. Repeating the process for letters U and I, the measured parameters were $\alpha_{U,I} = 42.22°$ and $e_{U,I} = 0.003$.

Figures 16.3(b)–N.3(d) show pseudocolor plots for the normalized Stokes parameters. These images exhibit a clear uniformity within the different parts of the object. The spatial distributions for α and e were calculated from the Stokes parameters through conventional expressions (see, for example, [24]). The mean values of the ellipse parameters for each part of the object were ($\langle \alpha_J \rangle = 2.5° \pm 1.4°$, $\langle e_J \rangle = -0.08 \pm 0.02$) and ($\langle \alpha_{U,I} \rangle = 43.6° \pm 1.1°$, $\langle e_{U,I} \rangle = -0.01 \pm 0.04$). The assigned uncertainties were the standard deviations of each distribution. These results were in good agreement with the values previously measured by the SP. The major discrepancy was found for $\alpha_J (\sim 6°)$, which only represents $\sim 3\%$ of the total range of values (from $-90°$ to $90°$) that can be taken by the azimuth.

These results demonstrate the possibility of performing spatially resolved Stokes polarimetry with the aid of CS. In particular, the system described here converts a commercial beam SP into a polarimetric imager. Although this system is based on liquid crystal elements, the method is valid for other types of polarimeters, provided that the selected device is itself spatially homogeneous, and the relationship between the measured signals and the Stokes parameters is linear, as in Eq. (16.3). Concerning the acquisition process, a TNLCD is used to project the intensity patterns over the object. Another possibility is to employ SLMs insensitive to polarization, like a DMD, as is done in the optical systems described in the following sections. The combination of DMDs with fast SPs may lead to the design of PI systems working at very high frequencies (~ 1 KHz), opening the door to near-real-time applications.

16.4 Single-Pixel Multispectral Imaging

Multispectral imaging (MI) is a useful optical technique that provides two-dimensional images of an object for a set of specific wavelengths within a selected spectral range [1]. Dispersive elements (prisms or gratings), filter wheels, or tunable band-pass filters, are typical components used in MI systems to acquire image spectral content [25]. Multispectral imaging provides both spatial and spectral information of an object and represents a powerful analysis tool in different scientific fields as medicine [26], pharmaceutics [27], astronomy [28], and agriculture [29]. In industry, new techniques have emerged that use VIS and NIR imaging to make quality and safety control, for example, in the detection of surface properties on fruits [30].

The second optical system described in this chapter is a CS imaging system able to provide spatially resolved information about the spectrum of the light reflected by an object [31]. A fiber spectrometer with no spatial resolution is used as a single-pixel detector. Now, the key element of the system that makes possible the CS acquisition process, is a digital micromirror device (DMD). The modulator sequentially generates a set of binary intensity patterns that sample the image of the object under consideration. The acquired data is subsequently processed to obtain a multispectral data cube.

A layout of the spectral camera is shown in Fig. 16.4(a). A white-light source illuminates a sample and a CCD camera lens images the object on a DMD, which is a reflective spatial light modulator that selectively redirects parts of an input light beam [32]. The DMD consists of an array of electronically controlled micromirrors, positioned over a CMOS memory cell, which can rotate about a hinge, as is schematically depicted in Fig. 16.4(b). The angular position of each micromirror admits two possible states ($+12°$ and $-12°$ respect to a common direction), depending on the binary state (logic 0 or 1) of the corresponding CMOS memory cell contents. As a consequence, light can be reflected at two angles depending on the signal applied to the mirror. The DMD used in this system is a Texas Instrument device (DLP Discovery 4100) with a resolution of 1920×1080 micromirrors, the panel size of the display is $0.95''$, the mirror pitch is $10.8\,\mu m$ and the fill factor is greater than 0.91. The optical axis of the "optical system 1" forms an angle with respect to the orthogonal direction to the DMD panel that approximately corresponds to twice the tilt angle of the device mirrors ($24°$). As is shown in Fig. 16.4(c), a micromirror oriented at $+12°$ orthogonally reflects the light, appearing as a bright pixel (ON state). In their turn, micromirrors oriented at $-12°$ work as dark pixels (OFF state). The light emerging from the bright pixels of the DMD is collected by "the optical system 2" (see Fig. 16.4a). This lens system couples the light into a silica multimode fiber with a diameter of $1000\,\mu m$, which is connected to a commercial concave grating spectrometer (Black Comet CXR-SR from Stellarnet). The spectral range of the fiber ranges from $220-1100\,nm$. The wavelength resolution of this spectrometer is $8\,nm$ (with a slit of $200\,\mu m$) and the maximum *signal-to-noise ratio* (SNR) is 1000:1. Technical specifications for this setup are summarized in Table 16.2.

An example of spectral image with resolution 256×256 pixels was performed with a sample object composed of an unripe cherry tomato together with a red one. The illumination source was a xenon white light lamp. The Walsh–Hadamard patterns addressed to the DMD had $N = 65536$ unit cells. Each unit cell was composed of 2×2 DMD pixels. With this resolution, the number of measurements was chosen to be $M = 6561$, which corresponds to a compression rate of 10:1 ($M \approx 0.1N$). The integration time of each spectrometer measurement was $300\,ms$.

In order to determine the object spectral reflectance, a spectrum was taken from a white reference (Spectralon diffuse 99% reflectance target from Labsphere, Inc.) to normalize the measured spectra during the CS acquisition process. In the case of plants, the chlorophyll content in leaves can be recovered from the spectral reflectance [33,34]. The data collected by the spectrometer for wavelengths lower than $500\,nm$, clearly affected by noise, imposed an inferior boundary to the usable spectral range. The results of the CS reconstruction for 15 spectral channels are shown in Fig. 16.5 (Plate 26). The selected central wavelengths λ_0 in the visible spectrum (VIS) range from $510-680\,nm$. The bandwidth of each spectrum channel was $10\,nm$ ($\lambda_0 \pm 5\,nm$). The recovered images were pseudo-colored and the color assignment (the wavelength to RGB transform) was carried out with the aid of standard XYZ Color-Matching Functions [35]. The CS algorithm provided an acceptable reconstruction in the near-infrared spectrum, around $860\,nm$. This is presented by means of a gray-level image. Figure 16.5 also includes a RGB image of the object.

The quality of the images obtained with the single-pixel camera in Fig. 16.4(a) is evaluated by performing a multispectral imaging experiment, by sending Walsh–Hadamard patterns of

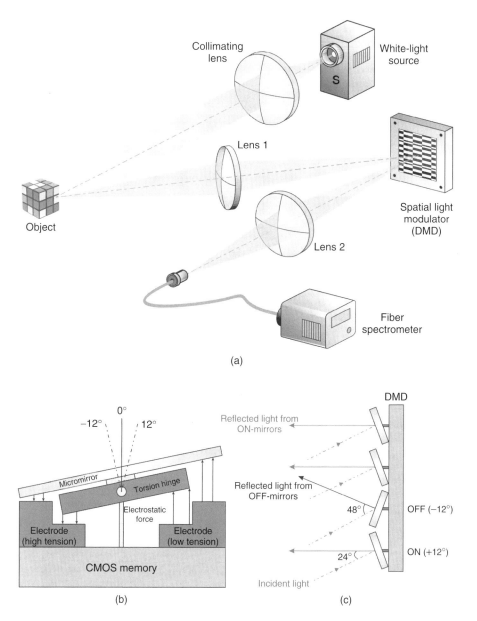

Figure 16.4 (a) Optical system for multispectral imaging using a single-pixel detector, (b) Individual DMD micromirror showed in the transverse view indicating its two possible orientations. (c) Working mode scheme of the DMD

Table 16.2   Technical specifications of the multispectral camera

Wavelength range	505–865 nm
Number of channels	15
Image resolution	256 × 256 pixels
Compression ratio	10:1
Integration time	300 ms
Pixel pitch	21.6 μm

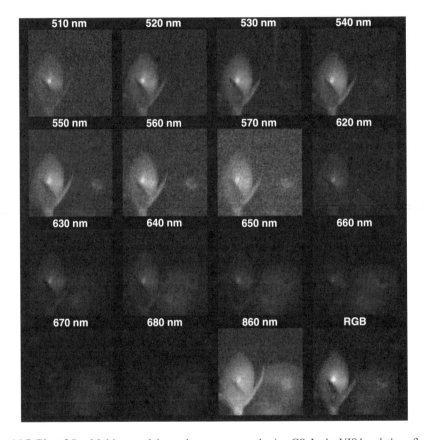

Figure 16.5 (Plate 26)   Multispectral data cube reconstructed using CS. In the VIS band, the reflectance for each spectral channel is represented by means of a 256 × 256 pseudo-color image. In the NIR band we show a gray-scale representation. A colorful image of the scene made up from the conventional RGB channels is also included. *Source*: F. Soldevila, E. Irles, V. Durán, P. Clemente, M. Fernández-Alonso, E. Tajahuerce, and J. Lancis 2013, Figure 4. Reproduced with permission from Springer. *See plate section for the color version*

64×64 unit cells ($N = 4096$) to the modulator, each one composed of 8×8 DMD pixels. In this case, the sample scene is constituted by two small square color objects. The number of measurements was $M = 4096$ ($M = N$), which allowed us to fulfill the Nyquist criterion. Eight central wavelengths, λ_0, were selected in the visible spectrum. The bandwidth of the corresponding spectral channels was 20 nm ($\lambda_0 \pm 10$ nm). Aside from the channels at the boundaries of the spectral range under consideration, the values of λ_0 correspond to peak emissions of commercial light-emitting diodes. The object spectral reflectance was determined again by means of the previously used white reference. The integration time of the spectrometer was set at 300 ms.

For each spectral channel, the off-line CS algorithm was resolved with the complete set of measurements. After a suitable filtering, the recovered matrix served as a reference (lossless) image $I_{ref}(i,j)$, where (i,j) indicates the location of an arbitrary image pixel. The reconstruction process was then repeated using decreasing fractions of the total number of pixels. In particular, the value of M was varied from 5 to 90% of N, and the fidelity of the reconstructed images was estimated by calculating the *mean square error* (MSE), given by

$$MSE = \frac{1}{N} \sum_i \sum_j \left[I(i,j) - I_{ref}(i,j) \right]^2, \tag{16.4}$$

where $I(i,j)$ is the noisy image obtained for a given value of M. The *peak signal-to-noise ratio* (PSNR) was used to evaluate the quality of the reconstruction. It is defined as the ratio between the maximum possible power of a signal and the power of the noise that affects the fidelity of its representation. In mathematical terms, [36]

$$PSNR = 10 \log \left(\frac{I_{max}^2}{MSE} \right) = 20 \log(I_{max}) - 10 \log(MSE) \tag{16.5}$$

Here, I_{max} is the maximum pixel value of the reference image. For each spectral channel, the reference images were represented by 2^8 gray-levels, so $I_{max} = 255$. Figures 16.6(a) and (b) show the curves for the MSE and the PSNR versus M for the different values of λ_0. Both figures point out that the image quality improves as the number of measurements grows and approximates to the Nyquist limit. However, it should be noted that the slope of both curves becomes visibly smoother for all spectral channels when $M \geq 0.4N$. In the case, for instance, of $\lambda_0 = 610$ nm, $MSE = 0.13I_{max}^2$ and $PSNR = 28.72$ dB for $M = 0.4N$ (the PSNR for $0.9N$ is only somewhat greater, 29.10 dB).

Although this single-pixel camera acquires sequentially the spatial information of the input object, it allows collection of all the spectral content at once, in contrast to those cameras based on tunable band-pass filters, which perform a wavelength sweep to measure the spectral information. In addition, the number of channels, their spectral resolution, and the total wavelength range of the single-pixel system are those provided by the spectrometer working as detector. This fact makes possible to exploit the high performance of commercially available devices. Thus, the spectral system can, in principle, cover the whole VIS spectrum and part of the NIR range (up to 1.1 microns), while conventional multispectral systems require pixelated sensors specifically designed for the infrared range (like InGaAs cameras).

Apart from the detector, the illumination is another key element to ensure a minimum signal along the selected spectral range. The use of a high power Xe arc lamp provides a continuous

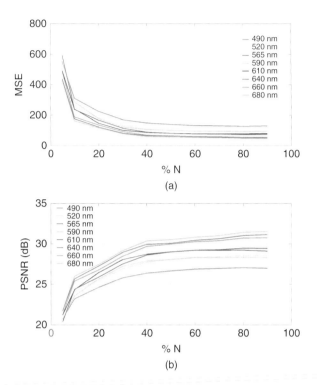

Figure 16.6 (a) MSE and (b) PSNR of the recovered images versus the number of measurements. Each curve corresponds to a spectral channel. *Source*: F. Soldevila, E. Irles, V. Durán, P. Clemente, M. Fernández-Alonso, E. Tajahuerce, and J. Lancis 2013, Figure 2. Reproduced with permission from Springer

and roughly uniform spectrum across the VIS region. However, the decreasing source irradiance at the "blue" side of the VIS spectrum, as well as the low reflectance of samples at that region, limits the spectral range to wavelengths higher than 500 nm.

This single-pixel multispectral camera presents a trade-off between image resolution and acquisition time. Increasing the illumination level or considering lower integration times (by a reduction of the spectral resolution) can make the acquisition time to drop by at least one order of magnitude. A comparable trade-off can be found in cameras based on acousto-optic or liquid crystal tunable filters, where the higher spectral resolution (number of channels), the longer acquisition time, with a strong dependence on the exposure time of the pixelated sensor used as a detector. A hyperspectral camera (i.e., a camera with more than 100 spectral channels) can take a few minutes in acquiring a data cube for image resolutions similar to those presented in this work [37].

16.5 Single-Pixel Spectropolarimetric Imaging

In certain applications, MI can be improved by adding spatially resolved information about the light polarization. Multispectral polarimetric imaging facilitates the analysis and identification of soils [38], plants [39], and surfaces contaminated with chemical agents [40]. In the field of biomedical optics, multispectral polarimetric imaging has been applied to the characterization of human colon cancer [41] or the pathological analysis of skin [42]. In many cases, polarimetric analysis can be performed by just including a linear polarizer in an imaging system to record images for various selected orientations of its transmission axis [42,43]. A simple configuration that includes two orthogonal polarizers integrated in a spectral system has been used for noninvasively imaging of the microcirculation through mucus membranes and on the surface of solid organs [43]. An illustrative example of a spectral camera with polarimetric capability is a system that combines an acousto-optic tunable filter with a liquid-crystal based polarization analyzer [44].

In this section we describe two different optical architectures for spectropolarimetric imaging. In the first one, polarimetry is performed by using a rotating linear analyzer in front of the detector, which leads to a linear polarization spectral imager. In the second one, the optical system is constituted by a fixed polarizer and two voltage-controlled variable retarders to spatially resolve the circular polarization component of light. In this way, this single-pixel multispectral system works as an imaging full-Stokes meter for each spectral channel.

16.5.1   Multispectral Linear Polarimetric Camera

A scheme of the multispectral linear polarimetric camera is depicted in Fig. 16.7. The optical system is similar to that described in the previous section, see Fig. 16.4(a), but now includes a linear polarizer. The sample scene is constituted by two square capacitors with a width of 7 mm. A xenon white light lamp is used again as illumination source. The light emerging from each

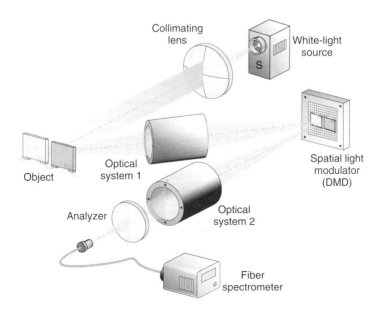

Figure 16.7   Optical system to obtain polarimetric multispectral imaging by using a single-pixel detector

Table 16.3 Technical specifications of the multispectral linear polarimetric camera

Wavelength range	470–700 nm
Number of channels	8
Image resolution	128×128 pixels
Compression ratio	5:1
Integration time	500 ms
Pixel pitch	43.2 μm

element of the scene had a different linear polarization. This spatial distribution of polarization was achieved by means of a linear polarizer located after the object, which had its area split in two parts, each of which with its transmission axis oriented at orthogonal directions (0 and 90°, respectively). The resolution of the patterns generated by the DMD was 128×128 unit cells ($N = 16384$) composed of 4×4 DMD pixels. The number of measurements was $M = 3249$, which corresponds to ~20% of N (i.e., a compression rate of 5:1). The integration time of the spectrometer was fixed to 500 ms. Eight central wavelengths λ_0 were selected in the VIS spectrum. The bandwidth of the channels was 20 nm ($\lambda_0 \pm 10$ nm). For each channel, four orientations of the polarization analyzer were sequentially considered in separated measurement series. The technical specifications of this camera are outlined in Table 16.3.

Figure 16.8 (Plate 27) shows the image reconstructions with the optical system in Fig. 16.7. Each column of the figure corresponds to a spectral channel and each row shows the results for a given orientation of the analyzer. A colorful image of the object is also shown (see Plate 27 in the plate section). This RGB image was made up from the data taken for the second configuration of the analyzer (45°). The result for 680 nm is presented by means of a gray-level image due to its proximity to the near infrared range.

The polarizer included in the single-pixel optical system in Fig. 16.7 limits the total spectral range, since the optical behavior of polarizing films is wavelength dependent. As a consequence, the upper boundary of the spectral range is ~700 nm. However, the use of high grade crystalline polarizers can solve this limitation.

16.5.2 Multispectral Full-Stokes Imaging Polarimeter

In principle, it is possible to obtain information about the spatial distribution of the Stokes parameters of light, S_i ($i = 0, \dots, 3$) from polarimetric images recovered for each spectral channel. In the previous optical system, as a linear polarizer is used as analyzer, the spatial distribution of S_0, S_1, and S_2 can be straightforwardly derived. However, a full Stokes polarimeter requires adding at least a linear retarder.

The scheme of a full-Stokes polarimeter is shown in Fig. 16.9. A white light beam generated by a xenon lamp is collimated by a lens and illuminates a sample object, whose image is formed on a digital micromirror device (DMD) by a pair of lenses. The light emerging from the DMD is guided to a single-pixel detector with the aid of a fourth lens. In order to achieve both polarimetric and spectral information, the single-pixel detector consists of two *liquid crystal variable retarders* (LCVR) (liquid crystal variable retarder from Meadowlark) with

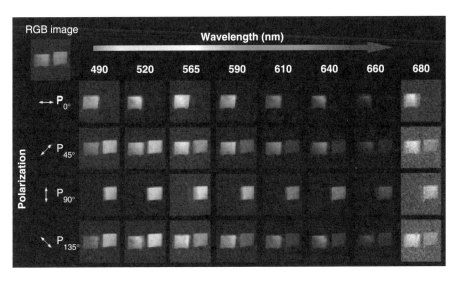

Figure 16.8 (Plate 27) Multispectral image cube reconstructed by CS algorithm for four different configurations of the polarization analyzer. The RGB image of the object is also included. In the VIS spectrum all channels are represented by pseudo-color images and a gray-scale representation is used for the wavelength closer to the NIR spectrum. *Source*: F. Soldevila, E. Irles, V. Durán, P. Clemente, M. Fernández-Alonso, E. Tajahuerce, and J. Lancis 2013, Figure 5. Reproduced with permission from Springer. *See plate section for the color version*

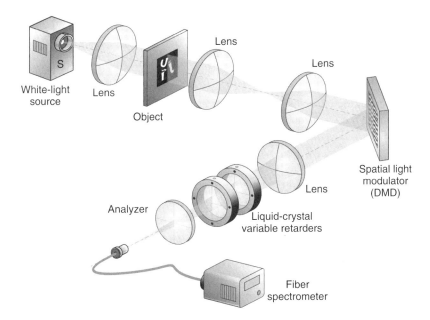

Figure 16.9 Scheme of the multispectral full-Stokes imaging polarimeter

their slow axis oriented at 45° and 0°, followed by a linear analyzer with its transmission axis oriented at 45° and a commercial fiber spectrometer (Black Comet CXRSR from StellarNet). Each LCVR is precalibrated to introduce controlled retardances in each chromatic channel of interest. The commercial fiber spectrometer is the same used in the preceding section.

By acquiring four images for different retardances of the LCVRs, it is possible to compute the Stokes parameters of each pixel of the scene. The CS algorithm provides an intensity map for the scene, which corresponds to the spatial distribution of the Stokes parameter S_0'. By using Stokes–Mueller calculus, it is possible to relate, for each pixel, the value of the Stokes vector with the measured irradiance S_0'. From the Mueller matrix expressions of a retarder wave plate and a linear polarizer, it is possible to show that the relationship between the recovered irradiance and the original Stokes parameters is

$$S_0'(2\delta_1, 2\delta_2) = \frac{1}{2}S_0 + \frac{1}{2}\sin(2\delta_1)\sin(2\delta_2)S_1$$

$$+ \frac{1}{2}\cos(2\delta_2)S_2 - \frac{1}{2}\cos(2\delta_1)\sin(2\delta_2)S_3, \tag{16.6}$$

where $2\delta_1$ and $2\delta_2$ are the phase retardances introduced, respectively, by the two LCVRs. Equation 16.6 establishes an undetermined system with four unknown quantities (the Stokes parameters of the incident light). In order to solve that system, a minimum of four pairs of phase retardances must be applied to the two LCVRs. After the off-line reconstructions, the Stokes vector in each point of the scene is given by $S_0' = M \cdot S$, where

$$M = \frac{1}{2}\begin{pmatrix} 1 & \sin(2\delta_1^{(1)})\sin(2\delta_2^{(1)}) & \cos(2\delta_2^{(1)}) & -\cos(2\delta_1^{(1)})\sin(2\delta_2^{(1)}) \\ 1 & \sin(2\delta_1^{(2)})\sin(2\delta_2^{(2)}) & \cos(2\delta_2^{(2)}) & -\cos(2\delta_1^{(2)})\sin(2\delta_2^{(2)}) \\ 1 & \sin(2\delta_1^{(3)})\sin(2\delta_2^{(3)}) & \cos(2\delta_2^{(3)}) & -\cos(2\delta_1^{(3)})\sin(2\delta_2^{(3)}) \\ 1 & \sin(2\delta_1^{(4)})\sin(2\delta_2^{(4)}) & \cos(2\delta_2^{(4)}) & -\cos(2\delta_1^{(4)})\sin(2\delta_2^{(4)}) \end{pmatrix}. \tag{16.7}$$

The subscripts in the elements of **M** relate to each one of the LCVRs and the superscripts denote each one of the four acquisitions. The solution of this linear system provides the spatial distribution of the Stokes parameters.

As a direct application of the single-pixel spectral Stokes polarimeter, a photoelasticity measurement on a piece of polystyrene was carried out. In the process of fabrication, the piece of polystyrene is shaped in a certain form. Due to this, the material presents stresses that cause a spatial distribution of birefringence. This distribution can be seen when the piece is placed between crossed linear polarizers and illuminated with white light, as can be seen in Fig. 16.10.

The Walsh–Hadamard patterns addressed to the DMD had a resolution of 128×128 unit cells ($N = 16384$). Each unit cell was composed of 4×4 DMD pixels. The number of measurements was chosen to be $M = 3249$, which corresponds to $\sim 20\%$ of N (a measurement rate of 5:1). The integration time of each spectrometer measurement was set to 20 ms. These technical specifications of the camera shown in Fig. 16.9 are summarized in Table 16.4. In Fig. 16.11 (Plate 28) we show the experimental results of the distribution of the normalized Stokes parameters for eight chromatic channels, each one with 20 nm width ($\lambda_0 \pm 10$ nm). To simplify data display, image reconstructions are arranged in a table. Each column corresponds

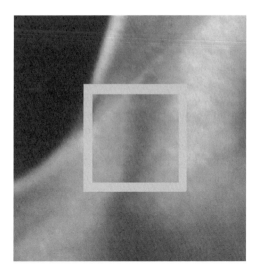

Figure 16.10   Color picture of the polystyrene sample. It is placed between two crossed linear polarizers and illuminated with white light. Color fringes are a consequence of the different states of polarization produced by the stress in the piece. The square indicates the region of interest imaged by the spectral camera

Table 16.4   Technical specifications of the multispectral full-Stokes imaging polarimeter

Wavelength range	450–730 nm
Number of channels	8
Image resolution	128×128 pixels
Compression ratio	5:1
Integration time	20 ms
Pixel pitch	43.2 μm

Figure 16.11 (Plate 28)   Spatial distribution of the Stokes parameters of the polystyrene piece. Each distribution is represented by a pseudo-colored 128×128 pixels picture. The values range from -1 (blue) to 1 (red). *See plate section for the color version*

to a spectral channel and each row shows the spatial distribution of a normalized Stokes parameter.

To simplify data display, image reconstructions are arranged in a table. Each column corresponds to a spectral channel and each row shows the spatial distribution of a normalized Stokes parameter.

As can be seen in Fig. 16.11, the expected fringe distribution of the Stokes parameters is recovered, if we compare this result with that in Fig. 16.10. For wavelengths near the IR, reconstructions are a bit noisy. This is caused by the low amount of light the source emits in this zone of the spectrum, which causes the reconstructions on these channels to have lower SNR. This problem can be solved by increasing the integration time of the spectrometer, but this makes measurement times much greater and the reconstructions on the channels in the visible region of the spectrum do not improve their SNRs. Using a light source with a flatter spectrum will solve the quality drop near the IR region.

16.6   Conclusion

We have described several multi-dimensional single-pixel imaging techniques providing the spatial distribution of multiple optical properties of an input scene. In all cases, the key element of the optical system is a SLM that sequentially generates a set of intensity light patterns to sample the input scene. In this way, it is possible to apply the theory of compressive sampling to data acquired with a single-pixel sensor. In particular, we have described a single-pixel hyperspectral imaging polarimeter. This system is able to provide spatially resolved measurements of Stokes parameters for different spectral channels. In this case the spatial light modulator is a digital micromirror device, and the sensor is composed by polarizing elements followed by a commercial fiber spectrometer. Experimental results for color objects with an inhomogeneous polarization distribution show the ability of the method to measure the spatial distribution of the Stokes parameters for multiple spectral components.

Acknowledgments

This work has been partly funded by the Spanish Ministry of Education (project FIS2010-15746) and the Excellence Net from the Generalitat Valenciana about Medical Imaging (project ISIC/2012/013). Also funding from Generalitat Valenciana through Prometeo Excellence Programme (project PROMETEO/2012/021) is acknowledged.

References

[1] Brady D., *Optical Imaging and Spectroscopy*, 1st edn. John Wiley & Sons, Ltd, 2009.
[2] Weissleder R. and M. J. Pittet, "Imaging in the era of molecular oncology," *Nature*, vol. **452**, no. 7187, pp. 580–589, Apr. 2008.
[3] Tyo J. S., D. L. Goldstein, D. B. Chenault, and J. A. Shaw, "Review of passive imaging polarimetry for remote sensing applications.," *Appl. Opt.*, vol. **45**, no. 22, pp. 5453–5469, Aug. 2006.
[4] Brady D. J., M. E. Gehm, R. A. Stack, D. L. Marks, D. S. Kittle, D. R. Golish, *et al.*, "Multiscale gigapixel photography," *Nature*, vol. **486**, no. 7403, pp. 386–389, Jun. 2012.
[5] Geelen B., N. Tack, and A. Lambrechts, "A snapshot multispectral imager with integrated tiled filters and optical duplication," pp. 861314–861313, 2013.

[6] Zhao X. and F. Boussaid, "Thin photo-patterned micropolarizer array for CMOS image sensors," *Photonics Technol.*, vol. **21**, no. 12, pp. 805–807, 2009.

[7] Candès E. and M. Wakin, "An introduction to compressive sampling," *Signal Process. Mag. IEEE*, no. March 2008, pp. 21–30, 2008.

[8] Pratt W., J. Kane, and H. Andrews, "Hadamard transform image coding," *Proc. IEEE*, vol. **57**, no. 1, 1969.

[9] Magalhães F., F. M. Araújo, M. V. Correia, M. Abolbashari, and F. Farahi, "Active illumination single-pixel camera based on compressive sensing," *Appl. Opt.*, vol. **50**, no. 4, pp. 405–414, Feb. 2011.

[10] Duarte M. and M. Davenport, "Single-pixel imaging via compressive sampling," *Signal Process.*, March 2008, pp. 83–91, 2008.

[11] Solomon J. E., "Polarization imaging," *Appl. Opt.*, vol. **20**, no. 9, pp. 1537–1544, May 1981.

[12] Oldenbourg R., "A new view on polarization microscopy," *Nature*, vol. **381**, no. 6585, pp. 811–82, Jun. 1996.

[13] Terrier P., V. Devlaminck, and J. M. Charbois, "Segmentation of rough surfaces using a polarization imaging system," *J. Opt. Soc. Am. A. Opt. Image Sci. Vis.*, vol. **25**, no. 2, pp. 423–430, Mar. 2008.

[14] Demos S. G. and R. R. Alfano, "Optical polarization imaging," *Appl. Opt.*, vol. **36**, no. 1, pp. 150–155, Jan. 1997.

[15] Baba J. S., J.-R. Chung, A. H. DeLaughter, B. D. Cameron, and G. L. Coté, "Development and calibration of an automated Mueller matrix polarization imaging system," *J. Biomed. Opt.*, vol. **7**, no. 3, pp. 341–349, Jul. 2002.

[16] Laude-Boulesteix B., A. De Martino, B. Drévillon, and L. Schwartz, "Mueller polarimetric imaging system with liquid crystals," *Appl. Opt.*, vol. **43**, no. 14, pp. 2824–2832, May 2004.

[17] de Boer J. F. and T. E. Milner, "Review of polarization sensitive optical coherence tomography and Stokes vector determination," *J. Biomed. Opt.*, vol. **7**, no. 3, pp. 359–371, Jul. 2002.

[18] Song H., Y. Zhao, X. Qi, Y. T. Chui, and S. A. Burns, "Stokes vector analysis of adaptive optics images of the retina," *Opt. Lett.*, vol. **33**, no. 2, pp. 137–139, Jan. 2008.

[19] Durán V., P. Clemente, M. Fernández-Alonso, E. Tajahuerce, and J. Lancis, "Single-pixel polarimetric imaging," *Opt. Lett.*, vol. **37**, no. 5, pp. 824–826, Mar. 2012.

[20] Sabatke D. S., M. R. Descour, E. L. Dereniak, W. C. Sweatt, S. A. Kemme, and G. S. Phipps, "Optimization of retardance for a complete Stokes polarimeter.," *Opt. Lett.*, vol. **25**, no. 11, pp. 802–4, Jun. 2000.

[21] Meadowlark Optics. *Liquid Crystal Polarimeter User Manual*. Available at www.meadowlark .com/store/PMI_Users_Manual_2.10.pdf, 2012.

[22] Davis S., R. Uberna, and R. Herke, "Retardance sweep polarimeter and method," *US Pat. 6,744,509*, vol. **2**, no. 12, 2004.

[23] Durán V., J. Lancis, E. Tajahuerce, and Z. Jaroszewicz, "Cell parameter determination of a twisted-nematic liquid crystal display by single-wavelength polarimetry," *J. Appl. Phys.*, vol. **97**, no. 4, p. 043101, 2005.

[24] Brosseau C., *Fundamentals of Polarized Light: A Statistical Optics Approach*, 1st edn. John Wiley & Sons, Inc., 1998.

[25] Boreman G. D., "Classification of imaging spectrometers for remote sensing applications," *Opt. Eng.*, vol. **44**, no. 1, p. 013602, Jan. 2005.

[26] Stamatas G. N., M. Southall, and N. Kollias, "In vivo monitoring of cutaneous edema using spectral imaging in the visible and near infrared," *J. Invest. Dermatol.*, vol. **126**, no. 8, pp. 1753–60, Aug. 2006.

[27] Hamilton S. J. and R. A. Lodder, "Hyperspectral imaging technology for pharmaceutical analysis," in *Proc. SPIE 4626, Biomedical Nanotechnology Architectures and Applications*, pp. 136–147, 2002.

[28] Scholl J. F., E. K. Hege, M. Hart, D. O'Connell, and E. L. Dereniak, "Flash hyperspectral imaging of non-stellar astronomical objects," in *Proc. SPIE 7075, Mathematics of*

Data/Image Pattern Recognition, Compression, and Encryption with Applications XI, vol. **7075**, p. 70750H–70750H–12, 2008.

[29] Dale L. M., A. Thewis, C. Boudry, I. Rotar, P. Dardenne, V. Baeten, and J. A. F. Pierna, "Hyperspectral imaging applications in agriculture and agro-food product quality and safety control: a review," *Appl. Spectrosc. Rev.*, vol. **48**, no. 2, pp. 142–159, Mar. 2013.

[30] Mehl P. M., Y.-R. Chen, M. S. Kim, and D. E. Chan, "Development of hyperspectral imaging technique for the detection of apple surface defects and contaminations," *J. Food Eng.*, vol. **61**, no. 1, pp. 67–81, Jan. 2004.

[31] Soldevila F., E. Irles, V. Durán, P. Clemente, M. Fernández-Alonso, E. Tajahuerce, and J. Lancis, "Single-pixel polarimetric imaging spectrometer by compressive sensing," *Appl. Phys. B*, vol. **113**, no. 4, pp. 551–558, 2013.

[32] Sampsell J. B., "Digital micromirror device and its application to projection displays," *J. Vac. Sci. Technol. B Microelectron. Nanom. Struct.*, vol. **12**, no. 6, p. 3242, Nov. 1994.

[33] Vila-Francés J., J. Calpe-Maravilla, J. Muñoz-Mari, L. Gómez-Chova, J. Amorós-López, E. Ribes-Gómez, and V. Durán-Bosch, "Configurable-bandwidth imaging spectrometer based on an acousto-optic tunable filter," *Rev. Sci. Instrum.*, vol. **77**, no. 7, p. 073108, 2006.

[34] Zou X., J. Shi, L. Hao, J. Zhao, H. Mao, Z. Chen, *et al.*, "*In vivo* noninvasive detection of chlorophyll distribution in cucumber (Cucumis sativus) leaves by indices based on hyperspectral imaging," *Anal. Chim. Acta*, vol. **706**, no. 1, pp. 105–112, Nov. 2011.

[35] Mather J., "Spectral and XYZ color functions," 2005. [Online]. Available at: www.mathworks.com/matlabcentral/fileexchange/7021-spectral-and-xyz-color-functions (accessed December 6, 2013).

[36] Pratt W. K., *Digital Image Processing*, 4th edn. John Wiley & Sons, Inc., 2007.

[37] Zuzak K. J., M. D. Schaeberle, E. N. Lewis, and I. W. Levin, "Visible reflectance hyperspectral imaging: characterization of a noninvasive, *in vivo* system for determining tissue perfusion," *Anal. Chem.*, vol. **74**, no. 9, pp. 2021–2028, May 2002.

[38] Coulson K. L., "Effects of reflection properties of natural surfaces in aerial reconnaissance," *Appl. Opt.*, vol. **5**, no. 6, pp. 905–917, Jun. 1966.

[39] Vanderbilt V. C., L. Grant, L. L. Biehl, and B. F. Robinson, "Specular, diffuse, and polarized light scattered by two wheat canopies," *Appl. Opt.*, vol. **24**, no. 15, pp. 2408–2418, Aug. 1985.

[40] Haugland S. M., E. Bahar, and A. H. Carrieri, "Identification of contaminant coatings over rough surfaces using polarized infrared scattering," *Appl. Opt.*, vol. **31**, no. 19, pp. 3847–3852, Jul. 1992.

[41] Pierangelo A., A. Benali, M.-R. Antonelli, T. Novikova, P. Validire, B. Gayet, and A. De Martino, "*Ex-vivo* characterization of human colon cancer by Mueller polarimetric imaging," *Opt. Express*, vol. **19**, no. 2, pp. 1582–1593, Jan. 2011.

[42] Zhao Y., L. Zhang, and Q. Pan, "Spectropolarimetric imaging for pathological analysis of skin," *Appl. Opt.*, vol. **48**, no. 10, pp. D236–246, Apr. 2009.

[43] Groner W., J. W. Winkelman, A. G. Harris, C. Ince, G. J. Bouma, K. Messmer, and R. G. Nadeau, "Orthogonal polarization spectral imaging: a new method for study of the microcirculation," *Nat. Med.*, vol. **10**, no. 10, pp. 1209–1212, 1999.

[44] Gupta N. and D. R. Suhre, "Acousto-optic tunable filter imaging spectrometer with full Stokes polarimetric capability," *Appl. Opt.*, vol. **46**, no. 14, pp. 2632–2637, May 2007.

17

Passive Polarimetric Imaging

Daniel A. LeMaster and Michael T. Eismann
Air Force Research Laboratory, USA

17.1   Introduction

The polarization properties of emitted, reflected, and scattered light are useful tools in remote sensing. These properties are inferred from a collection of radiometric measurements made by a device called a polarimeter. An imaging polarimeter extends this collection of measurements over a two-dimensional array of samples.

Imaging polarimetry has applications extending from medical diagnostics to astronomy. This chapter is devoted to the specific topic of passive broadband electro-optical and infrared Stokes imaging polarimetry for surveillance and reconnaissance in natural environments. Within this scope, there are numerous applications including clutter suppression and contrast enhancement [1–6], image segmentation [7], material characterization [8–10], shape extraction [11], and imaging through scattering media [12, 13]. Additional applications exists and, even among the applications specified, this list is reductive.

Though much of the material presented here is independent of any particular sensor, the scope of this chapter is limited in its discussion of architectures for polarization imagers. Sections 17.5 and 17.6 concentrate on modulated polarimeter technology. For instance, rotating analyzer and microgrid polarimeters are considered from both a data reduction matrix point of view and as linear systems using Fourier analysis. These architectures are most familiar to the authors though perhaps other architectures are better for certain applications.

There are several other very worthwhile resources from eminent authors that the reader is encouraged to explore. Reference [14] is probably the earliest book on the topic of imaging polarimetry and it contains an extensive survey of the work done in the 1980s and prior decades. More recently, [15] is highly recommend as a survey of passive imaging polarimetry up until 2006 and especially for its analysis of imaging polarimeter architectures. A very good treatment of reflective and emissive polarimetric phenomenology and radiative transfer is found in [16]. Excellent and often cited references on polarization science and polarimetry (though not polarization imaging) include [17] and [18]. Our goals in this chapter are to

Multi-dimensional Imaging, First Edition. Edited by Bahram Javidi, Enrique Tajahuerce and Pedro Andrés.
© 2014 John Wiley & Sons, Ltd. Published 2014 by John Wiley & Sons, Ltd.

concisely cover all of the important topics required for working in this field, expand on some underrepresented points, and to highlight what is new over the last several years.

17.2   Representations of Polarized Light

The purpose of this section is to discuss the nature of polarization and to solidify the physical concepts of polarized, partially polarized, and unpolarized light. It will quickly become clear that a representation based on electric fields will not be up to the task. The electric field representation will be traded for the Stokes parameters that represent polarized light in terms of measurable radiometric quantities. The interaction of polarized light with matter is then described in terms of Mueller matrices. This Stokes–Mueller formalism will be used throughout the chapter.

17.2.1   Optical Electro-Magnetic Fields

At any fixed point in free space, an optical electro-magnetic field may be described by a pair of orthogonal electric field components of the form,

$$E_x(t) = Re \left[A_x(t) e^{i\delta_x(t)} e^{2\pi i v t} \right] \tag{17.1}$$

where $A_x(t)$ is the amplitude of the field at time t, $\delta_x(t)$ is its phase, and v is the frequency of field oscillation (or the center frequency for polychromatic radiation). Strictly speaking, Eq. 17.1 is valid when the field's spectral bandwidth, Δv, is much smaller than v [19]. This "narrowband" restriction will be removed later on but, for now, it is useful for illustrating an important point. Once a coordinate system is established, the field components are labeled E_h for the horizontally oriented component and E_v for the vertically oriented component. E_h and E_v are perpendicular to the direction of propagation.

For monochromatic radiation, $A_x(t)$ and $\delta_x(t)$ are constants with time and, when $E_h(t)$ and $E_v(t)$ are plotted parametrically, they form the familiar polarization ellipse [20],

$$\left(\frac{E_h(t)}{A_h} \right)^2 + \left(\frac{E_v(t)}{A_v} \right)^2 - 2 \frac{E_h(t)E_v(t)}{A_h A_v} \cos(\delta_v - \delta_h) = \sin^2(\delta_v - \delta_h). \tag{17.2}$$

The time dependence of the amplitude and phase has been intentionally dropped to emphasize that these terms are constant. As shown in Fig. 17.1, this ellipse can also be described by an orientation angle, ψ, and an ellipticity angle, χ,

$$\psi = \frac{1}{2} \arctan \left[\frac{2A_h A_v}{A_h^2 - A_v^2} \cos(\delta_v - \delta_h) \right], \tag{17.3}$$

and

$$\chi = \frac{1}{2} \arcsin \left[\frac{2A_h A_v}{A_h^2 + A_v^2} \sin(\delta_v - \delta_h) \right]. \tag{17.4}$$

ψ is restricted to the range $(0, \pi)$ and the range of χ is $\left(-\frac{\pi}{4}, \frac{\pi}{4} \right)$. ψ defines the direction in which the EM field will exert maximum force and χ describes how the direction of the force evolves with time. The polarization state is said to be "right-handed" when $\chi > 0$ or

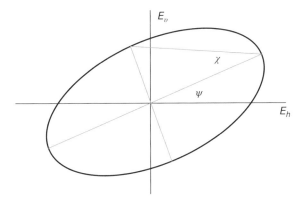

Figure 17.1 Polarization ellipse for monochromatic radiation

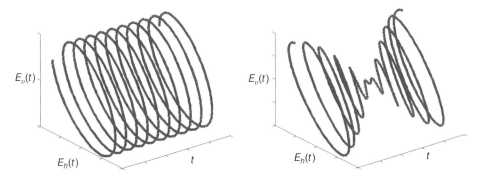

Figure 17.2 Time varying EM fields. Monochromatic radiation (*left*) and a mixture of frequencies with time varying phase (*right*)

"left-handed" when $\chi < 0$. This handedness refers to the direction through which the field orientation rotates as seen by the observer. In cases where $\chi = 0$, the ellipse collapses into linear polarization states.

The EM fields encountered in passive polarimetric imaging are polychromatic and generally have amplitude and phase terms that vary with time. Figure 17.2 shows a monochromatic EM field that may be fully described by the polarization ellipse on the left and a field that is a mixture of two frequencies with a linearly time vary phase on the right. A single polarization ellipse does not characterize this latter type of EM field as it evolves. While ψ and χ can be retained in some sense, this example demonstrates the need to adopt a more robust way to characterize polarization. The next section will provide one method for achieving this goal, the Stokes parameters.

17.2.2 Stokes Parameters and Mueller Matrices

Stokes parameters and Mueller matrices are used widely for polarization problems in remote sensing. The Stokes parameters are useful because they efficiently represent any kind of EM field described by Eq. 17.1 over realistic measurement intervals and may be inferred from simple linear combinations of directly measurable radiometric quantities. The interactions of Stokes parameters with the physical world are modeled through Mueller matrix transformations. Derivation of the Stokes parameters from electromagnetic principles can be found in [17]. For the present work, the Stokes–Mueller formalism is introduced only to the extent that it will be useful for solving problems later on.

The Stokes parameters fully describe polarization using the second-order statistics of the field. In terms of Eq. 17.1 the Stokes parameters are defined as,

$$S_0 = \langle \tilde{E}_h(t)\tilde{E}_h^*(t) \rangle + \langle \tilde{E}_v(t)\tilde{E}_v^*(t) \rangle \tag{17.5}$$

$$S_1 = \langle \tilde{E}_h(t)\tilde{E}_h^*(t) \rangle - \langle \tilde{E}_v(t)\tilde{E}_v^*(t) \rangle \tag{17.6}$$

$$S_2 = \langle \tilde{E}_h(t)\tilde{E}_v^*(t) \rangle + \langle \tilde{E}_v(t)\tilde{E}_h^*(t) \rangle \tag{17.7}$$

$$S_3 = i\left(\langle \tilde{E}_h(t)\tilde{E}_v^*(t) \rangle - \langle \tilde{E}_v(t)\tilde{E}_h^*(t) \rangle \right) \tag{17.8}$$

where $\langle . \rangle$ is the time average over some measurement interval (an interval that is long compared to the fundamental period of the EM field) and $\tilde{E}_x(t)$ is the complex analytic portion of the optical EM field, that is, $E_x(t) = Re[\tilde{E}_x(t)]$. Note that no assumption has been made regarding spectral bandwidth or about any component of $E_x(t)$ that varies over the measurement interval.

The Stokes parameters may be cast into other useful physical quantities. The first parameter, S_0, is proportional to the magnitude of the Poynting vector. S_0 is always positive and

$$S_0^2 \geq S_1^2 + S_2^2 + S_3^2 \tag{17.9}$$

The remaining parameters are interpreted as:

$$S_1 = PS_0 \cos 2\chi \cos 2\psi \tag{17.10}$$

$$S_2 = PS_0 \cos 2\chi \sin 2\psi \tag{17.11}$$

$$S_3 = PS_0 \sin 2\chi \tag{17.12}$$

where, in additional to time-averaged orientation and ellipticity, a new term, P is defined to be the *Degree of Polarization* (DOP) on the interval $(0, 1)$. In terms of the Stokes parameters,

$$P = \frac{\sqrt{S_1^2 + S_2^2 + S_3^2}}{S_0} \tag{17.13}$$

The term "unpolarized light" is often used to describe the $P = 0$ case but, strictly speaking, $P = 0$ is a case of no preferred polarization over the measurement interval. This case is also sometimes referred to as "randomly polarized". When $P = 1$, the radiation is said to be "fully polarized" and anything in between is said to be "partially polarized". A commonly used and closely related term is the *Degree of Linear Polarization* (DOLP),

$$P = \frac{\sqrt{S_1^2 + S_2^2}}{S_0}. \tag{17.14}$$

Parameters S_1 and S_2 describe the time average orientation direction while S_3 is affiliated with the time averaged ellipticity. When only S_0 and one other parameter are non-zero:

- S_1 by itself describes horizontal ($S_1 > 0$) or vertical ($S_1 < 0$) states
- S_2 by itself describes +45° ($S_2 > 0$) or −45° ($S_2 < 0$) states
- S_3 by itself describes right circular ($S_3 > 0$) or left circular ($S_3 < 0$) states.

The interaction of the Stokes parameters with their environment is modeled using Mueller matrices, M, and Stokes vectors, S,

$$\mathbf{S}^{(out)} = M\mathbf{S}^{(in)} = \begin{bmatrix} m_{00} & m_{01} & m_{02} & m_{03} \\ m_{10} & m_{11} & m_{12} & m_{13} \\ m_{20} & m_{21} & m_{22} & m_{23} \\ m_{30} & m_{31} & m_{32} & m_{33} \end{bmatrix} \begin{bmatrix} S_0 \\ S_1 \\ S_2 \\ S_3 \end{bmatrix}, \tag{17.15}$$

where superscripts *(in)* and *(out)* are used to indicate the original and transformed Stokes vectors. The set of all Stokes vectors does not meet the requirement to form a vector space [21] (e.g., since S_0 is always positive, this set contains no inverse elements). Nonetheless the term "Stokes vector" is in consistent widespread use.

For incoherent radiation, Stokes parameters from two (or more) sources combine via vector addition (i.e., element-wise addition). The Stokes–Mueller formalism may also be used to solve some problems dealing in coherent radiation but, to sum coherent radiation, the Jones calculus [17] must be used instead.

Each element in an optical system has a corresponding Mueller matrix. The cumulative effects of these individual Mueller matrices are found through matrix multiplication. For example, if Stokes vector $\mathbf{S}^{(in)}$ interacts with optical element 1 before element 2 then

$$\mathbf{S}^{(out)} = M_2 M_1 \mathbf{S}^{(in)}. \tag{17.16}$$

This matrix product does not commute.

The example Mueller matrices in Eqs. 17.17–17.19 will be useful for describing the polarimeters in Section 17.5. M_D is the Mueller matrix of a linear diattenuator,

$$M_D(p_h, p_v) = \frac{1}{2} \begin{bmatrix} p_h^2 + p_v^2 & p_h^2 - p_v^2 & 0 & 0 \\ p_h^2 - p_v^2 & p_h^2 + p_v^2 & 0 & 0 \\ 0 & 0 & 2p_h p_v & 0 \\ 0 & 0 & 0 & 2p_h p_v \end{bmatrix} \tag{17.17}$$

where p_h and p_v are electric field amplitude attenuation coefficients for polarization in the horizontal and vertical directions. A linear diattenuator is often referred to as a polarizer.

Next is the retarder,

$$M_W(\delta) = \begin{bmatrix} 1 & 0 & 0 & 0 \\ 0 & 1 & 0 & 0 \\ 0 & 0 & \cos\delta & \sin\delta \\ 0 & 0 & -\sin\delta & \cos\delta \end{bmatrix} \tag{17.18}$$

where δ is additional phase delay imposed on the h component of the optical electric field relative to the v component. An example of a device with a fixed retardance at a given wavelength is a wave plate. Other devices, such as liquid crystals, may be operated in a variable retardance mode.

Matrix M_R rotates a Stokes vector through angle θ about the optical axis.

$$M_R(\theta) = \begin{bmatrix} 1 & 0 & 0 & 0 \\ 0 & \cos 2\theta & \sin 2\theta & 0 \\ 0 & -\sin 2\theta & \cos 2\theta & 0 \\ 0 & 0 & 0 & 1 \end{bmatrix} \tag{17.19}$$

This Mueller matrix is particularly useful for interactions with elements that are more simply described in rotated coordinate system. For instance, the Mueller matrix of a linear diattenuator that is rotated by θ from the horizontal is given by

$$M_R(-\theta)M_D(p_h, p_v)M_R(\theta) \tag{17.20}$$

17.2.3   The Poincaré Sphere

The Poincaré sphere is a useful tool for visualizing how polarization states change as system parameters vary. As shown in Fig. 17.3, the Poincaré coordinate system is composed of the Stokes parameters S_1, S_2, and S_3. Stokes vectors plotted on the sphere are normalized so that $S_0 = 1$ and only fully polarized light ($P = 1$) is represented on this surface. The longitude of a point on the sphere is twice the orientation angle as measured from the positive S_1 axis. Similarly, latitude is given by twice the ellipticity angle as measured from the S_1, S_2 plane. An application of the Poincaré sphere is found later in Section 17.5.2.

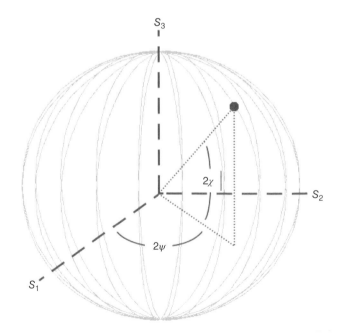

Figure 17.3   A fully polarized Stokes vector is a point on the Poincaré sphere

17.3 Polarized Reflection and Emission

Different materials and surfaces reflect and emit light in different ways. These differences are responsible for the polarization contrast exploited in imaging polarimetry. The purpose of this section is to introduce these concepts using the Stokes–Mueller calculus and to draw some specific conclusions on how to exploit these phenomena.

17.3.1 Reflection

At wavelengths shorter than $3\,\mu m$, an object's polarization signature is dominated by reflection and scattering at or near its surface. Surface reflections are often modeled by Fresnel's laws applied in a statistical sense to the various microscopic facets of the target surface. Light that penetrates the surface is either absorbed or scattered in the bulk of the material. Some of this bulk scattered light is transmitted back out in what is often a depolarizing process. The combined effects of surface and bulk reflections are modeled by the *Bidirectional Reflectance Distribution Function* (BRDF) Mueller matrix. The element in the xth row and y^{th} column in this Mueller matrix is defined to be:

$$f_{xy}(\theta_i, \phi_i, \theta_r, \phi_r, \lambda) = \frac{dL_x(\theta_r, \phi_r, \lambda)}{dE_y(\theta_i, \phi_i, \lambda)} \qquad (17.21)$$

where L_x is the radiance in Stokes parameter x reflected into direction (θ_r, ϕ_r) given an input irradiance, E_y, in Stokes parameter y incident from direction (θ_i, ϕ_i) at wavelength λ. By convention, angles θ_i and θ_r are the elevation angles of the incident and reflected light measured from the target surface normal direction. Angles ϕ_i and ϕ_r are the azimuth angles of the incident and reflected light. The BRDF elements have units of inverse steradians. The target reflected Stokes vector, $\mathbf{S}^{(r)}$, is calculated from the BRDF and the incident Stokes vectors, $\mathbf{S}^{(i)}$, arriving from around the hemisphere

$$\mathbf{S}^{(r)}(\theta_r, \phi_r, \lambda) = \int_0^{2\pi} \int_0^{\frac{\pi}{2}} f(\theta_i, \phi_i, \theta_r, \phi_r, \lambda) \mathbf{S}^{(i)}(\theta_i, \phi_i, \lambda) \cos\theta_i \sin\theta_i d\theta_i d\phi_i. \qquad (17.22)$$

In this equation, all Stokes vector components are in units of radiance.

Polarimetric BRDFs have been proposed for a variety of applications. Here, we will consider a simplified empirical BRDF model based on the microfacet model [22] with an added term for diffuse depolarization to demonstrate a few salient points,

$$f = d + M_{fr}(\beta, \hat{n}) \frac{o(\alpha, b, \sigma)}{4\cos(\theta_i)\cos(\theta_r)} \qquad (17.23)$$

where d is a depolarizing Mueller matrix that lumps together all bulk and diffuse surface reflections. In d, the only nonzero element is $m_{00} = d$. M_{fr} is the Mueller matrix for Fresnel reflection and o is the microfacet orientation distribution function. Explicit dependence on the incident and reflected angles has been suppressed for clarity. Each of these terms, their meanings, and their associated parameters are described next.

The Fresnel reflection Mueller matrix is

$$M_{fr}(\beta, \hat{n}) = \frac{1}{2} \begin{bmatrix} r_s r_s^* + r_p r_p^* & r_s r_s^* - r_p r_p^* & 0 & 0 \\ r_s r_s^* - r_p r_p^* & r_s r_s^* + r_p r_p^* & 0 & 0 \\ 0 & 0 & 2\mathrm{Re}(r_s r_p *) & 2\mathrm{Im}(r_s r_p *) \\ 0 & 0 & -2\mathrm{Im}(r_s r_p *) & 2\mathrm{Re}(r_s r_p *) \end{bmatrix} \qquad (17.24)$$

with Fresnel amplitude reflection coefficients [23]

$$r_s = \frac{\cos \beta - \sqrt{\hat{n}^2 - \sin^2 \beta}}{\cos \beta + \sqrt{\hat{n}^2 - \sin^2 \beta}} \qquad (17.25)$$

$$r_p = \frac{\hat{n}^2 \cos \beta - \sqrt{\hat{n}^2 - \sin^2 \beta}}{\hat{n}^2 \cos \beta + \sqrt{\hat{n}^2 - \sin^2 \beta}} \qquad (17.26)$$

where $\hat{n}$ as the relative refractive index, $\hat{n} = \frac{n_2}{n_1}$. Index n_2 is the reflecting media and n_1 is the index of the surrounding atmosphere. These refractive indices are, in general, complex. The subscript s refers to the amplitude reflection coefficient for the electric field component perpendicular to the plane of incidence (the plane containing the incident and reflected beams). In other words, the s direction is parallel to the surface. Subscript p is the reflection coefficient for the electric field component that is perpendicular to the s direction. It is important to note that the Fresnel Mueller matrix and, by extension, the BRDF is defined with respect to the orientation of the reflecting surface.

Several useful points need to be made about the physics of Fresnel reflections. First,

$$r_s r_s^* \geq r_p r_p^*, \qquad (17.27)$$

so randomly polarized light (e.g., from the Sun) tends to become s-polarized upon reflection. In the reference frame of the surface, this is also the positive S_1 direction. Second, Fresnel reflections from non-absorbing materials (real refractive index) tend to be less reflective but more strongly polarizing. Conversely, strongly absorbing materials (large imaginary refractive index component) are strongly reflective but less polarizing. Finally, the extent to which reflected light is polarized is heavily dependent on the incidence angle β. All of these points are demonstrated in Fig. 17.4 for silicon dioxide ($\hat{n} = 1.46$) and gold ($\hat{n} = 0.36 + 2.77i$) at 500 nm wavelength.

The Fresnel reflection Mueller matrix also shows that, for isotropic materials, a single reflection will never convert randomly polarized radiation into S_3. This "first surface" reflection dominates many polarization signatures found in nature. As a result, it is often not worth the cost or added complexity required to build an imaging polarimeter that is sensitive to elliptical polarization states.

The microfacet orientation distribution function, o, describes the fractional amount of light that is reflected into direction (θ_r, ϕ_r) from direction (θ_i, ϕ_i) due to facets oriented with an angular surface normal deviation of α from the average local surface normal. The relationships between all of the angles involved in the definition of the BRDF are

$$\cos \alpha = \frac{\cos \theta_i + \cos \theta_r}{2 \cos \beta} \qquad (17.28)$$

$$\cos(2\beta) = \cos \theta_i \cos \theta_r + \sin \theta_i \sin \theta_r \cos(\phi_i - \phi_r). \qquad (17.29)$$

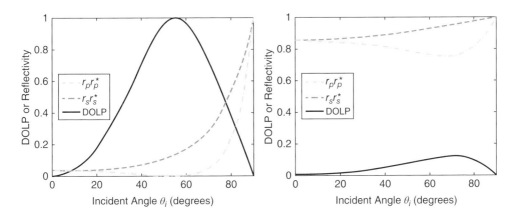

Figure 17.4 Fresnel reflection and DOLP: (*left*) silicon dioxide, (*right*) gold

The microfacet distribution function is assumed to be Gaussian in this model and parameterized by the standard deviation, σ, of the local surface slope, $\tan \alpha$, and a weight, b, which determines the relative strength of the Fresnel component over the diffuse component, *d*.

$$o(\alpha, b, \sigma) = \frac{b}{2\pi\sigma^2\cos^3\alpha} \exp \left(\frac{-\tan^2\alpha}{2\sigma^2} \right) \tag{17.30}$$

The combined effects of the BRDF model parameters are shown in Fig. 17.5 for surfaces illuminated with randomly polarized light in the forward scattering plane ($|\phi_r - \phi_i| = 180°$). On the left is a near specular surface reflection ($\sigma = 0.05$, $d = 10^{-5}$, and $b = 0.2$) off of silicon dioxide from 30° incidence. The largest reflection occurs in the target average specular direction (shown as $-30°$). The degree of linear polarization (DOLP) in this direction is about 0.4, which is consistent with a 30° incident angle in Fig. 17.4. DOLP continues to increase as microfacet reflection angle increases but the maximum achievable DOLP is tempered by the grazing angle dominance of the depolarizing diffuse reflection component.

A roughened silicon dioxide surface ($\sigma = 0.1$, $d = 0.03$, and $b = 0.2$) is shown on the right of Fig. 17.5. In this case, there is significant scattering in all directions but polarized reflections are relegated to near the target average specular direction and further muted by the influence of the large diffuse reflection component.

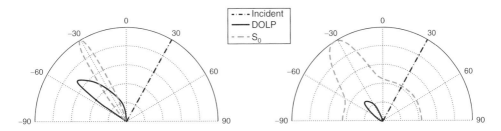

Figure 17.5 Two example cross sections of total radiance (normalized) and degree of linear polarization: (*left*) a near specular surface, (*right*) a roughened surface. In both cases, S_0 peak normalized

This example demonstrates why it is generally inadvisable to collect polarimetric imagery in an orientation with the Sun (or other major illumination source) behind the camera. All of the polarization response is confined to the forward scattering direction. This example also demonstrates why collection of polarization from an airborne platform is best done at a slant angle (where the polarization response will be highest) rather than from nadir (where the image resolution will be highest). This is an important trade off to consider in collection planning. Note that this example does not address reflection contributions from multiple or distributed sources (e.g., skylight), which is a key part of the utility of Eq. 17.22.

Note that the reflected light will be polarized parallel to the microfacet surface defined by the incident and reflected angles. The plane of incidence (which contains the incident, microfacet normal, and reflected directions) is rotated by an angle η with respect to the plane defined by the average target surface normal and the direction of reflection. For an observer using the average target surface normal coordinate system for both the incident and reflected Stokes vectors, the apparent BRDF is

$$f^{global}(\theta_i, \phi_i, \theta_r, \phi_r, \lambda) = M_R(\eta)f(\theta_i, \phi_i, \theta_r, \phi_r, \lambda)M_R(-\eta) \qquad (17.31)$$

where M_R is the rotation Mueller matrix from Eq. 17.19. In this coordinate system, a material that obeys a Fresnel microfacet model will shift some reflected light between S_1 and S_2 when illuminated by an unpolarized source and $|\eta| > 0$. There is no geometry where incident unpolarized light will be reflected into S_3. Because α increases as $|\eta|$ increases, the microfacet distribution function will weight the increasing $|\eta|$ contributions to total radiance less heavily.

Figure 17.6 shows a NIR band Stokes images of an object (a black backpack) camouflaged in S_0 by its proximity to the shadows of the trees. It has a low overall reflectivity, which explains why it blends with the shadows, but the sunlight that it reflects is quite clearly polarized. The natural surroundings exhibit only weakly polarized reflections in S_1 and so the target stands out. From the shadow angle, it is clear that the camera is not in the principal plane of the target ($|\eta| > 0$) and so there is a significant S_2 component to the target signature as well.

In this example, a contrast stretch is applied to each Stokes image individually. To provide a better impression of how much polarization content there is in this scene, a DOLP image is provided in Fig. 17.7. This DOLP image is scaled between 0 (randomly polarized) and the peak polarization value of the backpack, about 11%. As a reference point, the next most polarized feature in the scene is the sun-illuminated grass with a DOLP of about 1% or less.

Figure 17.6   NIR polarimetric imagery of a hidden object. S_0 (*left*), S_1 (*middle*), S_2 (*right*)

Figure 17.7   NIR DOLP image of the hidden object in Fig. 17.6

17.3.2   Emission

Objects at terrestrial temperatures spontaneously emit measurable radiation at wavelengths of 3 μm and longer. The amount of radiation emitted at a given wavelength depends on the object's temperature, T, and a object dependent property called emissivity. Like BRDF, emissivity is determined by material type, surface conditions, and viewing orientation. The Stokes vector for emission from a surface is

$$\mathbf{S}^{(e)}(\lambda, T, \theta, \phi) = L_{BB}(T, \lambda)\epsilon(\theta, \phi, \lambda) \tag{17.32}$$

where L_{BB} is the Planck blackbody radiation equation and ϵ is the emissivity vector.

For opaque materials, randomly polarized emission occurring within approximately one skin depth of the surface will become partially polarized upon transmission across the surface boundary [24]. Because of this, the emissivity vector ϵ is the first column of the Mueller matrix representing this boundary transmission.

$$\epsilon(\theta, \phi, \lambda) = \mathbf{M}_e(\theta, \phi, \lambda)\begin{bmatrix}1\\0\\0\\0\end{bmatrix} \tag{17.33}$$

where $\mathbf{M}_e$ is the Mueller matrix for emission.

Two additional facts allow the emissivity vector to be conveniently described in terms of the polarimetric BRDF [25]. All materials exchange perfect blackbody radiation with their surroundings when the entire system is in a state of thermal equilibrium. Furthermore, perfect blackbody radiation is randomly polarized. Therefore, in a chamber at thermal equilibrium, the following must be true at any temperature:

$$
\begin{bmatrix} 1 \\ 0 \\ 0 \\ 0 \end{bmatrix} = \epsilon(\theta, \phi, \lambda) + \rho(\theta, \phi, \lambda) \begin{bmatrix} 1 \\ 0 \\ 0 \\ 0 \end{bmatrix} \tag{17.34}
$$

where $\rho(\theta, \phi, \lambda)$ is the *Hemispherical Directional Reflectivity* (HDR) Mueller matrix with entries

$$
\rho_{xy}(\theta, \phi, \lambda) = \int_0^{2\pi} \int_0^{\frac{\pi}{2}} f_{xy}(\theta_i, \phi_i, \theta, \phi, \lambda) \cos \theta_i \sin \theta_i d\theta_i d\phi_i. \tag{17.35}
$$

This expression of HDR describes the fraction of total light reflected in direction (θ, ϕ) when the target is illuminated uniformly from all directions. The meaning of Eq. 17.34 is clear–the Stokes vector at any point inside the cavity is randomly polarized (left-hand side); therefore, the sum of all emitted and reflected light must also be randomly polarized for every surface (right-hand side). This is Kirchoff's law, generalized for polarization.

Regardless of the exact polarization BRDF model used,

$$
\rho_{30}(\theta, \phi, \lambda) = 0 \tag{17.36}
$$

because $f_{30} = 0$ whenever Fresnel reflection is the primary mechanism for polarization (very few naturally occurring exceptions come to mind). Additionally,

$$
\rho_{20}(\theta, \phi, \lambda) = 0 \tag{17.37}
$$

because of the rotational symmetry of the microfacet distribution. This outcome is consistently observed in nature as well. A complete definition of the emissivity vector now emerges:

$$
\epsilon(\theta, \phi, \lambda) = \begin{bmatrix} 1 - \rho_{00}(\theta, \phi, \lambda) \\ -\rho_{10}(\theta, \phi, \lambda) \\ 0 \\ 0 \end{bmatrix}. \tag{17.38}
$$

Thus emitted radiation is polarized in the direction of the average target surface normal (i.e., p-polarized). This is the $-S_1$ direction in the coordinate system defined previously for target surfaces. Equation 17.38 is useful in polarimetric remote sensing because it is also a very good approximation of natural emission behavior outside of the theoretical chamber at thermal equilibrium.

Figure 17.8 is an example of emission polarization using the BRDF model in Eq. 17.23 to calculate HDR with parameters $\hat{n} = 1.8 + 1.0i$, $d = 10^{-5}$, $\sigma = 0.05$, and $b = 1$ (i.e., a relatively smooth absorbing material). The dashed curve shows the total emitted radiation normalized by $L_{bb}(T, \lambda)$. Total emission has a broad peak in the surface normal direction ($0°$) and then falls off gradually with increasing angle. Polarized emission, represented by degree of linear polarization, is zero in the surface normal direction and then increases as the viewing angle increases. This pattern is rotationally symmetric with azimuth and is typical of many processed surfaces and materials. Just like in the reflection case, polarized emission is best viewed at a slanted angle.

All objects simultaneously emit and reflect radiation. To gain an appreciation for the magnitude of these combined effects on polarization signature, consider Fig. 17.9. Here, the material

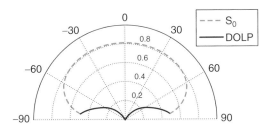

Figure 17.8 An example cross section of total radiance (normalized) and degree of linear polarization for polarized emission

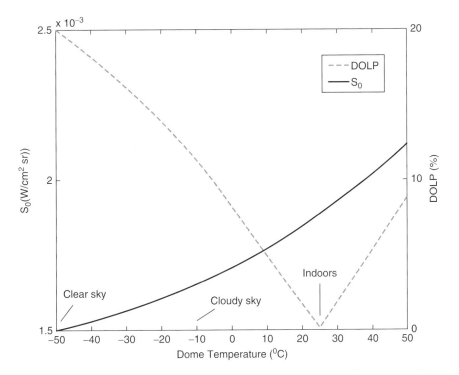

Figure 17.9 Total radiance and DOLP as a function of apparent background temperature. The target temperature is 25°C and observation angle is 60° away from the target surface normal

used in Fig. 17.8 is observed in the LWIR (8 to 10 µm) under a uniform blackbody hemisphere at a look angle of 60°. As the hemisphere warms up, the total apparent target radiance (reflected and emitted) increases and the apparent degree of polarization decreases. The degree of polarization trend reverses once the hemisphere becomes warmer than the target. Past the equilibrium temperature, reflection dominates and the apparent target polarization switches from p-polarized to s-polarized. Atmospheric effects will be dealt with in the following section but some useful apparent temperatures are labeled on the plot for reference. Among other

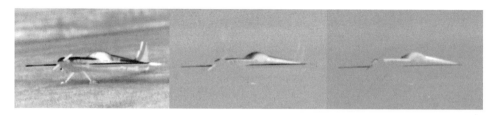

Figure 17.10   LWIR polarimetric imagery of a remote control aircraft. S_0 (*left*), S_1 (*middle*), S_2 (*right*)

things, this plot demonstrates why indoor emissive band polarization imaging is typically muted. Additional discussion on this topic is found in [26].

Any example of a LWIR polarimetric image is shown in Fig. 17.10. The S_1 and S_2 images are scaled so that the darkest regions have the most negative values and the lightest regions are the most positive values. S_0 values are always positive. On the top half of the aircraft, the horizontal surfaces show a preference for the negative S_1 direction and the vertical surfaces show a preference for the positive S_1 direction. Emission on the canopy smoothly transitions from positive to negative S_1 following the curvature. These three observations are consistent with emission from these surfaces being p-polarized (i.e., perpendicular to the surface). On the undercarriage, none of the surfaces show strong polarization because the emitted radiation is mixed with s-polarized reflections of the warm ground. The net result is effectively zero polarization. Note that the ground beyond the aircraft is largely randomly polarized. This is true of many natural backgrounds.

There is polarization content in both S_1 and S_2 because the aircraft surfaces are rotated somewhat with respect to the horizontal and vertical axes of the camera reference frame. It is important to remember that the s and p directions are defined with respect to the surface, not with respect to the camera.

17.4   Atmospheric Contributions to Polarimetric Signatures

The reflection and emission signatures of materials will be altered by the atmosphere. In many cases, these changes are substantial and highly variable. This variability is both temporal, in the sense that atmospheric conditions constantly change, and spatial because each part of the sky affects the target signature in a different way depending on the target BRDF. The radiative transfer equations that describe the polarization effects of the atmosphere are provided in exquisite detail in [16]. In practice, implementing these equations requires extensive measurements and modeling. Many of the required parameters cannot be measured directly without access to the target site.

The following discussion will be divided into polarimetric effects in the reflective bands ($\lambda < 3$ μm) and the emissive bands ($\lambda > 3$ μm). This is a distinction that arises because the remotely sensed radiation at these shorter wavelengths is dominated by solar illumination and scattering, while thermal emission from objects becomes a major component at wavelengths above 3 μm. These spectral regions are further sub-divided by the atmospheric windows; that is, the spectral bands for which the atmosphere has sufficiently small absorption that light reflected or emitted from an object can adequately transmit through it. Figure 17.11 (Plate 29) illustrates the ground

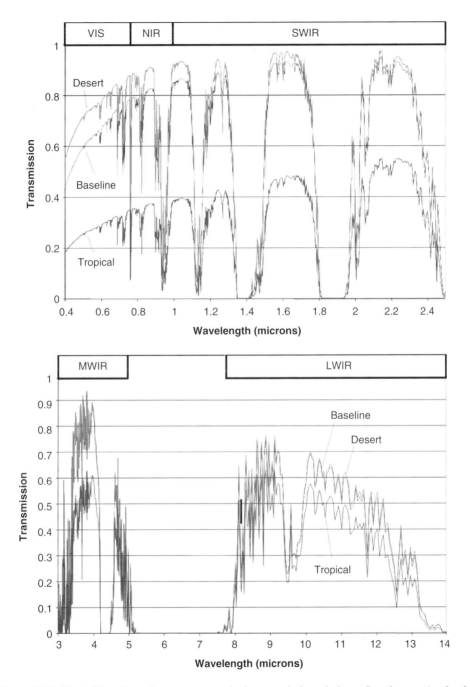

Figure 17.11 (Plate 29) Ground to space atmospheric transmission windows. *See plate section for the color version*

to space transmission over a vertical path in both the reflective (0.4–2.5 μm) and emissive (3–14 μm) spectral regions for some nominal meteorological conditions [27]. The various visible, near infrared (NIR), short-wave infrared (SWIR), mid-wave infrared (MWIR), and long-wave (LWIR) spectral windows are depicted in the transmission plots.

17.4.1   Reflective Bands

In the visible, NIR, and SWIR bands, the atmosphere contributes to the apparent polarization signature in the following ways:

1. solar illumination is absorbed or scattered before reaching the target,
2. target reflected photons are scattered out of the path or absorbed,
3. photons reflected from the target surroundings are scattered into the path,
4. photons from the Sun are scattered into the path.

Scattering has a substantial impact on polarization sensing through an atmospheric path. Scattering models are typically based on Mie calculations [28] from empirical measurements of the compositional characteristics and size distributions of scattering media at various altitudes, and are often characterized in terms of their scattering coefficient and phase function, both of which can be highly wavelength dependent. Figure 17.12 provides the spectral distribution of some example scattering coefficients at two altitudes for mid-latitude summer, 23 km visibility, rural scattering conditions based on the MODTRAN scattering database. In the reflective spectral region, the strong wavelength dependence of the Rayleigh scattering component is evident. At longer wavelengths, the scattering coefficients are lower and exhibit a resonance near 9 μm as the wavelength matches the peak in the size distribution of the larger but less concentrated aerosol particulates. The phase function is modeled by a Henyey–Greenstein function based on an asymmetry parameter that varies somewhat modestly with altitude and conditions.

Rayleigh scattering creates a strong band of polarized sky radiance. The Mueller matrix for Rayleigh scattering [20] is given by

$$
M_{ray} = K \begin{bmatrix} 1 + \cos^2\theta & -\sin^2\theta & 0 & 0 \\ -\sin^2\theta & 1 + \cos^2\theta & 0 & 0 \\ 0 & 0 & 2\cos\theta & 0 \\ 0 & 0 & 0 & 2\cos\theta \end{bmatrix} \tag{17.39}
$$

where, in this case, angle θ is defined with respect to the direction of propagation prior to the scattering event. The scattering parameter, K, defines the strength of the scattering interaction and scales with wavelength by λ^{-4}. Scattering Mueller matrices are defined in reference to a plane which contains both the incident and scattered light directions. By convention, polarization in this plane is taken to be the positive S_1 direction [23]. When unpolarized light (such as light from the Sun) is Rayleigh scattered, peak polarization occurs at $\theta = 90^0$ and is oriented perpendicularly to the scattering plane (i.e., $-S_1$ direction). Under clear sky conditions, a band of sky polarization builds up around this peak as shown in Fig. 17.13. (The dark band across this image is the sun occulter. The white regions immediately around the sun are due to image

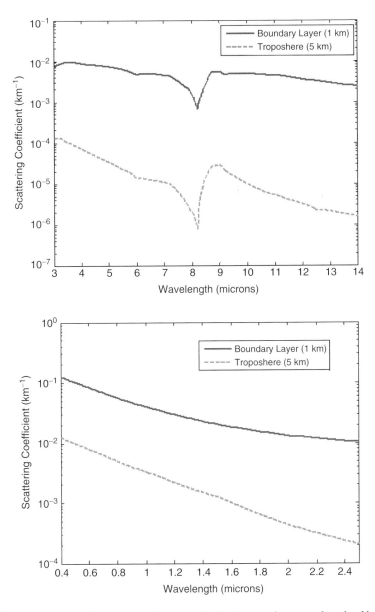

Figure 17.12 Scattering coefficients as a function of altitude at various wavelengths. Note the scale change between plots

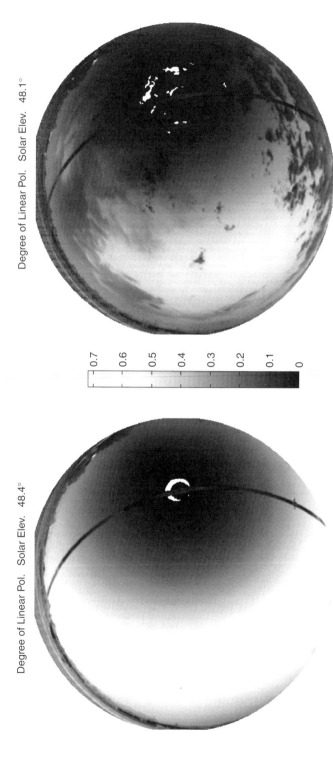

Figure 17.13 Full-sky polarization images at 450 nm: clear skies (*left*) and cloudy skies (*right*) above Montana State University in June, 2012. *Source:* Courtesy of Professor J. Shaw and Dr. N. Pust, Montana State University

saturation, not actual sky polarization.) Multiple scattering events, each with a different scattering plane, collude to reduce the apparent sky polarization at the ground to something less than the Rayleigh peak (around 70% in this case).

The overall effect of sky polarization will vary depending on the sun-target-sensor geometry. In [29], for a diffuse painted metal target looking generally toward the specular direction of the Sun, direct solar illumination dominated the S_1 signature while the Sun is high but clear sky Rayleigh scattering dominated near sunrise and sunset when the specular contribution of the Sun is weakest. Results for S_2 were less conclusive due to complicating factors. For a more specular target, the duration of solar dominance will be reduced. In cases where there is no direct solar illumination, such as targets in shadows [30], sky/Rayleigh scattering polarization will play a significant role throughout the day.

A pure Rayleigh scattering model is only valid in cases where the scatterers are isotropic and much smaller than λ. Experimental evidence shows that the composition of clouds, aerosols, and surface reflectance all contribute significantly to the polarization character of the sky [31, 32]. Examples of sky polarization under the influence of these complicating factors is shown as measured in Fig. 17.13 (*right*) for clouds and Fig. 17.14 from a validated clear sky model that includes aerosol and ground reflection effects [33].

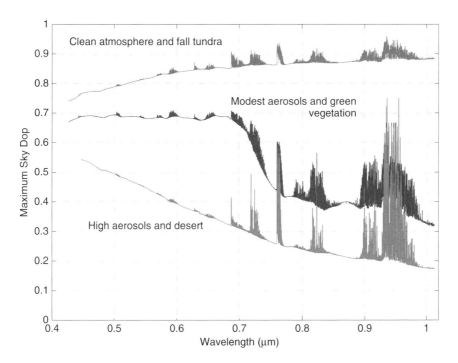

Figure 17.14 Examples of peak sky polarization calculated for various measured aerosol and background conditions in [31]. *Source*: Courtesy of Professor. J. Shaw and Dr. N. Pust, Montana State University

Figure 17.12 suggests diminished sky polarization in the SWIR band. On a clear day, SWIR sky polarization will follow the banding pattern exhibited at shorter wavelengths but with less overall radiance and degree of polarization. Researchers have reported SWIR (1.54 μm) sky degree of polarization varying between approximately 20 and 38% over a 10.5^o Rayleigh scattering arc in a clear summer sky [34].

17.4.2 Emissive Bands

As wavelength increases in the MWIR and LWIR bands, the driving influence of the atmosphere on polarization signatures shifts from scattering to emission. Referring back to Fig. 17.12, the continued overall decline in scattering throughout these bands is evident. In addition to the lingering effects of scattering, the atmosphere affects apparent target signatures through emission of randomly polarized light in the following ways:

1. target reflection of atmospheric emissions and clouds,
2. emission along the path.

In the first case, atmospheric emissions are effectively unpolarized but this radiation becomes partially polarized upon reflection. The reflected light is s-polarized (parallel to the surface) and adds incoherently to the p-polarized (perpendicular to the surface) emission. The combination of s and p polarized radiation results in a reduced apparent degree of polarization. This is part of the reason why humidity is major source of polarization signature variability, especially in the LWIR band (water vapor content also affects absorption). These interactions and others were modeled [35] and measured [36] in detail by Shaw for the case of water bodies.

The role of clouds is highly variable in the emissive bands. In addition to acting as an illumination source for target reflected photons (as discussed previously), clouds also affect the rates of heating and cooling of objects throughout the scene. These combined effects were demonstrated by Felton *et al.* [37] in a study of MWIR and LWIR polarization contrast over several diurnal cycles. They found that LWIR S_1 contrast for a tank hull target is reduced in the presence of clouds but MWIR S_1 increases under the same conditions. LWIR S_1 contrast was reduced to the noise level under totally overcast conditions. In this test, the tank hull temperature was allowed to float with the air temperature. If the tank were operating then the heated surfaces would likely show additional thermal and polarization contrast.

Emission along the path is randomly polarized and will reduce the apparent degree of polarization of the sensed radiation. While the contribution of path emission is insignificant over short distances, it can be fairly substantial over longer propagation paths. Figure 17.15 (Plate 30) illustrates the path radiance over the 3 to 14 μm spectral range over a zenith ground-to-space path for three meteorological conditions. The path radiance is compared to 230 K and 280 K blackbody spectral radiance functions to provide a relative sense of the magnitude of the path radiance [27]. For example, the path radiance is on the order of 30% of the sensed radiation at a 10 μm wavelength for this propagation path.

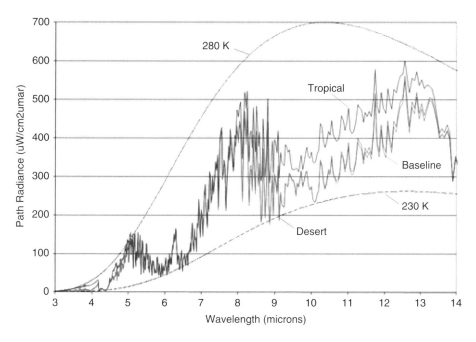

Figure 17.15 (Plate 30) Examples of path emission along a ground to space line of sight as a function of range and atmospheric conditions. *See plate section for the color version*

17.5   Data Reduction Matrix Analysis of Modulated Polarimeters

17.5.1   Important Equations

The *Data Reduction Matrix* (DRM) [18] is a tool for mapping radiometrically calibrated measurements from a polarimeter into Stokes parameters. The DRM is also a tool for modeling calibration errors and choosing optimal sets of measurements. These concepts will be introduced briefly and then applied to the analysis of real systems. For an imaging system, there may be a single DRM for all detectors or it may vary between detector elements. The spatial variability of the DRM will also be addressed in the examples.

Radiation described by Stokes vector $\mathbf{S}$ is incident upon a polarization analyzer defined by Mueller matrix $\boldsymbol{M}_i$. After passing through the analyzer, the following relationship holds between the measured signal I_i and the incident Stokes vector:

$$I_i = \mathbf{A}_i\mathbf{S} + n_i \qquad (17.40)$$

$\mathbf{A}_i$ is a row vector given by the first row of $\boldsymbol{M}_i$ and n_i is noise in the measurement. Recall that the first row of $\boldsymbol{M}_i$ maps the incident Stokes vector into the total irradiance on the detector. For N measurements arranged in a vector $\mathbf{I}$, a polarimeter measurement matrix is defined as

$$W = \begin{bmatrix} \mathbf{A}_1 \\ \vdots \\ \mathbf{A}_N \end{bmatrix} \tag{17.41}$$

and the least-squares estimate of **S** is

$$\hat{\mathbf{S}} = W^+ \mathbf{I} \tag{17.42}$$

where the data reduction matrix, W^+, is the pseudo-inverse of W. When W is overdetermined, that is, when there are more irradiance measurements than Stokes parameters, the Moore–Penrose pseudo-inverse is

$$W^+ = \left(W^T W\right)^{-1} W^T. \tag{17.43}$$

In an optimal configuration (described next), $W^T W$ is a diagonal matrix that makes this pseudoinverse calculation straightforward.

Stokes reconstruction errors occur because of noise and calibration errors in the data reduction matrix [38, 39]. The reconstruction error is the difference between the estimated and measured Stokes vectors

$$\boldsymbol{\epsilon} = \hat{\mathbf{S}} - \mathbf{S}. \tag{17.44}$$

For noise, the error is simply

$$\boldsymbol{\epsilon} = W^+ \mathbf{n} \tag{17.45}$$

where **n** is the vector of noise samples. This type of error is minimized by minimizing the condition number of W. For calibration matrix errors of the form,

$$W_{error} = W + \boldsymbol{\Delta} \tag{17.46}$$

the Stokes reconstruction error is

$$\boldsymbol{\epsilon} = W^+ \boldsymbol{\Delta} \mathbf{S}. \tag{17.47}$$

In these two equations, W is the true calibration matrix and $\boldsymbol{\Delta}$ is an additive offset containing all of the calibration error terms. For calibration errors, the total Stokes reconstruction error depends on the input polarization state.

17.5.2 Example Stokes Polarimeters

The work of the last few sections can be brought together through a series of examples. A straightforward implementation of a full Stokes polarimeter consisting of a fixed analyzer (linear diattenuator) and a rotating retarder is shown in Fig. 17.16.

The Mueller matrix of this system is given by

$$M(\theta, \delta) = M_D(p_h, p_v) M_R(-\theta) M_W(\delta) M_R(\theta) \tag{17.48}$$

and the ith measurement vector is given by

$$\mathbf{A}_i = \frac{1}{2} \left[1 \;\; \cos^2 2\theta_i + \cos\delta \sin^2 2\theta_i \;\; (1 - \cos\delta)\cos 2\theta_i \sin 2\theta_i \;\; -\sin\delta \sin 2\theta_i \right] \tag{17.49}$$

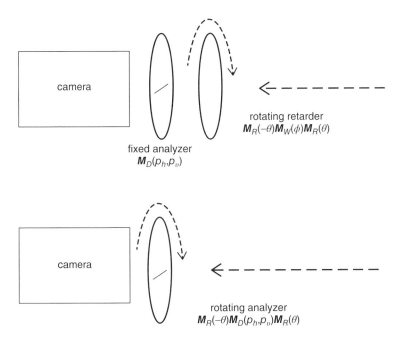

Figure 17.16   Diagrams of the rotating retarder (*top*) and rotating analyzer (*bottom*) polarimeters

for the case where the diattenuator is ideal with a horizontal preference for transmission ($p_h = 1$ and $p_v = 0$). The rotating retarder polarimeter can be analyzed on the Poincaré sphere. Figure 17.17 illustrates how choice of retardance, δ, affects the analyzer states available to the polarimeter. Each plotted point corresponds to the Stokes vector that is transmitted with maximum irradiance through the system at retarder orientation angle θ_i as it rotates through 180°. In the $\delta = 90^\circ$ case, the curve has components along each of the Stokes parameter directions. Consequently, there are many possible combinations of measurements made along this curve that will result in reconstruction of the full Stokes vector using Eq. 17.42. On the other hand, the $\delta = 180^\circ$ curve has zero length in the S_3 direction for all θ_i and there is no measurement combination that will result in reconstruction of S_3.

Clearly, some choices of δ are superior to others if reconstruction of the full Stokes vector is the goal. After δ is selected, some set of retarder orientation angles must also be selected to build up W. The best choices will be those that minimize errors in the presence of noise across the reconstructed Stokes parameters. This minimum error situation is found by selecting the retardance and measurement orientation angles that minimize the Frobenius norm condition number of W

$$c = ||W||_F ||W^+||_F \qquad (17.50)$$

such that

$$||W||_F = \sqrt{\sum_{i=1}^{m} \sum_{j=1}^{n} |w_{ij}|^2} \qquad (17.51)$$

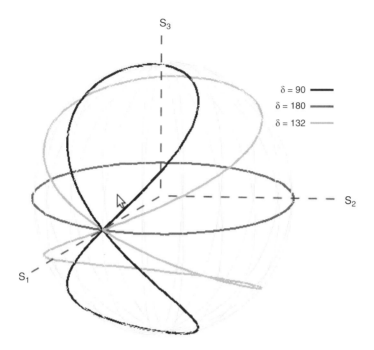

Figure 17.17   Variations on the rotating retarder polarimeter plotted on the Poincaré sphere

where W is taken to be an $m \times n$ matrix with elements w_{ij}. In [40] it is shown that the minimum condition number criterion for the rotating retarder polarimeter is achieved at $\delta = 132°$ (this curve is also shown in Fig. 17.17). Assuming four measurements are made to retrieve each Stokes vector, the optimal values for θ_i are $\pm 15.1°$ and $\pm 51.6°$.

As shown in Section 17.3, little is sacrificed by ignoring S_3 in passive polarimetric imaging. For this reason, linear Stokes polarimeters, that is, systems that determine S_0, S_1, and S_2 exclusively, are in widespread use. The rotating retarder polarimeter with $\delta = 180°$ is one example of a linear Stokes polarimeter that comes with several advantages but large broadband retarders are expensive in the infrared bands. An alternative way of constructing such a system is to place a rotating linear diattenuator/analyzer at the aperture of the imaging system as show in Figure 17.16 (*bottom*). For this rotating analyzer case, the throughput of the system is sensitive to the Mueller matrix of the optics between the analyzer and the detector array. A procedure for including the polarizing effects of the focusing optics will be presented in Section 17.7.

Both the rotating retarder and rotating analyzer schemes are examples of *division-of-time* (DoT) polarimeters. Another option is to spatially vary the measurement analyzers in a repeating pattern at the detector level as shown in Fig. 17.18. This is called a division-of-focal plane (DoFP) or microgrid polarimeter. The term "microgrid" refers to the wire grid polarizers that are bonded to the detector array to achieve polarization sensitivity.

The following example applies to either rotating analyzer or microgrid imaging polarimeters. In the rotating analyzer case, the full DRM is built up over time for every pixel. For microgrid systems, the DRM at each pixel can be built up simultaneously after some form of interpolation

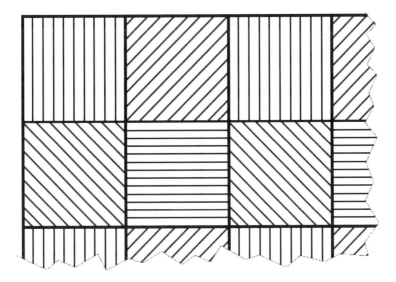

Figure 17.18 A microgrid polarizer array. Thin lines represent direction of the thin wires that prevent light polarized in this direction to pass

using adjacent pixels (however, a better option will be discussed in Section 17.6). The Mueller matrix for this imaging system is given by:

$$M(\theta) = M_R(-\theta)M_D(p_h, p_v)M_R(\theta). \tag{17.52}$$

For the sake of simplicity, the Mueller matrix of the focusing optics for this imaging system is assumed to be the identity matrix. The resulting analyzer vector is:

$$\mathbf{A}_i = \frac{1}{2}\left[p_h^2 + p_v^2 \ \ (p_h^2 - p_v^2)\cos 2\theta_i \ \ (p_h^2 - p_v^2)\sin 2\theta_i\right] \tag{17.53}$$

This analyzer vector has been truncated to three elements because reconstruction of S_3 is not a concern. Of course, $\mathbf{A}_i$ can now only operate on three element Stokes vectors. The energy in any unaccounted for S_3 component is still retained in S_0. For instance, right-hand circular polarized light would be seen by this polarimeter as unpolarized light whether or not there is a zero-valued fourth element in this analyzer vector. This simplification sacrifices nothing and will make calculation of the DRM easier later on.

To find three Stokes parameters, the number of measurements N must be greater than or equal to 3. The analyzer orientation angles are selected to minimize the condition number of W thereby optimizing performance in the presence of noise. The current system is only sensitive to linear polarization states so the minimum condition number is found by spreading the choices for θ_i out evenly over all possible values of the orientation angle ψ. For instance, for a three measurement system, the separation between each θ_i should be 60°. For a four channel system (like the microgrid case), this spacing is 45°. In practice, more that three rotating analyzer measurements may be made to reduce noise. Each of these configurations have a condition number of $\sqrt{2}$.

The example of the four channel linear Stokes polarimeter is useful because it can represent either the microgrid or rotating analyzer cases and the pseudoinverse of W may be calculated by hand. The analyzer angles are optimally spaced at 0, 45, 90, and 135 degrees. Without loss of generality, assume that the neutral density transmission loss in the analyzer, $p_h^2 + p_v^2 = 1$. The true transmission loss can be multiplied back in after analysis of the system is complete. The resulting analyzer matrix and DRM are:

$$W = \frac{1}{2} \begin{bmatrix} 1 & D & 0 \\ 1 & 0 & D \\ 1 & -D & 0 \\ 1 & 0 & -D \end{bmatrix} \tag{17.54}$$

$$W^+ = \begin{bmatrix} \frac{1}{2} & \frac{1}{2} & \frac{1}{2} & \frac{1}{2} \\ D^{-1} & 0 & -D^{-1} & 0 \\ 0 & D^{-1} & 0 & -D^{-1} \end{bmatrix} \tag{17.55}$$

where D is the linear diattenuation,

$$D = \frac{p_h^2 - p_v^2}{p_h^2 + p_v^2}. \tag{17.56}$$

Diattenuation expresses how well an analyzer extinguishes light that is polarized perpendicular to its preferred transmission axis. For rotating analyzer systems, diattenuation can be close to 1 (nearly ideal). Microgrid polarimeters have lower effective diattenuation values because of optical (diffraction) and electronic crosstalk between adjacent detectors. The following example examines the effect of diattenuation on Stokes reconstruction errors.

Using Eq. 17.45, the reconstruction error due to noise is

$$\epsilon = W^+ n = \begin{bmatrix} \frac{1}{2}(n_1 + n_2 + n_3 + n_4) \\ D^{-1}(n_1 - n_3) \\ D^{-1}(n_2 - n_4) \end{bmatrix}. \tag{17.57}$$

Thus the reconstruction of S_0 is unaffected by diattenuation but the noise in S_1 and S_2 increases as diattenuation decreases.

If diattenuation is ignored and the analyzers are assumed to be ideal in the calibration matrix then a different kind of Stokes reconstruction error occurs. To find this error, the calibration matrix of the idealized four channel polarimeter ($D = 1$) is decomposed according to Eq. 17.46

$$W_{ideal} = W + \Delta = \frac{1}{2} \begin{bmatrix} 1 & D & 0 \\ 1 & 0 & D \\ 1 & -D & 0 \\ 1 & 0 & -D \end{bmatrix} + \frac{1}{2} \begin{bmatrix} 0 & 1-D & 0 \\ 0 & 0 & 1-D \\ 0 & D-1 & 0 \\ 0 & 0 & D-1 \end{bmatrix}. \tag{17.58}$$

The reconstruction error that results is given by Eq. 17.47,

$$\epsilon = W^+ \Delta S = (D^{-1} - 1) \begin{bmatrix} 0 \\ S_1 \\ S_2 \end{bmatrix}. \tag{17.59}$$

In other words, S_1 and S_2 are overestimated if diattenuation is not included in the DRM. This example is intended to be illustrative but imperfect diattenuation is only one of many ways in which polarimeter calibration may be non-ideal. In general, every optical element in the polarimeter contributes some diattenuation, retardance, depolarization, and misalignment [18].

17.6 Fourier Domain Analysis of Modulated Polarimeters

The applications of imaging polarimetry include many cases where the intended target varies appreciably (temporally and/or spatially) during the measurement interval. The purpose of this section is to address Stokes reconstruction for dynamic scenes. This discussion will unfold via the examples of the rotating analyzer and microgrid polarimeters described in Section 17.5. The tools developed below may then be readily extended by the reader to other modulated polarimeter designs.

17.6.1 Rotating Analyzer

The analyzer vector of the rotating analyzer polarimeter is described by Eq. 17.53. Assume ideal polarization optics, that is, $D = 1$. From Eq. 17.40, the irradiance at an arbitrary detector at time t is given by

$$I(t) = \frac{1}{2}[S_0(t) + S_1(t)\cos 2\theta(t) + S_2(t)\sin 2\theta(t)]. \tag{17.60}$$

As the analyzer rotates,

$$\theta(t) = 2\pi f_R t \tag{17.61}$$

where f_R is the rotation frequency of the analyzer. The detector array samples this signal with frequency f_S which is assumed to be greater than f_R. To avoid unnecessary details, the sample duration is taken to be effectively instantaneous and $I(t)$ is bandlimited. The nth sample of this signal is

$$I(n) = \frac{1}{2}[S_0(n) + S_1(n)\cos(\alpha n) + S_2(n)\sin(\alpha n)] \tag{17.62}$$

where

$$\alpha = 4\pi\frac{f_R}{f_S} \tag{17.63}$$

and the *Discrete-Time Fourier Transform* (DTFT) of 17.62 is,

$$\tilde{I}(\omega) = \frac{1}{2}\left\{\tilde{S}_0(\omega) + \pi\delta(\omega - \alpha)\left[\tilde{S}_1(\omega) + i\tilde{S}_2(\omega)\right] + \pi\delta(\omega + \alpha)\left[\tilde{S}_1(\omega) - i\tilde{S}_2(\omega)\right]\right\} \tag{17.64}$$

where ω is the angular frequency.

As a brief review, the DTFT is periodic over increments of 2π radians. Integer multiples of the folding frequency, $\omega = \pi$, correspond to the maximum analog frequency, $f_S/2$, that may be accurately reconstructed by the sampled data ($f_S/2$ is also referred to as the Nyquist frequency). Spectral content beyond the Nyquist frequency is aliased into lower frequency portions of the spectrum resulting in undesirable artifacts in the reconstructed data. The S_1 and S_2 components of 17.64, which contain all of the polarization information in the sampled signal, are modulated out to a $\pm\alpha$ sideband while the total intensity component, S_0, is the baseband. Spectral content near the center of the baseband and sidebands represent signal components that vary relatively slowly while more rapid signal variations are captured at frequencies away from the band centers.

The frequency range occupied by the signal is referred to as its bandwidth. A sketch of the magnitude of this spectrum when sampled for maximum available bandwidth is shown in Fig. 17.19. It is possible that the bandwidth of the S_0 component will be different than the S_1 and S_2 bandwidths. To guarantee that a signal with analog bandwidth B can be reconstructed properly,

$$\frac{2\pi}{f_S}B \le \alpha \tag{17.65}$$

and

$$3\alpha \le 2\pi. \tag{17.66}$$

Together, these requirements ensure that three full bandwidths (one for the baseband and two for the sidebands) may be contained without overlap within the DTFT spectrum. As shown by [41], these requirements become

$$f_S \ge 6f_R \tag{17.67}$$

and

$$B \le 2f_R \tag{17.68}$$

One key result of this analysis is that the temporal bandwidth capacity of division-of-time polarimeters is always less than that of the underlying imaging system. In other words, sensitivity to polarization is achieved by sacrificing temporal resolution.

This Fourier analysis also shows how Stokes parameter reconstruction can be viewed as a combination of modulation and low pass filtering operations. The DRM in Section 17.5 represents one particular choice of low pass filter, a rectangular window in time.

17.6.2 Microgrid Polarimeters

Fourier analysis of the microgrid polarimeter was first carried out in [42]. Referring back to Fig. 17.18, each pixel in a raw microgrid intensity image samples a point in the scene as projected through a polarization analyzer. Ideally, the analyzer at each sampling point is identical

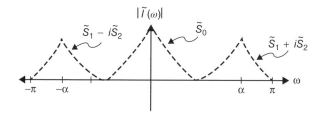

Figure 17.19   Magnitude of the discrete time rotating analyzer spectrum with annotations for the Stokes components

except for rotation in a repeating pattern (0°, 45°, 90°, and 135°) as shown in Fig. 17.18. The sampled intensity at each point (m, n) in a raw microgrid image is given by

$$I(m, n) = \frac{1}{2} S_0(m, n) + \frac{1}{4} \cos(\pi m)[S_1(m, n) + S_2(m, n)]$$

$$+ \frac{1}{4} \cos(\pi n)[S_1(m, n) - S_2(m, n)] \qquad (17.69)$$

where each of the $S_x(m, n)$ terms represent one of the Stokes parameter images as it varies in space. The DSFT of I is given by:

$$\tilde{I}(\xi, \eta) = \frac{1}{2} \tilde{S}_0(\xi, \eta) + \frac{\pi}{4} \left[\tilde{S}_1 (\xi - \pi, \eta) + \tilde{S}_2(\xi - \pi, \eta) \right]$$

$$+ \frac{\pi}{4} \left[\tilde{S}_1 (\xi, \eta - \pi) - \tilde{S}_2(\xi, \eta - \pi) \right] \qquad (17.70)$$

where each of ξ and η are real-valued spatial frequency coordinates defined over one period of the spectrum, that is, $-\pi \leq \xi, \eta \leq \pi$. An illustration of the microgrid DSFT is shown in Fig. 17.20.

Similar to the rotating analyzer case, the S_1 and S_2 components of the microgrid spectrum are modulated out into sidebands while the S_0 signal remains at baseband. Tyo *et al.* [42] established a sufficient condition to avoid aliasing in terms of a band radius in ξ and η space:

$$r_{S_0} + r_{S_1 \pm S_2} < \pi \qquad (17.71)$$

where each r_x is the radius, in radians per sample, of the smallest circle that encloses the band limits on S_0 or $S_1 \pm S_2$. In almost every case, the bandwidth occupied by S_0 is larger than the $S_1 \pm S_2$ bandwidth.

This bandwidth condition can be difficult to meet. Given a detector spacing of d_x, the Nyquist frequency of the microgrid imager is

$$f_x = \frac{1}{2d_x} \qquad (17.72)$$

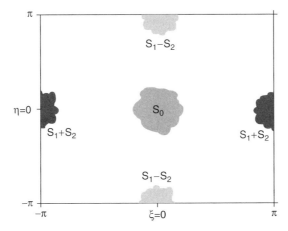

Figure 17.20 Magnitude of the microgrid analyzer spectrum with annotations for the Stokes components

Note that d_x is the detector spacing, not the spacing between like-oriented analyzer samples. The ultimate limit on the image bandwidth is imposed by the optical system:

$$f_o = \frac{1}{\lambda F_\#} \tag{17.73}$$

where λ is the operating wavelength and $F_\#$ is the f-number of the optical system [43]. If $f_x < f_o$ then aliasing and spectral mixing between the S_0 baseband and the S_1, S_2 sidebands is going to occur. Though detector sizes continue to decrease, aliasing is common problem for microgrid systems in the infrared bands.

Microgrid Stokes image reconstruction under these conditions has been attempted using like-polarization interpolation [44], Fourier domain processing [42, 45], bilateral filtering [46], and multi-frame super-resolution [47].

17.6.3   Band-Limited Stokes Reconstruction

Stokes parameter estimation using the data reduction matrix is highly susceptible to reconstruction errors due to changes in the input signal. In other words, the DRM works best when the signal presented to the aperture is approximately constant. Lacasse *et al.* [41] provide an alternative to the DRM that can exploit time/space varying signals for modulated polarimeters up to the bandwidth limits presented in the previous sections. For a division-of-time polarimeter, the reconstructed Stokes vector at sample n is

$$\hat{\mathbf{S}}(n) = \mathbf{Z}^{-1}(n) \left[w * \mathbf{A}^T I \right] (n) \tag{17.74}$$

where

$$\mathbf{Z}(n) = \left[w * \mathbf{A}^T \mathbf{A} \right] (n). \tag{17.75}$$

In these equations, $w(n)$ is a filter window; $\mathbf{A}(n)$ is the analyzer row vector at this reconstruction location; $I(n)$ is the measured irradiance; and $*$ is the discrete convolution operator. The window function is the low pass filter used to weight the measurements contributing to the Stokes reconstruction at n. Matrix $\mathbf{Z}(n)$ separates out the Stokes parameters from the products of the demodulation process in $[w * \mathbf{A}^T I](n)$.

Equation 17.74 is equivalent to the traditional DRM when the window function, $w(n)$, is rectangular [48]

$$w_r(n) = \begin{cases} 1 & \text{if } \quad n < N - 1 \\ 0 & \quad \text{otherwise.} \end{cases} \tag{17.76}$$

The length, N, of the reconstruction window determines how fast the polarimeter responds to signal variations, how accurately the signal will be reconstructed, and how much cross-spectral contamination there will be between the baseband and sidebands in the reconstruction process. As an example, Fig. 17.21 shows the filter response for two different rectangular windows applied to a rotating analyzer polarimeter that has been sampled according to Eqs. 17.67 and 17.68. The three sample window has a broad frequency response but it also has a significant response outside of the maximum available reconstruction bandwidth. Except for the case of a constant signal (where the signal spectrum consists of delta functions at 0 and $\pm\alpha$ radians/sample), this filter guarantees some cross-spectral contamination. When the window is made

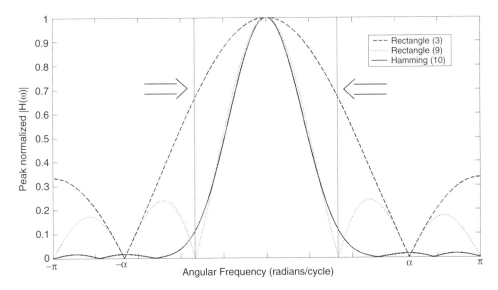

Figure 17.21 Frequency response to various Rotating Analyzer polarimeter reconstruction filters. Maximum reconstruction bandwidth is indicated by arrows on the figure

three times longer, this out of band response is diminished at the cost of diminished high frequency response *within* the passband.

Equation 17.74 allows the reconstruction window to be tailored for improved performance. For real-time systems, a Hamming window is one option that offers an improvement over the rectangular DRM window. The Hamming window of length N is

$$w_h(n) = a - (1 - a) * \cos\left(\frac{2\pi(n)}{N-1}\right) \tag{17.77}$$

where $a = 0.54$. A 10 sample Hamming window is also shown in Fig. 17.21. It has an in-band frequency response that is almost identical to the nine sample rectangle window while providing nearly complete out of band rejection.

It is also possible to approximate a filter with DSFT bandwidth α and unit magnitude inside the pass-band (i.e., a rectangular frequency response) using a truncated sinc style window. However, this window will result in substantial reconstruction artifacts if the input measurements are not truly band-limited, which is often the case in real systems.

In the example here, n is a one-dimensional series of samples in a division-of-time polarimeter but this method extends readily to the two-dimensional microgrid case. Unlike the methods listed at the end of Section 17.6.2, this method also allows for non-ideal and spatially varying analyzer vectors.

17.7 Radiometric and Polarimetric Calibration

Reliable imaging polarimetry requires sensor calibration. This calibration is composed of two parts: a mapping of sensor digital counts into radiometric units and then an accurate mapping of radiometric values into polarimetric quantities. The procedures for conducting radiometric and polarimetric calibration are described in this section.

17.7.1 Radiometric Non-Uniformity Correction

Each detector in an infrared imaging array responds to the arrival of photons in a slightly different way. Read-out electronics also affect how this detected signal is interpreted and recorded. Additionally, vignetting may affect apparent scene radiance at the edge of the image even in well-designed optical systems. These non-uniformities in detector and read-out response degrade the image on both small (e.g., pixel by pixel) and large (rows or regions of pixels) scales. The polarimetric data reduction matrix makes no accommodation for this issue and, therefore, these detector variations must be eliminated as a preprocessing step.

For InSb and HgCdTe imaging arrays, $L(x, y)$, the radiometrically corrected response for a detector in array position (x, y), is well modeled as using a piecewise linear fit. For imaging polarimeters without rotating optics, this relationship is given by

$$L(x, y) = m_j(x, y)C(x, y) + b_j(x, y) \tag{17.78}$$

where $C(x, y)$ is the raw digital counts from the camera, $m_j(x, y)$ is the slope of the detector response and $b_j(x, y)$ is the offset. The subscript j represents the subrange of count values over which each b_j and m_j are valid. These subranges are generally set globally for the entire array. The width of these subranges must be established experimentally. Each pair of b_j and m_j are readily determined by recording the response of the system at a minimum of two known radiometric levels within each count subrange and then solving the resulting system of equations. In general, a new set of calibration coefficients is required whenever the camera is turned off, when integration time or gain is changed, or after the camera response drifts during periods of extended use.

For imaging polarimeters with moving parts, a radiometric calibration is also required at each analyzer position. In this case,

$$L(x, y) = m_{ij}(x, y)C(x, y) + b_{ij}(x, y) \tag{17.79}$$

where the additional subscript i now corresponds to the analyzer orientation θ_i.

For imagers in the MWIR and LWIR atmospheric transmission bands, a blackbody radiator is used for radiometric input. At shorter wavelengths, a integrating sphere is often used though other less expensive diffuse sources can be purchased or built. In all cases, the radiating surface must produce uniform radiance over an area at least as large as the aperture of the imager. During calibration, this radiating surface is placed as close the camera aperture as possible.

Detector response will drift during periods of continuous use. Periodic re-calibration is necessary, especially for HgCdTe arrays, which places limits on sensor availability. Radiometric calibration time can be extended significantly through the use of scene-based non-uniformity correction. Two recent papers [49, 50] address this issue specifically for the case of microgrid polarimetric imagery.

17.7.2   Polarimetric Calibration

The purpose of polarimetric calibration is to ensure that the data reduction matrix faithfully reconstructs Stokes vectors throughout the scene. Unlike radiometric calibration, polarimetric calibration is only necessary when some change occurs affecting the optical system. For instance, changes to lenses, bandpass filters, or any of the polarization optics are all reasons to conduct a polarimetric calibration.

Persons, *et al.* [51], describe a laboratory setup for polarization calibration as shown in Fig. 17.22. A transmissive polarization target is placed in front of an incoherent illumination source so that it can be imaged by the camera under test. The target is designed so that the transmitted Stokes vector can be changed according to the needs of the test. The polarization target is tilted slightly so that stray light reflected from its surface and into the camera is held fixed by an external reference. After setup is complete, the data collection proceeds as follows

1. The illumination source is set to temperature T_1
2. Measurements $m = 1$ to M are made
   (a) The polarization reference is set to generate Stokes vector $\mathbf{S}_m(T_x)$
   (b) The camera records the response vector $\mathbf{L}_m(T_x)$. The response vector consists of radiance measurements from each analyzer state
3. The illumination source is set to temperature T_2
4. Step 2 is repeated

The estimate of the measurement matrix, $\hat{\mathbf{W}}$, is then given by

$$\hat{\mathbf{W}} = [L(T_1) - L(T_2)][S(T_1) - S(T_2)]^+ \tag{17.80}$$

where

$$S(T_x) = \begin{bmatrix} \mathbf{S}_1(T_x) & \dots & \mathbf{S}_M(T_x) \end{bmatrix} \tag{17.81}$$

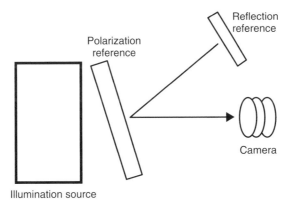

Figure 17.22   Laboratory setup for polarization calibration

and

$$L(T_\mathrm{x}) = \begin{bmatrix} \mathbf{L}_1(T_\mathrm{x}) & \ldots & \mathbf{L}_M(T_\mathrm{x}) \end{bmatrix}. \tag{17.82}$$

The purpose of the signal subtraction (and of using the reflection reference) is to remove any unmodeled contributions from polarization target emission or reflections from the target's surroundings.

Obviously, this calibration procedure is only valid if the Stokes vectors produced by the polarization reference are known precisely, at least to a constant radiometric scale factor. As an example, the polarization reference for calibrating a linear Stokes polarimeter may be a rotating polarizer (linear diattenuator) in a rotation stage. The diattenuation of this device (see Eq. 17.56) must be taken into account. Additionally, the spectral content of the illumination source should be consistent with real-world operations. For polarimetric imagers in the emissive bands, the reference source is a blackbody operated at temperatures inside of the camera's radiometric range. It should be clear from the discussion in Section 17.5 that the Stokes vectors in Eq. 17.81 should be spread out as far apart as possible on the accessible parts Poincaré sphere.

17.8   Polarimetric Target Detection

Sometimes it is desired to accentuate objects in a scene that are anomalous to the Stokes image background. We refer to these objects of interest as targets, and this process of accentuating them as target detection. The optimum detection strategy is the Neyman–Pearson test that forms a detection statistic as a likelihood ratio of an alternative (target present) hypothesis and a null (target absent) hypothesis [52]. These likelihood functions represent the known or assumed statistical models for the target and background. In an anomaly detector, no information is known or assumed for the target, and the detector is completely specified by the statistical model of the background. A common model for vector data is the multivariate normal model, which leads to the Mahalanobis distance or multivariate energy detector. This, however, is not usually a good choice for multispectral or polarimetric imagery because the background is non-stationary and not well represented by global normal statistics.

To address the non-stationarity problem associated with multivariate imagery, Reed and (Xiaoli) Yu developed a spatially adaptive anomaly detector based on local normal statistics that has exhibited good performance for multispectral imaging sensors [53]. This RX algorithm forms a pixel-by-pixel detection statistic given by

$$r(\mathbf{x}) = (\mathbf{x} - \mathbf{m}_b)^T C_b^{-1}(\mathbf{x} - \mathbf{m}_b) \tag{17.83}$$

where $\mathbf{x}$ is the vector data for the pixel under test (or sample mean over a target region), $\mathbf{m}_b$ is the local sample mean vector for a background region around the pixel under test, and C_b is the local sample covariance matrix for the background region as depicted in Fig. 17.23. The size of the target region is selected to match the expected target size, and the size of the background region is a compromise between covariance matrix estimation accuracy (large region) and spatial adaptivity (small region). Sometimes a guard region is placed around the target region to avoid corrupting the covariance matrix with target data. As the pixel under

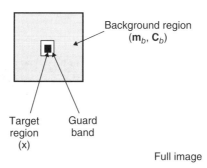

Figure 17.23   Regions of interest in the RX algorithm

test moves across the image, the local target mean vector and background covariance matrix need to be recomputed, producing a computational cost to the spatial adaptivity. Targets are detected by thresholding the detection statistic at a level that provides the desired false alarm rate. Pixels with detection statistic values above threshold are declared as targets.

The benefit of the spatial adaptivity of the RX algorithm is that it supports good detection performance even in the presence of inhomogeneous background clutter, which is very important when applied to multispectral imagery. When applied to polarimetric imagery, this spatial adaptivity might not seem as important for the theoretical case in which the background is truly unpolarized. In this case, the target can be detected simply by thresholding the degree of polarization image. However, there can be polarized content in the background clutter, or even variations in the S_1 and S_2 Stokes images due to miscalibration and other sensor artifacts, that result in inhomogeneous clutter in these components. In such cases, one can use the RX algorithm simply by defining $\mathbf{x}$ in the detection statistic as the entire Stokes vector or the two linear components depending on what target characteristics are considered anomalous. If the entire Stokes vector is used, the algorithm will detect objects that are anomalously intense or polarized. By removing the S_0 component in the RX implementation, the influence of intensity is removed.

As an example, the RX algorithm was applied to the polarimetric dataset in Fig. 17.6 with the results displayed in Fig. 17.24. The image on the left shows the degree of linear polarization, which indicates the highly polarized object, but also shows some variation in the background polarization. The RX algorithm was applied in this case only to the S_1 and S_2 components of the polarimetric image since the S_0 component was so heavily dominated by the background clutter. A single pixel target window was used with a 49×49 pixel background window and 19×19 pixel guard band (nominally the target size). The RX detection statistic image is shown on the right where it is scaled at one-fourth the peak response. In this case, the detection statistic at the center of the target was over a factor of two higher than anywhere else in the image, so the target could be detected with no false alarms if a threshold was used at this level. With the given scaling, it is evident that there are some bright single pixel responses throughout the image that would represent false alarms if not filtered out by either a high enough detection statistic threshold or spatial filtering to remove isolated single-pixel detections.

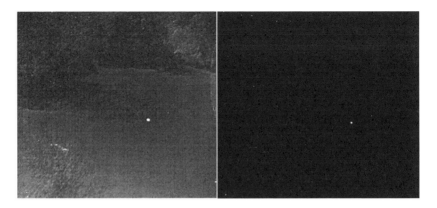

Figure 17.24   Comparison of a DOLP image (*left*) and RX detection map (*right*)

References

[1] Rogne T., F. Smith, and J. Rice, "Passive target detection using polarized components of infrared signatures," *Proceedings of the SPIE 1317*, 1990.

[2] Howe J., M. Miller, R. Blumer, T. Petty, M. Stevens, D. Teale, and M. Smith, "Polarization sensing for target acquisition and mine detection," in *Proceedings SPIE*, vol. 4133, 2000.

[3] Sadjadi F. and C. Chun, "Automatic detection of small objects from their infrared state-of-polarization vectors," *Optics Letters*, vol. **28**, no. 7, pp. 531–533, 2003.

[4] Gurton K., M. Felton, R. Mack, D. LeMaster, C. Farlow, M. Kudenov, and L. Pezzaniti, "MidIR and LWIR polarimetric sensor comparison study," in *Proceedings SPIE 7672*, 2010.

[5] Ratliff B., D. LeMaster, R. Mack, P. Villeneuve, J. Weinheimer, and J. Middendorf, "Detection and tracking of RC model aircraft in LWIR microgrid polarimeter data," in *Proceedings SPIE*, vol. 8160, p. 1, 2011.

[6] Goudail F. and J. S. Tyo, "When is polarimetric imaging preferable to intensity imaging for target detection?," *Journal of the Optical Society of America A*, vol. **28**, pp. 46–53, Jan 2011.

[7] Thilak V., D. Voelz, and C. Creusere, "Image segmentation from multi-look passive polarimetric imagery," in *Proceedings SPIE*, vol. 6682, 2007.

[8] Wolff L., "Polarization-based material classification from specular reflection," *Pattern Analysis and Machine Intelligence, IEEE Transactions on*, vol. **12**, no. 11, pp. 1059–1071, 1990.

[9] Thilak V., D. Voelz, and C. Creusere, "Polarization-based index of refraction and reflection angle estimation for remote sensing applications," *Applied Optics*, vol. **46**, no. 30, pp. 7527–7536, 2007.

[10] Hyde IV M., J. Schmidt, M. Havrilla, and S. Cain, "Determining the complex index of refraction of an unknown object using turbulence-degraded polarimetric imagery," *Optical Engineering*, vol. **49**, no. 12, pp. 126201–126201, 2010.

[11] Koshikawa K., "A polarimetric approach to shape understanding of glossy objects," in *Proceedings of the 6th International Joint Conference on Artificial Intelligence-Volume 1*, pp. 493–495, Morgan Kaufmann Publishers Inc., 1979.

[12] Tyo J., M. Rowe, E. Pugh Jr, N. Engheta, *et al.*, "Target detection in optically scattering media by polarization-difference imaging," *Applied Optics*, vol. **35**, no. 11, pp. 1855–1870, 1996.

[13] Chenault D. and J. Pezzaniti, "Polarization imaging through scattering media," *SPIE Proceedings 4133*, 2000.

[14] Egan W., *Photometry and polarization in remote sensing*. Elsevier, New York, NY, 1985.

[15] Tyo J. S., D. L. Goldstein, D. B. Chenault, and J. A. Shaw, "Review of passive imaging polarimetry for remote sensing applications," *Applied Optics*, vol. **45**, no. 22, pp. 5453–5469, 2006.

[16] Schott J., *Fundamentals of Polarimetric Remote Sensing*. SPIE Press, Bellingham, WA, 2009.

[17] Goldstein D., *Polarized Light, Revised and Expanded*. CRC Press, 2010.

[18] Chipman R., "Polarimetry," in *Handbook of Optics, Third Edition Volume I: Geometrical and Physical Optics, Polarized Light, Components and Instruments (set)*, 3rd edn, ch. 15, New York, NY: McGraw-Hill, 2010.

[19] Goodman J. W., *Statistical Optics*, New York, John Wiley & Sons, Inc., 1985.

[20] Collett E., *Field Guide to Polarization*. SPIE Press, Bellingham, WA, 2005.

[21] Hoffman K. and R. Kunze, *Linear Algebra*. Prentice-Hall, Englewood Cliffs, NJ, 1971.

[22] Priest R. G. and T. A. Gerner, "Polarimetric BRDF in the microfacet model: theory and measurements," tech. rep., DTIC Document, 2000.

[23] Pedrotti F. and L. Pedrotti, *Introduction to Optics*, 2nd edn, Prentice Hall, 1993.

[24] Sandus O., "A review of emission polarization," *Applied Optics*, vol. **4**, no. 12, pp. 1634–1642, 1965.

[25] Resnick A., C. Persons, and G. Lindquist, "Polarized emissivity and Kirchhoff's law," *Applied Optics*, vol. **38**, no. 8, pp. 1384–1387, 1999.

[26] Tyo J. S., B. M. Ratliff, J. K. Boger, W. T. Black, D. L. Bowers, and M. P. Fetrow, "The effects of thermal equilibrium and contrast in lwir polarimetric images," *Opt. Express*, vol. **15**, pp. 15161–15167, Nov 2007.

[27] Eismann M. T., *Hyperspectral Remote Sensing*. SPIE Press, Bellingham, WA, 2012.

[28] Bohren C. F. and D. R. Huffman, *Absorption and Scattering of Light by Small Particles*. Wiley-VCH Verlag GmbH & Co. KGaA, 2008.

[29] Pust N. J., J. A. Shaw, and A. R. Dahlberg, "Concurrent polarimetric measurements of painted metal and illuminating skylight compared with a microfacet model," *SPIE Proceedings 7461*, 2009.

[30] Lin S.-S., K. M. Yemelyanov, J. Edward N. Pugh, and N. Engheta, "Separation and contrast enhancement of overlapping cast shadow components using polarization," *Optics Express*, vol. **14**, pp. 7099–7108, Aug 2006.

[31] Pust N. J. and J. A. Shaw, "Wavelength dependence of the degree of polarization in cloud-free skies: simulations of real environments," *Optics Express*, vol. **20**, pp. 15559–15568, Jul 2012.

[32] Pust N. J., J. A. Shaw, and A. Dahlberg, "Visible-NIR imaging polarimetry of painted metal surfaces viewed under a variably cloudy atmosphere," *Proceedings of the SPIE 6972*, 2008.

[33] Pust N. J., A. R. Dahlberg, M. J. Thomas, and J. A. Shaw, "Comparison of full-sky polarization and radiance observations to radiative transfer simulations which employ AERONET products," *Optics Express*, vol. **19**, pp. 18602–18613, Sep 2011.

[34] Miller M., R. Blumer, and J. Howe, "Active and passive SWIR imaging polarimetry," *Proceedings of the SPIE 4481*, 2002.

[35] Shaw J. A., "Degree of linear polarization in spectral radiances from water-viewing infrared radiometers," *Applied Optics*, vol. **38**, no. 15, pp. 3157–3165, 1999.

[36] Shaw J. A., "Polarimetric measurements of long-wave infrared spectral radiance from water," *Applied Optics*, vol. **40**, no. 33, pp. 5985–5990, 2001.

[37] Felton M., K. P. Gurton, J. L. Pezzaniti, D. B. Chenault, and L. E. Roth, "Measured comparison of the crossover periods for mid- and long-wave IR (MWIR and LWIR) polarimetric and conventional thermal imagery," *Optics Express*, vol. **18**, pp. 15704–15713, Jul 2010.

[38] Tyo J., "Noise equalization in stokes parameter images obtained by use of variable-retardance polarimeters," *Optics Letters*, vol. **25**, no. 16, pp. 1198–1200, 2000.

[39] Tyo J., "Design of optimal polarimeters: maximization of signal-to-noise ratio and minimization of systematic error," *Applied Optics*, vol. **41**, no. 4, pp. 619–630, 2002.

[40] Ambirajan A. and D. Look Jr, "Optimum angles for a polarimeter: part I," *Optical Engineering*, vol. **34**, no. 6, pp. 1651–1655, 1995.

[41] LaCasse C., R. Chipman, and J. Tyo, "Band limited data reconstruction in modulated polarimeters," *Optics Express*, vol. **19**, no. 16, pp. 14976–14989, 2011.

[42] Tyo J. S., C. F. LaCasse, and B. M. Ratliff, "Total elimination of sampling errors in polarization imagery obtained with integrated microgrid polarimeters," *Optics Letters*, vol. **34**, no. 20, pp. 3187–3189, 2009.

[43] Goodman J., *Introduction to Fourier Optics*. McGraw-Hill Companies, 1988.

[44] Ratliff B. M., C. F. LaCasse, and J. S. Tyo, "Interpolation strategies for reducing IFOV artifacts in microgrid polarimeter imagery," *Optics Express*, vol. **17**, pp. 9112–9125, May 2009.

[45] LeMaster D., "Stokes image reconstruction for two-color microgrid polarization imaging systems," *Optics Express*, vol. **19**, no. 15, pp. 14604–14616, 2011.

[46] Ratliff B., C. LaCasse, and J. Tyo, "Adaptive strategy for demosaicing microgrid polarimeter imagery," in *Aerospace Conference, 2011 IEEE*, pp. 1–9, IEEE, 2011.

[47] Hardie R., D. LeMaster, and B. Ratliff, "Super-resolution for imagery from integrated microgrid polarimeters," *Optics Express*, vol. **19**, no. 14, pp. 12937–12960, 2011.

[48] LaCasse C., J. Tyo, and R. Chipman, "Role of the null space of the DRM in the performance of modulated polarimeters," *Optics Letters*, vol. **37**, no. 6, pp. 1097–1099, 2012.

[49] Black W. T., C. F. LaCasse IV, and J. S. Tyo, "Frequency-domain scene-based non-uniformity correction and application to microgrid polarimeters," in *Proceedings SPIE 8160*, 2011.

[50] Ratliff B. M. and D. A. LeMaster, "Adaptive scene-based correction algorithm for removal of residual fixed pattern noise in microgrid image data," in *Proceedings SPIE 8364*, 2012.

[51] Persons C., M. Jones, C. Farlow, L. Morell, M. Gulley, and K. Spradley, "A proposed standard method for polarimetric calibration and calibration verification," in *Proceedings of the SPIE*, vol. 6682, 2007.

[52] Scharf L., *Statistical Signal Processing: Detection, Estimation, and Time Series Analysis*. Addison – Wesley, Reading, MA, 1991.

[53] Reed I. S. and X. Yu, "Adaptive multiple-band CFAR detection of an optical pattern with unknown spectral distribution," *Acoustics, Speech and Signal Processing, IEEE Transactions on*, vol. **38**, no. 10, pp. 1760–1770, 1990.

Index

Multi-dimensional Imaging, First Edition. Edited by Bahram Javidi, Enrique Tajahuerce and Pedro Andrés.
© 2014 John Wiley & Sons, Ltd. Published 2014 by John Wiley & Sons, Ltd.